DISCRETE MATHEMATICS

DISCRETE MATHEMATICS

Martin Aigner

Translated by David Kramer

AMERICAN MATHEMATICAL SOCIETY

Originally published in the German language by
Friedr. Vieweg & Sohn Verlag, 65189 Wiesbaden, Germany,
as "Martin Aigner: Diskrete Mathematik. 5. Auflage (5th edition)".
© Friedr. Vieweg & Sohn Verlag|GWV Fachverlage GmbH,
Wiesbaden, 2004

Translated by David Kramer.

2000 *Mathematics Subject Classification*. Primary 05A10, 05A15, 05B05, 05B15, 05C70,
05C85, 06E30, 11A07, 90C05, 94B05.

For additional information and updates on this book, visit
www.ams.org/bookpages/discmat

Library of Congress Cataloging-in-Publication Data

Aigner, Martin, 1942–
    [Diskrete mathematik. English]
    Discrete mathematics / Martin Aigner.
        p. cm
    Includes bibliographical references and index.
    ISBN-13: 978-0-8218-4151-8 (alk. paper)
    1. Computer science—Mathematics. I. Title.

QA76.9.M35 A34   2007
004—dc22   2006052285

Contents

Prefaces

Preface to the English Translation

The book that you hold in your hands is based on the 5th edition of the German text "Diskrete Mathematik" published by Vieweg-Verlag. The translator, David Kramer, has done an admirable job; he not only perfectly conveyed the spirit of the original, but improved the text in several points. My sincere thanks go to him and to Ina Mette, who initiated this project. The intentions and scope of the book are explained extensively in the preface to the first edition. It is my hope that readers will learn about the most important developments of discrete mathematics, and at the same time enjoy the subject as an intuitively appealing mathematical pleasure.

Translator's Note

I would like to express my thanks to those who have made my work on this translation the great pleasure that it was: to Ina Mette, of the American Mathematical Society, who invited me to take on this project and who has supported my work over the past year; to Martin Aigner, who way beyond the call of duty read the entire translation, correcting a number of errors and answering my queries when I got stuck; to Arlene O'Sean for copyediting; and to Barbara Beeton and Stephen Moye for TeX-nical support.

Preface to the First Edition

Fifty years ago, there was no concept of "discrete mathematics," and even today it is not a commonplace in German-speaking countries. Courses in discrete mathematics are not universally available, and certainly not with a comprehensive list of topics (in contrast, for example, to the USA, where it has long since established a permanent place for itself). Mathematicians generally understand "discrete mathematics" to mean combinatorics or graph theory, while computer scientists think of discrete structures or Boolean algebras. This book therefore has the goal of presenting all the topics necessary for further study in these areas.

Discrete mathematics deals primarily with finite sets. And what can one study in relation to finite set? First of all, one can count them, which is the classical theme of combinatorics. In Part 1 we shall learn about all the most important ideas and methods of *counting*. Depending on the problem, one is generally presented with a simple structure on a collection of finite sets in the form of relations. Graphs, which have the greatest scope of application, are an example of this. These aspects will be presented in Part 2 under the rubric of *graphs and algorithms*. Finally, there is often an algebraic structure on finite sets (or one may supply one in a natural way). *Algebraic systems* are the content of Part 3.

These three points of view form the Ariadne's thread that runs throughout this book. Another aspect that permeates our presentation has to do with the notion of optimization. The development that has completely revolutionized combinatorics over the past fifty years, bringing about the field of discrete mathematics, was the search for efficient algorithms. It no longer sufficed to solve a problem theoretically; one now wished to construct an explicit solution, and if possible using a fast algorithm. It is certainly not by chance that this optimization point of view gained ascendance at the end of the 1940s, exactly at the time of the development of the first fast computers. Therefore, we shall place great emphasis in this book on the algorithmic point of view, above all in Part 2, as announced in the title of that part. Discrete mathematics today is a basic science for information theory, and the topics covered in this book should be of equal interest to mathematicians and computer scientists.

The three parts are organized in such a way that except for Chapters 1 and 6—which deal respectively with the fundamentals of counting and graphs and should be read by all readers—they can be studied independently of one another. All the material can be covered in a two-semester course, while Chapters 1–3, 6–8, and 13 could form the content for a one-semester course. It is usual in a preface to a textbook like this for the author to point out the importance of working the exercises. In a book on discrete mathematics one cannot stress too highly the value of the exercises, which should be clear from the fact that the exercises and solutions take up almost one-fourth of the book. Discrete mathematics deals above all with concrete problems, and without practice, one will be unable to solve them despite all one's theoretical knowledge. Furthermore, the exercises often suggest questions that lead more deeply into the subject. The exercises in each chapter are divided (by a horizontal line) into two parts. Those of the first part should be relatively easy to solve, while those of the second are more difficult. Many exercises contain hints, and those indicated with ▷ have a solution at the end of the book. Each part ends with a brief bibliography with some suggestions for further study.

The only prerequisites for this book are familiarity with mathematical fundamentals and in a number of places some knowledge of linear algebra and calculus at the undergraduate level. The notation used is generally standard, with perhaps the following exceptions:

$$A = \sum A_i \qquad \text{the set } A \text{ is the } \textit{disjoint union} \text{ of the } A_i,$$

$$A = \prod A_i \qquad \text{the set } A \text{ is the } \textit{Cartesian product} \text{ of the } A_i,$$

$$\binom{A}{k} \qquad \text{the family of all } k\text{-subsets of } A.$$

The advantage of this notation is that it carries over immediately to the size of the sets:

$$\left| \sum A_i \right| = \sum |A_i|, \quad \left| \prod A_i \right| = \prod |A_i|, \quad \left| \binom{A}{k} \right| = \binom{|A|}{k}.$$

If the sets A_i are not necessarily disjoint, then we set as usual $A = \bigcup A_i$. The elements of $\prod A_i = A_1 \times \cdots \times A_n$ are as usual all n-tuples $(a_1, \ldots, a_n)$, $a_i \in A$. A k-set consists of k elements, and $\mathcal{B}(S)$ is the family of all subsets of S. The notation $\lceil x \rceil$, $\lfloor x \rfloor$ for $x \in \mathbb{R}$ means x rounded up, respectively down, to the next integer. Finally $|S|$ denotes the number of elements of the set S.

This book is the result of a course that I have given for students in mathematics and computer science. I am grateful for the collaboration (and criticism) of these students. I offer particular thanks to my colleague G. Stroth and my students T. Biedl, A. Lawrenz, and H. Mielke, who read through the entire text and made improvements in many places. My wholehearted thanks go to E. Greene, S. Hoemke, and M. Barrett for their transformation of the manuscript into LATEX, and to Vieweg-Verlag for their helpful and friendly collaboration.

Berlin, Easter 1993   Martin Aigner

Preface to the Fifth Edition

In the last few years, discrete mathematics has established itself as a basic subject in mathematics and computer science. There is a more or less agreed-upon set of standard topics, and the connections to other areas, primarily theoretical computer science, have been made to the benefit of all parties. If it is not too presumptuous of me to put a favorable interpretation on the many complimentary remarks and suggestions about this book, then it, too, since its appearance a decade ago, has contributed a little to this happy development.

The present edition represents a thorough revision and expansion. There are two new chapters: one on counting patterns with symmetries, which offers access to the most elegant theorems of combinatorics; the other resulting from the split of the chapter on codes into one on coding and the other on cryptography, not least because of the great significance of these topics in discussions both inside and outside mathematics. And finally, there are one hundred new exercises to give the reader something to think about and to encourage him or her to further study.

Like the first edition, this one was carefully composed by Margrit Barrett in LATEX. I offer her my hearty thanks, and also Christoph Eyrich, who did the the final editing, and Frau Schmickler-Hirzebruch, of Vieweg-Verlag, for her friendly collaboration.

Berlin, December 2003   Martin Aigner

Part 1

Counting

Discrete mathematics is the study of finite sets, and we would like to begin by asking how many elements there are in a set described by certain conditions. For example, we may ask how many pairs of distinct integers can be formed from the elements of the set $\{1, 2, 3, 4\}$. The answer is 6, as is easily obtained. However, this answer is of little interest, since it tells us nothing about the number of pairs in the sets $\{1, 2, \ldots, 6\}$ and $\{1, 2, \ldots, 1000\}$. The question becomes interesting only when we ask for the number of pairs in the set $\{1, \ldots, n\}$ for *arbitrary* n.

A typical discrete counting problem is the following: Given an *infinite* family of *finite* sets S_n (where n runs through the index set I, for example the natural numbers), determine the *counting function* $f : I \to \mathbb{N}_0$, $f(n) = |S_n|$, $n \in I$ (where the notation $|A|$ means the number of elements in the set A). Usually the sets S_n are described by simple combinatorial conditions.

But first we must consider a more philosophical problem: What is meant by a "determination" of f? The most satisfying result is of course a *closed formula*. For example, if S_n is the set of permutations of an n-set (a set of n elements), then we have $f(n) = n!$, and such a formula will always be accepted as an adequate determination. However, in most cases such a formula is unobtainable. What is one to do in such cases?

Summation

Suppose we don't wish to count all the permutations of $\{1, \ldots, n\}$, but only the fixed-point-free ones, that is, those for which for all i, the index i never appears at the ith place. Let D_n denote the number of such permutations. For example, 231, 312 are the only fixed-point-free permutations for $n = 3$, and so $D_3 = 2$. We shall prove later that

$$D_n = n! \sum_{k=0}^{n} \frac{(-1)^k}{k!}$$

for all n. In this case, what we have obtained is a summation formula.

Recurrence

From combinatorial considerations there follows, as we shall see later, the relationship $D_n = (n-1)(D_{n-1} + D_{n-2})$ for $n \geq 3$. The general formula then follows from the initial values $D_1 = 0$, $D_2 = 1$. For example, we obtain $D_3 = 2$, $D_4 = 9$, $D_5 = 44$. Sometimes, such a *recurrence formula*, expressing a function for one value of n in terms of previous values, is to be preferred to a closed formula. Thus the *Fibonacci numbers* F_n are defined by $F_0 = 0$, $F_1 = 1$, $F_n = F_{n-1} + F_{n-2}$, $n \geq 2$. Later, we shall derive from this the formula

$$F_n = \frac{1}{\sqrt{5}} \left(\left(\frac{1 + \sqrt{5}}{2} \right)^n - \left(\frac{1 - \sqrt{5}}{2} \right)^n \right),$$

but most people (and most computers, due to the irrationality of $\sqrt{5}$) will prefer the recurrence formula.

Generating Functions

A method that has proved extremely fruitful consists in expressing the values $f(n)$ of a counting function as *coefficients* of a power series

$$F(z) = \sum_{n \geq 0} f(n) z^n \,.$$

Here $F(z)$ is called the *generating function* of the counting function f. For example, we might ask about the number of n-subsets of an r-set for fixed r, in which case $f(n) = \binom{r}{n}$ (the binomial coefficient), and we know from the binomial theorem that

$$\sum_{n \geq 0} \binom{r}{n} z^n = (1 + z)^r.$$

We shall see how surprisingly simple it is to derive identities involving binomial coefficients.

Asymptotic Analysis

In later chapters we shall study algorithms for solving a variety of problems. In addition to proving an algorithm correct, we will be interested in how fast it is; thus we ask about the *run time* of the algorithm. Frequently an algorithm is expressed as a recurrence. In sorting problems, for example, we shall meet the recurrence relation

$$f(n) = \frac{2}{n} \sum_{k=0}^{n-1} f(k) + an + b$$

with $a > 0$. In this case it is easy to obtain a solution, but in general, a determination of $f(n)$ can be extremely difficult. We shall then attempt to estimate $f(n)$ by more tractable functions $a(n)$ and $b(n)$ with $a(n) \leq f(n) \leq b(n)$ and be satisfied if we can solve the problem *asymptotically*, that is, if we can find a familiar function $g(n)$ (for example a polynomial or exponential function) that has the same *order of magnitude* as $f(n)$.

Chapter 1

Fundamentals

1.1. Elementary Counting Principles

We would like to present here some fundamental rules on which all counting is based. The first two rules (which are so obvious that they don't even have to be proved) are based on a classification of the elements of the set to be counted.

Summation Rule. *Let $S = \sum_{i=1}^{t} S_i$ be a disjoint union of sets. Then $|S| = \sum_{i=1}^{t} |S_i|$.*

In applications, this summation rule generally appears in the following form: We classify the elements of S according to properties E_i $(i = 1, \ldots, t)$, which are mutually exclusive, and set $S_i = \{\, x \in S : x \text{ has property } E_i \,\}$.

This summation rule forms the basis of most recurrence formulas. Let us consider the following example: For a set X of size n (an n-set), let $S = \binom{X}{k}$ denote the set of all k-subsets of X, that is, $|S| = \binom{n}{k}$. Let $a \in X$. We classify the k-subsets A according to whether $a \in A$ or $a \notin A$, setting $S_1 = \{A \in S : a \in A\}$, $S_2 = \{A \in S : a \notin A\}$. We obtain the sets in S_1 by combining all $(k-1)$-subsets of $X \setminus \{a\}$ with a, that is, $|S_1| = \binom{n-1}{k-1}$, and all sets of S_2 by taking all k-subsets of $X \setminus \{a\}$, that is, $|S_2| = \binom{n-1}{k}$. According to the summation rule, we obtain the fundamental recurrence formula for binomial coefficients:

$$\binom{n}{k} = \binom{n-1}{k-1} + \binom{n-1}{k} \quad (n \geq 1).$$

In Section 1.4 we will discuss the binomial coefficients in considerable detail.

Product Rule. *Let $S = S_1 \times S_2 \times \cdots \times S_t$ be a direct product of sets. Then $|S| = \prod_{i=1}^{t} |S_i|$.*

Suppose there are three air routes connecting Kabul and Damascus, and five routes connecting Damascus and Moscow. Then there are $15 = 3 \cdot 5$ routes from Kabul to Moscow via Damascus.

It is often useful to clarify the product rule using a *tree diagram*. Let a, b, c denote the routes from Kabul to Damascus, and $1, 2, 3, 4, 5$ the routes from Damascus to Moscow. Then the following diagram shows the 15 routes from Kabul to Moscow:

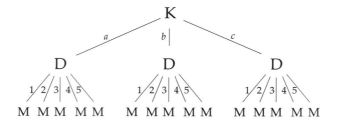

We shall call a sequence of zeros and ones a $(0,1)$-**word**, and the number of zeros and ones will be called the *length* of the word. How many different $(0,1)$-words of length n are there? For each place in the word there are two possibilities, and thus, according to the product rule, the answer is 2^n.

The next two rules compare two sets.

Rule of Equality. *If there is a bijection between two sets S and T, then $|S| = |T|$.*

A typical application of the rule of equality is this: We would like to count the elements of a set S. If we are able to construct a bijection between S and a set T, whose size we know, then we may conclude that $|S| = |T|$.

How many different subsets does an n-set X possess, for example the set $X = \{1, \ldots, n\}$? For each subset A we consider the **characteristic vector** $w(A) = a_1 a_2 \ldots a_n$ of A with $a_i = 1$ if $i \in A$, and $a_i = 0$ if $i \notin A$. Every vector $w(A)$ is therefore a $(0,1)$-word of length n, and one sees at once that the mapping w is a bijection between the set S of all subsets of $\{1, \ldots, n\}$ and the set T of all $(0,1)$-words of length n. But we already know the size of T, namely, $|T| = 2^n$, and thus from the rule of equality we conclude that $|S| = 2^n$ as well.

For our last rule we require a couple of definitions. An **incidence system** (S, T, I) consists of two sets S and T and a relation I (called the incidence) between the elements of S and T. If there is a relation $a \, I \, b$ between $a \in S$ and $b \in T$, then we call a and b *incident*. Otherwise, we say that they are nonincident. A well-known example can be taken from geometry: Let S be a set of points, T a set of lines, and $p \, I \, g$ means that the point p lies on the line g.

Rule of Double Counting. *Let (S, T, I) be an incidence system, and for $a \in S$ let $r(a)$ denote the number of elements of T incident to a, and analogously, $r(b)$ for $b \in T$ is the number of elements in S incident to b. Then the following relation holds:*

$$\sum_{a \in S} r(a) = \sum_{b \in T} r(b).$$

It becomes clear at once that this rule is valid if we consider the incidence system as a rectangular array. We number the elements of S and T thus: $S = \{a_1, \ldots, a_m\}$, $T = \{b_1, \ldots, b_n\}$. We now create an $m \times n$ matrix $M = (m_{ij})$, called the **incidence matrix**, by setting

$$m_{ij} = \begin{cases} 1 & \text{if } a_i \, I \, b_j, \\ 0 & \text{otherwise.} \end{cases}$$

The value of $r(a_i)$ is then precisely the number of ones in the ith row, and analogously, $r(b_j)$ is the number of ones in the jth column. The sum $\sum_{i=1}^{m} r(a_i)$ is then equal to the total number of ones (counted rowwise), while $\sum_{j=1}^{n} r(b_j)$ gives the same number (counted columnwise).

Example. As an example, let us consider the numbers from 1 to 8, with $S = T = \{1, \ldots, 8\}$, and say that $i \in S$, $j \in T$ are incident if i is a divisor of j, denoted by $i \mid j$. The associated incidence matrix then has the following form, where for the sake of clarity we have printed only the ones:

	1	2	3	4	5	6	7	8
1	1	1	1	1	1	1	1	1
2		1		1		1		1
3			1			1		
4				1				1
5					1			
6						1		
7							1	
8								1

The number of ones in column j is equal to the number of divisors of j, which we denote by $t(j)$, and so, for example, $t(6) = 4$, $t(7) = 2$. We now pose the question, how many divisors does a number between 1 and 8 have on *average*? That is, we would like to compute $\bar{t}(8) = \frac{1}{8}\sum_{j=1}^{8} t(j)$. In our example, $\bar{t}(8) = \frac{5}{2}$. From the table we obtain the following values:

n	1	2	3	4	5	6	7	8
$\bar{t}(n)$	1	$\frac{3}{2}$	$\frac{5}{3}$	2	2	$\frac{7}{3}$	$\frac{16}{7}$	$\frac{5}{2}$

How big, then, is $\bar{t}(n)$ for arbitrary n? At first glance, the prospects of answering this question seem hopeless. For prime numbers p, we have $t(p) = 2$, while for a power of two, the value $t(2^k) = k+1$ results. Nonetheless, let us attempt to apply the rule of double counting. Counting by columns, we obtain, as we have seen, $\sum_{j=1}^{n} t(j)$. How many ones are there in the ith row? Clearly, the ones represent the multiples of i, namely $1 \cdot i, 2 \cdot i, \ldots,$ and the last multiple less than or equal to n is $\lfloor \frac{n}{i} \rfloor \cdot i$, and so $r(i) = \lfloor \frac{n}{i} \rfloor$. Our rule therefore yields

$$\bar{t}(n) = \frac{1}{n} \sum_{j=1}^{n} t(j) = \frac{1}{n} \sum_{i=1}^{n} \left\lfloor \frac{n}{i} \right\rfloor \sim \frac{1}{n} \sum_{i=1}^{n} \frac{n}{i} = \sum_{i=1}^{n} \frac{1}{i},$$

where the error in passing from $\lfloor \frac{n}{i} \rfloor$ to $\frac{n}{i}$ for all i is less than 1, and is so in the sum as well. We shall often encounter the last value, $\sum_{i=1}^{n} \frac{1}{i}$. It is called the nth **harmonic number** H_n. From calculus we know that $H_n \sim \log n$ is about the size of the natural logarithm, and we thus obtain the astounding result that the divisor function, despite its irregularity, behaves completely regularly *on average*, namely, $\bar{t}(n) \sim \log n$.

1.2. The Fundamental Counting Coefficients

Some numbers, such as the binomial coefficients $\binom{n}{k}$, appear again and again. We would like to discuss these most important numbers in this section. By the end of the section it should have become clear why it is precisely these numbers that appear in so many counting problems.

The first concepts that we associate with sets are those of **subsets** and **partitions**. Let N be an n-set. Then as we know, $\binom{n}{k}$ denotes the number of k-subsets of N. The numbers $\binom{n}{k}$ are known as the **binomial coefficients**. Let us now look at all **set partitions** of n into k disjoint nonempty blocks, called k-partitions, and we will denote the number of such blocks by $S_{n,k}$. The numbers $S_{n,k}$ are called the **Stirling numbers of the second kind** (why of the second kind has a historical reason that will soon be made clear).

For example, $N = \{1, 2, 3, 4, 5\}$ possesses the following 2-partitions, where we have omitted the braces and commas:

$$
\begin{aligned}
12345 \quad = \quad & 1234+5, \quad 1235+4, \quad 1245+3, \\
& 1345+2, \quad 2345+1, \quad 123+45, \\
& 124+35, \quad 125+34, \quad 134+25, \\
& 135+24, \quad 145+23, \quad 234+15, \\
& 235+14, \quad 245+13, \quad 345+12.
\end{aligned}
$$

Therefore, $S_{5,2} = 15$.

In analogy to set partitions, we can also define **integer partitions**. Let $n \in \mathbb{N}$. Then $n = n_1 + n_2 + \cdots + n_k$, $n_i \geq 1$ $(\forall i)$, is called a k-partition of n. The number of them is denoted by $P_{n,k}$. Since the order of the n_i is arbitrary, we may assume that $n_1 \geq n_2 \geq \cdots \geq n_k$.

For $n = 8$ we obtain the following 4-partitions:

$$8 = 5+1+1+1, \; 4+2+1+1, \; 3+3+1+1, \; 3+2+2+1, \; 2+2+2+2.$$

Therefore, $P_{8,4} = 5$.

We have just observed that the order of summation in an integer partition is arbitrary. Therefore, we may speak of *unordered* integer partitions. Similarly, in enumerating subsets or set partitions, the order is arbitrary. We would now like to consider how many *ordered* k-subsets or k-partitions there are, where by ordered, we mean that the ordered subset $\{1, 2, 3\}$ is different from $\{3, 1, 2\}$ and from $\{3, 2, 1\}$, even though as ordinary sets they are equal. Likewise, the ordered set partitions $123 + 45$ and $45 + 123$ are different from each other, as are the integer partitions $3 + 3 + 1 + 1$ and $3 + 1 + 3 + 1$.

To compute the corresponding coefficients, we need the notion of permutations. Let N be an n-set, e.g., $N = \{1, 2, \ldots, n\}$. We consider words of length k all of whose entries, elements of N, are different. We call them k-**permutations** of N. For example, 1234 and 5612 are two 4-permutations of $N = \{1, 2, \ldots, 6\}$. How many such k-permutations are there? For the first entry we have n choices. Once the first element has been chosen, we have $n - 1$ choices for the second position. Once this element has been chosen, there remain $n - 2$ choices for the third place, and so on. The product rule then yields the following statement:

The number of k-permutations of an n-set $= n(n-1)\cdots(n-k+1)$.

In particular, for $k = n$, it follows that the number of n-permutations, that is, the number of permutations of the set N, is equal to $n! = n(n-1)\cdots 2 \cdot 1$.

The values $n(n - 1) \cdots (n - k + 1)$ appear so frequently in counting problems that they have acquired their own name:

$$n^{\underline{k}} := n(n-1)\cdots(n-k+1)$$

are called **falling factorials** (of n of length k). Analogously, we set

$$n^{\overline{k}} := n(n+1)\cdots(n+k-1)$$

and call $n^{\overline{k}}$ **rising factorials**.

And now back to our problem of counting ordered objects. For subsets and set partitions this is quite easy. Every k-subset results in $k!$ ordered k-subsets, and every k-set partition yields $k!$ ordered k-set partitions, since the distinct elements, respectively blocks, can be permuted in $k!$ ways. We thus obtain for these numbers

$$k! \binom{n}{k} \quad \text{and} \quad k! \, S_{n,k} \, .$$

It is now clear that the ordered k-subsets are nothing other than the k-permutations of N, and we therefore obtain for $\binom{n}{k}$ the usual formula

$$\binom{n}{k} = \frac{n(n-1)\cdots(n-k+1)}{k!} = \frac{n^{\underline{k}}}{k!} \, .$$

The counting of ordered integer partitions is a bit subtler, since the summands do not have to be *distinct*, and so some permutations will result in the same ordered partition. For example, from $3+1+1$ there are not $6 = 3!$ distinct ordered partitions, but only three, namely $3+1+1$, $1+3+1$, $1+1+3$. The following formula is a nice illustration of the rule of equality:

The number of ordered k-partitions of n is $\left(\begin{array}{c} n-1 \\ k-1 \end{array}\right).$

To prove this, we construct a bijection from the set S of all ordered k-partitions of n to the set T of all $(k-1)$-subsets in $\{1, 2, \ldots, n-1\}$. Let $n = n_1 + n_2 + \cdots + n_k \in S$. Then we define $f : S \to T$ by $f(n_1 + \cdots + n_k) = \{n_1, n_1 + n_2, \ldots, n_1 + \cdots + n_{k-1}\}$. Since $n_i \geq 1$, it follows that $1 \leq n_1 < n_1 + n_2 < \cdots < n_1 + \cdots + n_{k-1} \leq n - 1$; that is, $f(n_1 + \cdots + n_k) \in T$. The inverse map is $g(\{a_1 < a_2 < \cdots < a_{k-1}\}) = a_1 + (a_2 - a_1) + \cdots + (a_{k-1} - a_{k-2}) + (n - a_{k-1})$, and f, g are clearly inverses of each other. The result then follows from the rule of equality.

As an example, for $n = 6$, $k = 3$ we obtain the following $\binom{5}{2} = 10$ ordered 3-partitions of 6:

$$4+1+1, \ 1+4+1, \ 1+1+4, \ 3+2+1, \ 3+1+2, \ 2+3+1,$$
$$2+1+3, 1+3+2, \ 1+2+3, \ 2+2+2 \, .$$

Finally, we wish to introduce the notion of a **multiset**. Whenever we speak of a set, we always assume that all of its elements are *distinct*. In a multiset, we remove this restriction. Thus $M = \{1, 1, 2, 2, 3\}$ is a multiset over $\{1, 2, 3\}$, where 1 and 2 appear with *multiplicity* two, 3 with multiplicity 1. The size of a multiset is the total number of elements, counted with their multiplicities. In our example, $|M| = 5$. The following formula shows the significance of the rising factorials:

The number of k-multisets over an n-set $= \dfrac{n(n+1)\cdots(n+k-1)}{k!} = \dfrac{n^{\overline{k}}}{k!} \, .$

Again, the proof is given by the rule of equality. Let S be the set of all k-multisets over $\{1, 2, \ldots, n\}$, and T the set of all k-subsets of $\{1, 2, \ldots, n + k - 1\}$. Thus $|T| = \binom{n+k-1}{k} = \frac{n^{\overline{k}}}{k!}$. For $\{a_1 \leq a_2 \leq \cdots \leq a_k\} \in S$, we set $f(\{a_1 \leq \cdots \leq a_k\}) = \{a_1, a_2 + 1, a_3 + 2, \ldots, a_k + (k-1)\}$. We then have $1 \leq a_1 < a_2 + 1 < \cdots < a_k + (k-1) \leq n + (k-1)$, and so $f(\{a_1 \leq \cdots \leq a_k\}) \in T$. The inverse mapping is $g(\{b_1 < \cdots < b_k\}) = \{b_1 \leq b_2 - 1 \leq \cdots \leq b_k - (k-1)\}$, and the proof is complete.

Our fundamental counting coefficients appear quite naturally in counting mappings. Let us look at the mappings $f : N \to R$, where $|N| = n$, $|R| = r$. The *total number* of mappings is r^n, since for each element we have r possible images, so that with the product rule we obtain r^n. Likewise, the product rule yields $r(r-1)\cdots(r-n+1)$ for the number of *injective* mappings. And how about the *surjective* mappings? Each mapping f can be described by the set of preimages $\{f^{-1}(y) : y \in R\}$. For example, the mapping f given by

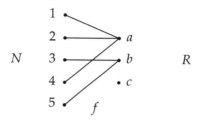

has the preimages $f^{-1}(a) = \{1, 2, 4\}$, $f^{-1}(b) = \{3, 5\}$, $f^{-1}(c) = \varnothing$. If in particular f is surjective, then the preimages form an *ordered r-partition* of N, and conversely, each such partition yields precisely one surjective mapping. To sum up, we have

$$|\text{map } (N, R)| = r^n,$$
$$|\text{inj } (N, R)| = r^{\underline{n}},$$
$$|\text{surj } (N, R)| = r! \, S_{n,r} \, .$$

Each mapping $f : N \to R$ has a *unique image* $A \subseteq R$, $A = \{f(x) : x \in N\}$, and f is surjective from N onto A. We therefore classify mappings according to their images, and the summation rule yields

$$r^n = |\text{map } (N, R)| = \sum_{A \subseteq R} |\text{surj } (N, A)| = \sum_{k=0}^{r} \sum_{|A|=k} |\text{surj } (N, A)|$$
$$= \sum_{k=0}^{r} \binom{r}{k} k! \, S_{n,k} = \sum_{k=0}^{r} S_{n,k} r^{\underline{k}},$$

and we obtain a formula that links the powers, falling factorials, and Stirling numbers:

(1.1)
$$r^n = \sum_{k=0}^{n} S_{n,k} r^{\underline{k}} \, .$$

We may terminate the summation at n, since clearly there are no k-partitions of N for $k > n$; that is, $S_{n,k} = 0$ holds for $k > n$.

It will help in remembering the counting coefficients if we think of the set N as a set of *balls*, R as a set of *boxes*, and a mapping $f : N \to R$ as *distributing* the balls among the boxes. The mapping is then injective if each box gets at most one ball, surjective if each box gets at least one ball. We assume that the balls are indistinguishable, but not the boxes. In that case, how many such distributions are there? In the injective case, we always choose n of the r boxes that contain a ball (which ball is irrelevant, since they are identical), and thus obtain exactly the n-subsets of R, of which there are $\binom{r}{n}$. If we permit arbitrary distributions, then we get precisely the n-multisets of R, of which there are $\frac{r^{\overline{n}}}{n!}$, as we have already calculated. And how about the surjective case? We know these distributions as well. Box i contains $n_i \geq 1$ balls, and thus altogether, $n = n_1 + \cdots + n_r$ is an ordered partition n, and the number of these is $\binom{n-1}{r-1}$.

If we now combine all cases according to whether balls and boxes are distinguishable, we obtain the following diagram, which provides an overview of all of our fundamental coefficients:

$|N| = n,\ |R| = r$	arbitrary	injective	surjective	bijective
N distinguishable R distinguishable	r^n	$r^{\underline{n}}$	$r!\,S_{n,r}$	$r! = n!$
N indistinguishable R distinguishable	$\dfrac{r^{\overline{n}}}{n!}$	$\dfrac{r^{\underline{n}}}{n!} = \binom{r}{n}$	$\binom{n-1}{r-1}$	1
N distinguishable R indistinguishable	$\displaystyle\sum_{k=1}^{r} S_{n,k}$	0 or 1	$S_{n,r}$	1
N indistinguishable R indistinguishable	$\displaystyle\sum_{k=1}^{r} P_{n,k}$	0 or 1	$P_{n,r}$	1

1.3. Permutations

Permutations of a set, such as the set $N = \{1, 2, \ldots, n\}$, can be represented in a number of ways. We may first consider a permutation π to be simply a bijective mapping $\pi = \left(\begin{smallmatrix} 1 & 2 & \cdots & n \\ \pi(1) & \pi(2) & \cdots & \pi(n) \end{smallmatrix}\right)$. If we fix the order of the domain set to be $1, 2, \ldots, n$, we can then write π uniquely as the *word* $\pi = \pi(1)\,\pi(2)\,\ldots\,\pi(n)$.

Every permutation π is equivalent to a set of *cycles*. For example, let

$$\pi = \begin{pmatrix} 1\,2\,3\,4\,5\,6\,7\,8\,9 \\ 5\,8\,3\,1\,9\,7\,6\,2\,4 \end{pmatrix}.$$

Then 1 is mapped to 5, 5 to 9, 9 to 4, and 4 to 1. The elements $(1,5,9,4)$ form a cycle. We see an analogous result for the remaining elements, and we thereby obtain the **cycle representation** of π, namely,

$$\pi = (1,5,9,4)(2,8)(3)(6,7).$$

The number of elements in a cycle is the *length* of the cycle. Cycles of length 1 are called **fixed points**. We make two observations: First, in a cycle representation, the *order* of the cycles is irrelevant. In our example we could as well write $\pi = (6,7)(1,5,9,4)(3)(2,8)$. It is still the same permutation. Second, we can represent an individual cycle beginning with an arbitrary element, after which the order of the remaining elements is fixed. For example, $(7,6)(9,4,1,5)(8,2)(3)$ is also a cycle representation of π.

For example, for $n = 3$ we obtain the six permutations written as words

123 132 213 231 312 321,

and in cycle representation

$$(1)(2)(3)\quad (1)(2,3)\quad (1,2)(3)\quad (1,2,3)\quad (1,3,2)\quad (1,3)(2).$$

In particular, the cycle representation of π yields a partition of N with the cycles as blocks. In analogy to sets, we define $s_{n,k}$ to be the number of permutations of $\{1,\ldots,n\}$ with k cycles, and we call $s_{n,k}$ the **Stirling numbers of the first kind**. As an example, we have $s_{n,1} = (n-1)!$, since in a cycle of length n we can take 1 as the initial element, and then permute the remaining elements arbitrarily. Another example is $s_{n,n-1} = \binom{n}{2}$, since a permutation with $n-1$ cycles consists of $n-2$ fixed points and one two-cycle, which clearly we may choose in $\binom{n}{2}$ ways. Clearly, from the definition we have

$$n! = \sum_{k=1}^{n} s_{n,k} \quad (n \geq 1).$$

For a permutation π we let $b_i(\pi)$ denote the number of cycles of length i, $i = 1,\ldots,n$, and $b(\pi)$ the total number of cycles. Thus

$$n = \sum_{i=1}^{n} i\, b_i(\pi), \qquad b(\pi) = \sum_{i=1}^{n} b_i(\pi).$$

The **type** of permutation π is the formal expression $t(\pi) = 1^{b_1(\pi)} \ldots n^{b_n(\pi)}$. In our example above, we have $t(\pi) = 1^1 2^2 4^1$. (We omit the numbers i with $b_i(\pi) = 0$.)

We observe at once that there are precisely as many possible types of permutation as there are integer partitions of n. For $n = 5$, for example, we obtain

Partition	Type
5	5^1
$4 + 1$	$1^1 4^1$
$3 + 2$	$2^1 3^1$
$3 + 1 + 1$	$1^2 3^1$
$2 + 2 + 1$	$1^1 2^2$
$2 + 1 + 1 + 1$	$1^3 2^1$
$1 + 1 + 1 + 1 + 1$	1^5

How many permutations are there of a given type $1^{b_1} 2^{b_2} \ldots n^{b_n}$? For the time being, we write empty cycles as

$$\underbrace{(.) \cdots (.)}_{b_1} \underbrace{(..) \cdots (..)}_{b_2} \underbrace{(\cdots) \cdots (\cdots)}_{b_3} \cdots$$

and fill in the places in the row with the $n!$ words. In this way we certainly obtain the permutations of the given type. In general, however, we will produce the same permutation multiple times. Since the *order* of cycles is irrelevant, we may permute the b_i cycles of length i as a unit, which yields a multiplicity of $b_1! b_2! \cdots b_n!$. Finally, we may fix the start element of a cycle, and so we obtain that cycle with multiplicity $1^{b_1} 2^{b_2} \cdots n^{b_n}$. The result: Let $\sum_{i=1}^n i b_i = n$:

The number of permutations of type $1^{b_1} 2^{b_2} \ldots n^{b_n}$ is $\dfrac{n!}{b_1! \cdots b_n! 1^{b_1} 2^{b_2} \cdots n^{b_n}}.$

In particular, we obtain

$$s_{n,k} = \sum_{(b_1, \ldots, b_n)} \frac{n!}{b_1! \cdots b_n! 1^{b_1} \cdots n^{b_n}} \quad \text{with} \quad \sum_{i=1}^n i b_i = n, \quad \sum_{i=1}^n b_i = k,$$

$$n! = \sum_{(b_1, \ldots, b_n)} \frac{n!}{b_1! \cdots b_n! 1^{b_1} \cdots n^{b_n}} \quad \text{with} \quad \sum_{i=1}^n i b_i = n.$$

For $n = 5$ we may now complete our list:

Number of Permutations	Stirling Numbers
24	$s_{5,1} = 24$
$\left.\begin{array}{c} 30 \\ 20 \end{array}\right\}$	$s_{5,2} = 50$
$\left.\begin{array}{c} 20 \\ 15 \end{array}\right\}$	$s_{5,3} = 35$
10	$s_{5,4} = 10$
1	$s_{5,5} = \ 1$

$$120 = 5!$$

We shall encounter permutations frequently, especially in sorting problems. If we consider a permutation $a_1, a_2, \ldots, a_n$ of $\{1, \ldots, n\}$ as a list, then we would like to bring the list back to its original ordering $1, 2, \ldots, n$ with the fewest possible number of exchanges. The exercises offer a first look at some of the problems that arise.

1.4. Recurrence Equations

For the binomial coefficients we have already obtained a satisfactory closed formula, $\binom{n}{k} = \frac{n(n-1)\cdots(n-k+1)}{k!}$, while for the Stirling numbers of the first kind $s_{n,k}$ we have a somewhat unwieldy summation formula (which causes difficulties due to the unknown number $P_{n,k}$ of summands).

For the numbers $S_{n,k}$, we currently have only the definition. Recurrence is going to give us a hand.

Binomial coefficients. We have

$$(1.2) \qquad \binom{n}{k} = \frac{n(n-1)\cdots(n-k+1)}{k!} = \frac{n^{\underline{k}}}{k!} \qquad (n \geq k \geq 0),$$

$$(1.3) \qquad \binom{n}{k} = \frac{n!}{k!(n-k)!} \qquad (n \geq k \geq 0),$$

and thus in particular,

$$(1.4) \qquad \binom{n}{k} = \binom{n}{n-k} \qquad (n \geq k \geq 0).$$

It is useful to define $\binom{n}{k}$ for negative numbers, indeed for arbitrary complex numbers n and for arbitrary integers k. We begin by setting $\binom{0}{0} = 1$. This makes sense because the empty set $\varnothing$ has precisely one 0-subset, namely $\varnothing$. We likewise set $n^{\underline{0}} = n^{\overline{0}} = 1$ for the falling and rising factorials, and $0! = 1$.

The expression $r^{\underline{k}} = r(r-1)\cdots(r-k+1)$ or $r^{\overline{k}} = r(r+1)\cdots(r+k-1)$ makes sense for arbitrary $r \in \mathbb{C}$; for example, $(-\frac{1}{2})^{\underline{3}} = (-\frac{1}{2})(-\frac{3}{2})(-\frac{5}{2}) =$

$-\frac{15}{8}$, $(-2)^{\overline{2}} = (-2)(-1) = 2$. For $k!$, however, we must assume $k \geq 0$, since the factorial function cannot be defined for $k < 0$ without some additional work. We therefore give the general definition for $r \in \mathbb{C}$:

$$(1.5) \qquad \binom{r}{k} = \begin{cases} \frac{r(r-1)\cdots(r-k+1)}{k!} = \frac{r^{\underline{k}}}{k!} & (k \geq 0), \\ 0 & (k < 0). \end{cases}$$

Recurrence. We have the formula

$$(1.6) \qquad \binom{r}{k} = \binom{r-1}{k-1} + \binom{r-1}{k} \qquad (r \in \mathbb{C}, k \in \mathbb{Z}).$$

This formula comes directly from (1.5). We give a second proof, which demonstrates the important "polynomial method." For $k < 0$, both sides of (1.6) are equal to 0, and for $k = 0$, both sides are equal to 1. Therefore, let $k \geq 1$. We know already that (1.6) is valid for all *natural* numbers r. If we replace r by a variable x, we obtain

$$\binom{x}{k} \overset{?}{=} \binom{x-1}{k-1} + \binom{x-1}{k}.$$

On both sides are polynomials in x over $\mathbb{C}$ of degree k, and we know that these two polynomials assume the same value for all natural numbers. We now invoke a theorem from algebra that polynomials of degree k that agree at $k+1$ points are identical. In this case they agree at infinitely many points, and so we indeed have the *polynomial equation*

$$(1.7) \qquad \binom{x}{k} = \binom{x-1}{k-1} + \binom{x-1}{k} \qquad (k \geq 1),$$

and therefore (1.6) is valid for all $x = r \in \mathbb{C}$.

We shall again call the polynomials $x^{\underline{k}} = x(x-1)\cdots(x-k+1)$ and $x^{\overline{k}} = x(x+1)\cdots(x+k-1)$ with $x^{\underline{0}} = x^{\overline{0}} = 1$ **falling** and **rising factorials**. Furthermore, we can also conclude (1.7) from the obviously valid equation $x^{\underline{k}} = x(x-1)^{\underline{k-1}} = (k+(x-k))(x-1)^{\underline{k-1}} = k(x-1)^{\underline{k-1}} + (x-1)^{\underline{k-1}}(x-k) = k(x-1)^{\underline{k-1}} + (x-1)^{\underline{k}}$ via division by $k!$.

The recurrence equation (1.6) yields, for $n, k \in \mathbb{N}_0$, **Pascal's triangle**

n \ k	0	1	2	3	4	5	6	7	$\cdots$
0	1								
1	1	1							
2	1	2	1						
3	1	3	3	1					
4	1	4	6	4	1				
5	1	5	10	10	5	1			
6	1	6	15	20	15	6	1		
7	1	7	21	35	35	21	7	1	

$$\binom{n}{k}$$

where the blank places in the table represent the value 0, since $\binom{n}{k} = 0$ for $n < k$. Discussion of the secrets and attractions of Pascal's triangle fills many a volume. We shall record here only three formulas. First is the row sum with index n, $\sum_{k=0}^{n} \binom{n}{k} = 2^n$, since we thus count precisely the subsets of an n-set. Let us now consider a column sum with index k up to row n, that is, $\sum_{m=0}^{n} \binom{m}{k}$. For $k = 2$, $n = 6$, we obtain $35 = \binom{7}{3}$, and for $k = 1, n = 5$, we have $15 = \binom{6}{2}$. In general,

$$(1.8) \qquad \sum_{m=0}^{n} \binom{m}{k} = \binom{n+1}{k+1} \qquad (n, k \geq 0).$$

For $n = 0$ this is certainly true, and by induction we obtain from (1.6)

$$\sum_{m=0}^{n+1} \binom{m}{k} = \sum_{m=0}^{n} \binom{m}{k} + \binom{n+1}{k} = \binom{n+1}{k+1} + \binom{n+1}{k} = \binom{n+2}{k+1}.$$

Finally, let us consider the diagonals from upper left to lower right, that is, the expression $\sum_{k=0}^{n} \binom{m+k}{k}$, where m denotes the starting row and n the ending row. In the triangle, the diagonal with $m = 3$, $n = 3$ is highlighted, and the sum is $35 = \binom{7}{3}$:

$$(1.9) \qquad \sum_{k=0}^{n} \binom{m+k}{k} = \binom{m+n+1}{n} \qquad (m, n \geq 0).$$

The proof is again carried out by induction. Moreover, (1.9) holds for all $m \in \mathbb{C}$.

Negation. We have the formula

$$(1.10) \qquad \binom{-r}{k} = (-1)^k \binom{r+k-1}{k} \qquad (r \in \mathbb{C}, \ k \in \mathbb{Z}).$$

We have as well

$$(-x)^{\underline{k}} = (-x)(-x-1)\cdots(-x-k+1) = (-1)^k \, x(x+1)\cdots(x+k-1),$$

and thus the general polynomial equation

(1.11) $$(-x)^{\underline{k}} = (-1)^k \, x^{\overline{k}} \, .$$

Division by $k!$ immediately yields (1.10). The formula of (1.11) is called the *reciprocity law* between the rising and falling factorials.

From (1.10) we may immediately derive a further astounding property of Pascal's triangle. Let us consider the *alternating sums* of a row, say of the seventh. We obtain 1, $1 - 7 = -6$, $1 - 7 + 21 = 15$, -20, 15, -6, 1, 0, and thus precisely the numbers in the row above, but with alternating signs. In fact, from (1.10) and (1.9) we see that

$$\sum_{k=0}^{m}(-1)^k \binom{n}{k} = \sum_{k=0}^{m}\binom{k - n - 1}{k} = \binom{m - n}{m} = (-1)^m \binom{n-1}{m} \, .$$

Binomial Theorem. By multiplying out the product on the left, we obtain

$$(x + y)^n = \sum_{k=0}^{n}\binom{n}{k} x^k \, y^{n-k} \qquad (n \geq 0) \, .$$

In particular, for $y = 1$, we have

$$(x + 1)^n = \sum_{k=0}^{n}\binom{n}{k} x^k \, .$$

If we set $x = 1$ and $x = -1$ in turn, we obtain the familiar formulas

(1.12) $$2^n = \sum_{k=0}^{n}\binom{n}{k} \quad \text{and} \quad 0 = \sum_{k=0}^{n}(-1)^k\binom{n}{k} \quad (n \geq 1) \, .$$

Finally, we would like to derive one of the most important formulas of all.

Vandermonde's identity. We have the identity

(1.13) $$\binom{x + y}{n} = \sum_{k=0}^{n}\binom{x}{k}\binom{y}{n - k} \qquad (n \geq 0) \, .$$

We prove the equation for natural numbers $x = r$, $y = s$. The rest follows from our polynomial method. Let R and S be disjoint sets with $|R| = r$, $|S| = s$. On the left we have $\binom{r+s}{n}$, that is, the number of all n-subsets of $R + S$. We now classify these subsets A according to their average $|A \cap R| = k$, $k = 0, \ldots, n$. If $|A \cap R| = k$, then we must have $|A \cap S| = n - k$; that is, there are exactly $\binom{r}{k}\binom{s}{n-k}$ k-subsets with $|A \cap R| = k$ (product rule). Applying the summation rule yields the result.

Stirling numbers. We first consider Stirling numbers of the second kind, $S_{n,k}$. As with binomial coefficients, we have $S_{n,k} = 0$ for $n < k$, since an

n-set permits at most an n-partition. We also set $S_{0,0} = 1$ and $S_{0,k} = 0$ for $k > 0$, and $S_{n,0} = 0$ for $n > 0$. We have the following recurrence equation:

$$(1.14) \qquad S_{n,k} = S_{n-1,k-1} + k\,S_{n-1,k} \qquad (n, k > 0)\,.$$

To prove this, we naturally use the summation rule. Let N be an n-set. We classify the k-partitions based on a fixed element $a \in N$. If $\{a\}$ forms a block by itself, then the remaining blocks form a $(k-1)$-partition of $N \smallsetminus \{a\}$. This yields the summand $S_{n-1,k-1}$. Otherwise, we remove a. In this case, $N \smallsetminus \{a\}$ is decomposed into k blocks, and we can insert a in any of these k blocks, and so we obtain $k\,S_{n-1,k}$ partitions in this second case.

Our recurrence yields the *Stirling triangle of the second kind*:

n \ k	0	1	2	3	4	5	6	7	$\cdots$
0	1								
1	0	1							
2	0	1	1						
3	0	1	3	1					
4	0	1	7	6	1				
5	0	1	15	25	10	1			
6	0	1	31	90	65	15	1		
7	0	1	63	301	350	140	21	1	

$S_{n,k}$

Some special values immediately drop out of this triangle: $S_{n,1} = 1$, $S_{n,2} = 2^{n-1} - 1$, $S_{n,n-1} = \binom{n}{2}$, $S_{n,n} = 1$. That $S_{n,n-1} = \binom{n}{2}$ is clear, since an $(n-1)$-partition consists of one pair and $n-2$ single elements. If we decompose N into two disjoint blocks, then these two sets $A, N \smallsetminus A$ are complements of each other, and $A \neq \varnothing, N$. Thus

$$S_{n,2} = \frac{2^n - 2}{2} = 2^{n-1} - 1.$$

Now on to the Stirling numbers $s_{n,k}$ of the first kind. As usual, we set $s_{0,0} = 1$, $s_{0,k} = 0$ $(k > 0)$, $s_{n,0} = 0$ $(n > 0)$. In this case, the recurrence equation is

$$(1.15) \qquad s_{n,k} = s_{n-1,k-1} + (n-1)s_{n-1,k} \qquad (n, k > 0)\,.$$

Once again, we classify the permutations of N with k cycles according to an element $a \in N$. There are $s_{n-1,k-1}$ such permutations containing a as a 1-cycle. Otherwise, $N \smallsetminus \{a\}$ decomposes into k cycles, and we can enter a in a cycle in front of any of the $n-1$ elements in $N \smallsetminus \{a\}$.

The initial values of Stirling's triangle of the first kind are as follows:

$n^{\ k}$	0	1	2	3	4	5	6	7
0	1							
1	0	1						
2	0	1	1					
3	0	2	3	1				
4	0	6	11	6	1			
5	0	24	50	35	10	1		
6	0	120	274	225	85	15	1	
7	0	720	1764	1624	735	175	21	1

$s_{n,k}$

We already know some of these values:

$$s_{n,1} = (n-1)!, \ s_{n,n-1} = \binom{n}{2}, \ s_{n,n} = 1.$$

To calculate $s_{n,2}$, we employ (1.15). Division by $(n-1)!$ yields

$$\frac{s_{n,2}}{(n-1)!} = \frac{(n-2)!}{(n-1)!} + \frac{(n-1)s_{n-1,2}}{(n-1)!} = \frac{s_{n-1,2}}{(n-2)!} + \frac{1}{n-1},$$

and so by iteration,

$$s_{n,2} = (n-1)! \left(\frac{1}{n-1} + \frac{1}{n-2} + \cdots + 1 \right) = (n-1)!H_{n-1},$$

where H_{n-1} denotes the familiar $(n-1)$st harmonic number.

Why, you may ask, are $s_{n,k}$ and $S_{n,k}$ called Stirling numbers of the first and second kinds? In Section 1.2 we showed that $r^n = \sum_{k=0}^{n} S_{n,k} r^{\underline{k}}$ for all $r \in \mathbb{N}$. With our trusty polynomial method we can thus obtain the polynomial equation

$$(1.16) \qquad\qquad x^n = \sum_{k=0}^{n} S_{n,k} x^{\underline{k}}.$$

Conversely, if we express the falling factorials $x^{\underline{n}}$ by powers x^k, then we maintain that

$$(1.17) \qquad\qquad x^{\underline{n}} = \sum_{k=0}^{n} (-1)^{n-k} s_{n,k} x^k .$$

For $n = 0$ this is clearly true. Induction now yields, with the help of (1.15),

$$x^{\underline{n}} = x^{\underline{n-1}}(x - n + 1) = \sum_{k=0}^{n-1} (-1)^{n-1-k} s_{n-1,k} x^k (x - n + 1)$$

$$= \sum_{k=0}^{n-1} (-1)^{n-1-k} s_{n-1,k} x^{k+1} + \sum_{k=0}^{n-1} (-1)^{n-k} (n-1) s_{n-1,k} x^k$$

$$= \sum_{k=0}^{n} (-1)^{n-k} (s_{n-1,k-1} + (n-1)s_{n-1,k}) x^k$$

$$= \sum_{k=0}^{n} (-1)^{n-k} s_{n,k} x^k .$$

This is the reason for the nomenclature first and second kind. The polynomial sequences (x^n) and $(x^{\underline{n}})$ can be uniquely expressed as linear combinations of each other, and the coefficients of $x^{\underline{n}}$ in such a representation expressed via x^k, and x^n via $x^{\underline{k}}$ (up to sign), are precisely the Stirling numbers of the first and second kinds. Later, we will take up again this idea when we derive general inversion formulas. Furthermore, in the literature, Stirling numbers of the first kind are also denoted by $(-1)^{n-k} s_{n,k}$, that is, with alternating signs.

1.5. Discrete Probability

Combinatorics and discrete probability theory were originally almost synonymous. One of the first textbooks on combinatorics, written by Whitworth in 1901, bore the title *Choice and Chance*. Everyone knows problems of the following sort: You are presented a bag containing four white and three blue balls. What is the probability that in drawing four balls from the bag you will select precisely two white balls and two blue balls? This is of course a combinatorial problem. Today, quite different questions play an important role. For transmitting information, we need $(0, 1)$-sequences that are as close as possible to being random. What is a "random" sequence? Or how does one generate a random number? Suppose we wish to test a hypothesis about permutations of length n. Since $n!$ is out of reach of the fastest computers for about $n = 20$, we "simulate" by selecting a permutation at "random." But what is a *random permutation*? Simply stated, the methods of discrete probability theory are counting arguments, which can be derived as well from combinatorics. But the concepts of probability theory frequently allow much more rapid and elegant access to the relevant results.

Probability theory begins with the idea of a **probability space** (Ω, p), which is a finite set, and a mapping $p : \Omega \to [0, 1]$ that assigns to every $\omega \in \Omega$ the **probability** $p(\omega)$, $0 \le p(\omega) \le 1$. Finally, we insist on the normalization $\sum_{\omega \in \Omega} p(\omega) = 1$.

A simple example is the throwing of a six-sided die. Here Ω is the set $\{1, 2, 3, 4, 5, 6\}$ and $p(\omega)$ gives the probability that the number ω will be thrown. For a fair (unbiased) die, $p(\omega) = \frac{1}{6}$ for all $\omega \in \Omega$. However, the die could as well be "loaded," say such that $p(1) = p(4) = \frac{1}{4}$, $p(2) = p(5) = p(6) = \frac{1}{10}$, $p(3) = \frac{1}{5}$. We call p the (probability) **distribution** on Ω. If we set $p_0 = \left(\frac{1}{6}, \frac{1}{6}, \frac{1}{6}, \frac{1}{6}, \frac{1}{6}, \frac{1}{6}\right)$, $p_1 = \left(\frac{1}{4}, \frac{1}{10}, \frac{1}{5}, \frac{1}{4}, \frac{1}{10}, \frac{1}{10}\right)$, then p_0, p_1 are two distributions.

An arbitrary subset $A \subseteq \Omega$ is called an **event**, and we define $p(A) = \sum_{\omega \in A} p(\omega)$. In particular, $p(\varnothing) = 0$, $p(\Omega) = 1$. In our experiment of dice throwing, we could ask, for example, for the probability that an even number is thrown. That is, we ask for the probability of the event $A = \{2, 4, 6\}$. For our two distributions, we obtain $p_0(A) = \frac{1}{2}$, $p_1(A) = \frac{9}{20}$; the loaded die disadvantages even numbers.

From the definition we immediately obtain

$$(1.18) \qquad A \subseteq B \Longrightarrow p(A) \le p(B),$$

$$(1.19) \qquad p(A \cup B) = p(A) + p(B) - p(A \cap B),$$

$$(1.20) \qquad p(A) = \sum_i p(A_i) \text{ for a partition } A = \sum_i A_i,$$

$$(1.21) \qquad p(\Omega \smallsetminus A) = 1 - p(A),$$

$$(1.22) \qquad p\left(\bigcup_i A_i\right) \le \sum_i p(A_i).$$

Formula (1.20) is of course nothing other than our summation rule. If all elements $\omega \in \Omega$ are of equal probability (*equiprobable*), $p(\omega) = \frac{1}{|\Omega|}$, we speak of a *uniform* probability space (Ω, p) and call p the **uniform distribution** on Ω. In this case we have also for an event that $p(A) = \frac{|A|}{|\Omega|}$, which is generally expressed as follows: $p(A)$ is the number of "favorable" outcomes ($\omega \in A$) divided by the number of "possible" outcomes ($\omega \in \Omega$).

Likewise, there is a product rule. Suppose we toss our die twice in succession and ask about the possible outcomes. The probability space is now $\Omega^2 = \{1, \ldots, 6\}^2$, and $p(\omega, \omega')$ is the probability that on the first throw we get ω, and on the second throw ω'. We say that the throws of the die are **independent** if $p(\omega, \omega') = p(\omega)\, p(\omega')$ for all $(\omega, \omega') \in \Omega^2$. For the event (A, B), we then have

$$p(A, B) = \sum_{(\omega, \omega')} p(\omega, \omega') = \sum_{(\omega, \omega') \in A \times B} p(\omega)\, p(\omega')$$

$$= \sum_{\omega \in A} p(\omega) \sum_{\omega' \in B} p(\omega') = p(A)\, p(B).$$

In general, we obtain the following analogue to the product rule: If the spaces $(\Omega_1, p_1), \ldots, (\Omega_m, p_m)$ are independent, then for the distribution p on $\Omega_1 \times \cdots \times \Omega_m$ and $A_1 \subseteq \Omega_1, \ldots, A_m \subseteq \Omega_m$, we have

$$(1.23) \qquad p(A_1, \ldots, A_m) = \prod_{i=1}^{m} p_i(A_i) \,.$$

For example, if the probability of an even number on the first throw is A and the probability of an odd number on the second throw is B, then for our distributions p_0, p_1, we have

$$p_{00} = p_0(A)\, p_0(B) = \frac{1}{2} \cdot \frac{1}{2} = \frac{1}{4},$$
$$p_{11} = p_1(A)\, p_1(B) = \frac{9}{20} \cdot \frac{11}{20} = \frac{99}{400} \,.$$

Let us first toss the fair die, and then the loaded one. We then obtain

$$p_{01} = p_0(A)\, p_1(B) = \frac{1}{2} \cdot \frac{11}{20} = \frac{11}{40} \,.$$

Probability theory becomes truly interesting when we assign values to the elements or events and then ask about the distribution of the values. A **random variable** X is a mapping $X : \Omega \to \mathbb{R}$. For example, in throwing a die two times, we can ask for the *sum* of the two tosses; that is, $X : \{1, \ldots, 6\}^2 \to T = \{2, 3, \ldots, 12\}$. In fact, X is a function, but the name random variable is in general use. Using X, we can now define a distribution on the image $T \subseteq \mathbb{R}$:

$$p_X(x) = p(X = x) := \sum_{X(\omega)=x} p(\omega) \qquad (x \in T) \,.$$

We have that $\sum_{x \in T} p_X(x) = \sum_x \sum_{X(\omega)=x} p(\omega) = 1$, and so p_X is in fact a distribution on T, the distribution *induced* by X. The value $p_X(x)$ thus gives precisely the probability that the random variable takes on the value x in a given experiment. If we consider problems involving only the random variable X, then we may work with the generally much smaller space (T, p_X) instead of the underlying space (Ω, p).

Let us examine the induced distributions $p_{00}(X = x)$ and $p_{11}(X = x)$ for our two dice:

s		2	3	4	5	6	7	8	9	10	11	12
$p_{00}(X = s)$	$\frac{1}{36}$ (	1	2	3	4	5	6	5	4	3	2	1)
$p_{11}(X = s)$	$\frac{1}{400}$ (	25	20	44	66	56	68	49	36	24	8	4)

If we have two random variables $X : \Omega \to T$, $Y : \Omega \to U$ with the induced distributions p_X and p_Y, then the *joint* distribution on $T \times U$ is defined by

$$p(x, y) = p(X = x \wedge Y = y) = \sum_{X(\omega)=x, Y(\omega)=y} p(\omega).$$

We call X and Y **independent** if

(1.24) $p(x, y) = p_X(x)\, p_Y(y)$ for all $(x, y) \in T \times U$.

It is immediately apparent how these concepts are to be generalized to m random variables: $X_1, \ldots, X_m$ are said to be independent if

$$p(X_1 = x_1 \wedge \cdots \wedge X_m = x_m) = \prod_{i=1}^{m} p(X_i = x_i) \quad \text{for } (x_1, \ldots, x_m).$$

Tossing a die twice with X_1 the first result and X_2 the second yields independent variables. If X_1 is the sum of the two throws and X_2 the product, then we would have no reason to expect independence, and indeed, we obtain for the smallest case that $x_1 = 2$, $x_2 = 1$, $p_{00}(X_1 = 2) = \frac{1}{36}$, $p_{00}(X_2 = 1) = \frac{1}{36}$, but

$$p_{00}(X_1 = 2 \wedge X_2 = 1) = \frac{1}{36} > p_{00}(X_1 = 2)\, p_{00}(X_2 = 1) = \frac{1}{36^2} \ .$$

To study the behavior of a random variable $X : \Omega \to T$, one makes use of certain measurements. The most important of these are the **expectation** EX and the **variance** VX.

The expectation tells us the value that the random variable assumes on *average*. We set

(1.25) $$EX := \sum_{\omega \in \Omega} p(\omega)\, X(\omega) \ .$$

If (Ω, p) is uniform, $|\Omega| = n$, then $EX = \frac{1}{n} \sum_{\omega \in \Omega} X(\omega)$ is simply the usual mean. With the help of the induced distribution p_X on T, we obtain

(1.26) $$EX = \sum_{x \in T} p_X(x)\, x \ ,$$

since clearly

$$\sum_{\omega \in \Omega} p(\omega)\, X(\omega) = \sum_{x \in T} \sum_{\omega : X(\omega)=x} p(\omega) x = \sum_{x \in T} p_X(x) x.$$

For our two dice, the expected value of the sum of two throws of one die each is $\frac{1}{36}(1 \cdot 2 + 2 \cdot 3 + \cdots + 1 \cdot 12) = 7$, and for the loaded die, it is 6.3. This small example shows already that we need some rules for the effective calculation of EX.

If X, Y are two random variables on Ω, then the sum $X + Y$ is also on Ω, and according to (1.25), we immediately obtain

$$(1.27) \qquad E(X + Y) = \sum_{\omega \in \Omega} p(\omega)\,(X(\omega) + Y(\omega)) = EX + EY .$$

Likewise, we have $E(\alpha X) = \alpha EX$ and $E(\alpha) = \alpha$ for a constant α. The expectation is therefore a *linear* function:

$$(1.28) \qquad E(\alpha_1 X_1 + \cdots + \alpha_m X_m) = \sum_{i=1}^{m} \alpha_i\, EX_i .$$

For our experiment with dice, we can say the following: The expectation of the sum of two tosses is equal to the sum of the expectations of the individual tosses. For example, if X_1 is the result of the first throw, and X_2 the result of the second, then $EX_1 = EX_2 = \frac{7}{2}$ for the fair die, and $EX_1 = EX_2 = 3.15$ for the loaded one. For the sum $X = X_1 + X_2$, we thus obtain $EX = EX_1 + EX_2 = 7$, respectively 6.3.

The *product* of two random variables is not so easily treated; nevertheless, we have the convenient formula

$$(1.29) \qquad E(XY) = (EX)(EY) \qquad \text{if } X, Y \text{ are independent.}$$

To prove this, we use (1.24) and (1.26). For $X : \Omega \to T$, $Y : \Omega \to U$ we have

$$\begin{aligned}
E(XY) &= \sum_{\omega \in \Omega} p(\omega)\, X(\omega)\, Y(\omega) = \sum_{(x,y) \in T \times U} p(X = x \land Y = y) \cdot x\,y \\
&= \sum_{(x,y) \in T \times U} p_X(x)\, p_Y(y)\, xy = \sum_{x \in T} p_X(x)x \cdot \sum_{y \in U} p_Y(y)y \\
&= (EX)(EY) .
\end{aligned}$$

The next important measurement is the variance of a random variable. Suppose we purchase five lottery tickets $\{L_1, \ldots, L_5\}$ each with the same probability $\frac{1}{5}$ of being a winner. The first time, we have payouts of $X : 0, 2, 5, 8, 85$ depending on the lottery, and the second time $Y : 18, 19, 20, 21, 22$. The expectation is the same in each case, $EX = EY = 20$, and yet the two payout results differ considerably. In the payout for Y, the payouts are grouped narrowly around the average EY, while in the case of X, they are widely scattered. It is precisely this "scattering" that is measured by the variance.

The variance VX is defined by

$$(1.30) \qquad VX = E\left((X - EX)^2\right) .$$

If we write $EX = \mu$ as usual, then $(X - \mu)^2$ is again a random variable, one that measures the squared distance from X to μ. The variance gives the magnitude of this expected distance. With the help of (1.28) we obtain

$$VX = E((X - \mu)^2) = E(X^2 - 2\mu X + \mu^2) = E(X^2) - 2\mu\,EX + E(\mu^2)$$
$$= E(X^2) - 2(EX)^2 + (EX)^2\,,$$

and therefore

(1.31) $$VX = E(X^2) - (EX)^2\,.$$

If X and Y are independent random variables, then in accord with (1.28) and (1.29),

$$V(X + Y) = E((X + Y)^2) - (E(X + Y))^2 = E(X^2) + 2E(XY)$$
$$+ E(Y^2) - (EX)^2 - 2(EX)(EY) - (EY)^2 = VX + VY\,.$$

Therefore, for two independent variables, the variance of a sum is equal to the sum of the variances.

What is the result of our two lottery drawings? In the case of X we obtain $VX = \frac{1}{5}(0^2 + 2^2 + 5^2 + 8^2 + 85^2) - 20^2 = 1063.6$, while $VY = 2$. The *standard deviation* $\sqrt{VX}$ is 32.6 in the first case, and 1.41 in the second. The first drawing is thus to be preferred by a gambler willing to take greater risks, the second by someone more cautious.

After all of these theoretical considerations, it is time for an example. Suppose we randomly choose a permutation π of length n. How many fixed points will π have? The underlying probability space is the set Ω of all $n!$ permutations, all with probability $\frac{1}{n!}$. The random variable in which we are interested is $F(\pi)$, the number of fixed points of π, and we would like to know the expectation EF. At first glance, this seems hopeless, since we do not yet have a formula for the number of permutations with k fixed points. But here theory comes to our aid.

Let $F_i : \Omega \to \{0, 1\}$ be the random variable with $F_i(\pi) = 1$ or 0, depending on whether π has a fixed point at the ith place. Clearly, F_i assumes the value 1 for $(n - 1)!$ permutations, and we obtain $EF_i = \frac{1}{n!}(n - 1)! = \frac{1}{n}$ for all i. Since clearly $F = F_1 + \cdots + F_n$, we conclude from (1.28) that

$$E(F) = \sum_{i=1}^{n} E(F_i) = 1\,.$$

That is, on average we may expect precisely one fixed point. The variance is not so immediately apparent, since the variables F_i are of course not independent. We have

$$E(F^2) = E\Big(\Big(\sum_{i=1}^{n} F_i\Big)^2\Big) = E\Big(\sum_{i=1}^{n}\sum_{j=1}^{n} F_i F_j\Big) = \sum_{i=1}^{n}\sum_{j=1}^{n} E(F_i F_j)$$

$$= \sum_{i=1}^{n} E(F_i^2) + 2 \sum_{1 \le i < j \le n} E(F_i F_j).$$

We now have $F_i^2 = F_i$, since $F_i = 1$ or 0, and therefore $E(F_i^2) = E(F_i) = \frac{1}{n}$. Moreover, $E(F_i F_j) = \sum\{p(\pi) : \pi \text{ has fixed points at } i \text{ and } j\} = \frac{(n-2)!}{n!} = \frac{1}{n(n-1)}$. It follows that $E(F^2) = 1 + \binom{n}{2}\frac{2}{n(n-1)} = 2$, and therefore

$$VF = E(F^2) - (EF)^2 = 2 - 1 = 1.$$

Our result is usually expressed as follows: A permutation has on average 1 ± 1 fixed points.

1.6. Existence Theorems

In most problems, we will not be able to compute the exact number of objects to be determined. We will have to be satisfied with *estimates* and theorems about *orders of magnitude*. We will have more to say about that in Chapter 5. A problem takes on an entirely different cast when we ask whether there *exists at all* an object satisfying certain conditions. We obtain an answer to this question if we succeed in actually *constructing* such an object or, conversely, proving the *impossibility of its existence*. Here we shall concentrate on the aspect of existence. Usually it will be too much effort to examine all possible objects to determine whether one of them satisfies the given conditions. We therefore need some sort of assertion that establishes such an object's existence *without* our having to search through all the objects, and indeed without our ever knowing which object it is that possesses the given properties.

Let us clarify this discussion with the help of an example. Let the numbers $a_1, a_2, \ldots, a_n$ be a collection of integers, not necessarily all distinct. Does there exist a subset of these numbers whose sum is a multiple of n? Since there are 2^n sums to consider, it will be impossible to try out all possible sums for large values of n. Nevertheless, can we assert the existence of such a sum? For small values of n such as $n = 2$, 3, 4, or 5, one can easily check that such a sum always exists. But is this true for arbitrary n?

The simplest method of solution, one that turns out to be extremely useful in applications, is called the *pigeonhole principle*.

The Pigeonhole Principle. *If n objects (pigeons) are distributed among r containers (pigeonholes), where $n > r$, then there must exist at least one container with more than one object in it.*

The assertion is obvious; there is nothing to prove. In the language of mappings, the pigeonhole principle states that if N and R are two sets with $|N| = n > r = |R|$ and if f is a mapping from N to R, then there exists $a \in R$ with $|f^{-1}(a)| \geq 2$. We can restate the pigeonhole principle in a sharper formulation:

Pigeonhole Principle: Sharper Version. *Let $f : N \to R$ with $|N| = n > r = |R|$. Then there exists $a \in R$ with $\left| f^{-1}(a) \right| \geq \left\lfloor \frac{n-1}{r} \right\rfloor + 1$.*

To see that this is true, observe that if we had $|f^{-1}(a)| \leq \left\lfloor \frac{n-1}{r} \right\rfloor$ for all $a \in R$, then we would have $n = \sum_{a \in R} |f^{-1}(a)| \leq r \left\lfloor \frac{n-1}{r} \right\rfloor < n$, which is obviously false.

With the pigeonhole principle we can easily solve our counting problem. We will show even more, namely, that among the sums $\sum_{i=k+1}^{\ell} a_i$ of successive numbers $a_{k+1}, a_{k+2}, \ldots, a_\ell$ there is already to be found a multiple of n. We set $N = \{0, a_1, a_1 + a_2, a_1 + a_2 + a_3, \ldots, a_1 + a_2 + \cdots + a_n\}$. If we divide an arbitrary integer m by n, we obtain as remainder one of the numbers $0, 1, \ldots, n-1$. We write $R = \{0, 1, 2, \ldots, n-1\}$ and define $f : N \to R$ by setting $f(m)$ equal to the remainder on division by n. Since $|N| = n + 1 > n = |R|$, it follows from the pigeonhole principle that there exist two sums $a_1 + \cdots + a_k$, $a_1 + \cdots + a_\ell$, $k < \ell$, that have the same remainder on division by n (where one of the two sums could be the empty sum, which is assumed equal to 0). Therefore, $\sum_{i=k+1}^{\ell} a_i = \sum_{i=1}^{\ell} a_i - \sum_{i=1}^{k} a_i$ has remainder 0, and so the sum is a multiple of n. We observe that the number n of summands is the least possible, since we can set $a_1 = a_2 = \cdots = a_{n-1} = 1$.

Another nice application of the pigeonhole principle is the following. Let $a_1, \ldots, a_{n^2+1}$ be a sequence of $n^2 + 1$ distinct real numbers. Then there exists either a *monotonically increasing* subsequence $a_{k_1} < a_{k_2} < \cdots < a_{k_{n+1}}$ ($k_1 < \cdots < k_{n+1}$) of length $n + 1$ or a *monotonically decreasing* subsequence $a_{\ell_1} > a_{\ell_2} > \cdots > a_{\ell_{n+1}}$ of length $n + 1$.

Here is takes a bit of skill in using the pigeonhole principle. We associate with a_i the integer t_i, which is the length of a longest monotonically increasing subsequence with initial element a_i; t_i is thus a number between 1 and $n^2 + 1$. If $t_i \geq n + 1$ holds for some i, then we have found a desired increasing sequence. Let us therefore assume that $t_i \leq n$ for all i. The mapping $f : a_i \mapsto t_i$ shows us, according to the sharper version of the pigeonhole principle, that there exists $s \in \{1, \ldots, n\}$ such that $\left\lfloor \frac{n^2}{n} \right\rfloor + 1 = n + 1$ numbers $a_{\ell_1}, a_{\ell_2}, \ldots, a_{\ell_{n+1}}$ ($\ell_1 < \ell_2 < \cdots < \ell_{n+1}$) all have maximal length s with initial term a_{ℓ_i}. We consider two adjacent terms

a_{ℓ_i}, $a_{\ell_{i+1}}$ of this partial sequence. If we had $a_{\ell_i} < a_{\ell_{i+1}}$, then there would exist an increasing subsequence $a_{\ell_{i+1}} < \cdots$ of length s and thus one of length $s + 1$ with initial term a_{ℓ_i}, in contradiction to $f(a_{\ell_i}) = s$. Thus the a_{ℓ_i} satisfy $a_{\ell_1} > a_{\ell_2} > \cdots > a_{\ell_{n+1}}$, and we have obtained our desired increasing sequence. The reader may easily determine that the assertion is no longer true for n^2, and so $n^2 + 1$ is again the best possible result.

Ramsey's Theorem. A far-reaching generalization of the pigeonhole principle was discovered by the logician Ramsey. In considering the pigeonhole principle, it can be useful to interpret the r pigeonholes as colors. A mapping $f : N \longrightarrow R$ is therefore a coloring of N, and the principle states that if there are more elements than colors, then for *every* coloring at least two elements must receive the same color.

We may make this more precise. Suppose we are given natural numbers $\ell_1, \ldots, \ell_r$ and an n-set N with $n \geq \ell_1 + \cdots + \ell_r - r + 1$. Then for *every* coloring of N there must exist a color i such that ℓ_i elements are colored with the color i. We call this the *Ramsey property* for $(\ell_1, \ldots, \ell_r)$. The pigeonhole principle relates to the case $\ell_1 = \cdots = \ell_r = 2$.

If an n-set has the Ramsey property, then so, of course, does every larger set. We are therefore interested in the smallest such n, and this is clearly $\ell_1 + \cdots + \ell_r - r + 1$, since for $m = \sum_{i=1}^r \ell_i - r = \sum_{i=1}^r (\ell_i - 1)$ one could have the coloring in which for every i exactly $\ell_i - 1$ elements are colored with i. Ramsey's theorem now asserts that there is an analogous result for colorings of h-sets ($h = 1$ is the pigeonhole principle). We shall demonstrate this only for $h = 2$, that is, pairs, and for two colors. The general case then follows easily (see Exercise 1.54). Here is a statement of Ramsey's theorem:

Let k and ℓ be natural numbers greater than or equal to 2. Then there exists a least integer $R(k, \ell)$, called the Ramsey number, such that the following holds: If N is an n-set with $n \geq R(k, \ell)$ and we color all pairs from N either red or blue, then there exists either a k-set in N whose pairs are all colored red or an ℓ-set whose pairs are all colored blue.

Clearly, we have $R(k, 2) = k$, since in a k-set either all pairs are colored red or one pair is colored blue ($\ell = 2$). Analogously, we have $R(2, \ell) = \ell$. We now use induction on $k + \ell$. We assume that $R(k - 1, \ell)$ and $R(k, \ell - 1)$ exist and we show that

$$R(k, \ell) \leq R(k - 1, \ell) + R(k, \ell - 1).$$

Suppose we are given the set N with $|N| = n = R(k - 1, \ell) + R(k, \ell - 1)$, whose pairs are arbitrarily colored red or blue. Let $a \in N$. Then $N \smallsetminus a$ decomposes into $R \cup B$, where $x \in R$ if $\{a, x\}$ is colored red, and $y \in B$ if $\{a, y\}$ is colored blue. Since

$$|R| + |B| = R(k - 1, \ell) + R(k, \ell - 1) - 1,$$

we must have either $|R| \geq R(k-1, \ell)$ or $|B| \geq R(k, \ell-1)$. Let us assume the first case. The second case proceeds similarly. By induction, there are in R either $k-1$ elements whose pairs are all colored red, in which case we have together with a our desired k-set, or there is an ℓ-set whose pairs are all colored blue, and again we are finished.

From the "Pascal" recurrence relation $R(k, \ell) \leq R(k-1, \ell) + R(k, \ell-1)$ and the initial conditions, we see at once that

$$R(k, \ell) \leq \binom{k+\ell-2}{k-1}.$$

For example, for the first case of interest we obtain $R(3, 3) \leq \binom{4}{2} = 6$, and 6 is also the exact value (why?).

Ramsey's theorem is mentioned in many books on mathematical puzzles with the following interpretation. Let N be a set of people. A red pair means that the pair are acquainted, and a blue pair means that they are unacquainted. The Ramsey number $R(3, 3) = 6$ indicates that in every group of six people, there are always either three who are mutually unacquainted or three who are mutually acquainted.

A quite different useful method is of a probability-theoretic nature. We define a probability space on our objects and show that the probability that an object satisfies given conditions is greater than zero. Then there must exist such an object.

As an illustration we consider the following coloring problem, which goes back to the famous Hungarian mathematician Paul Erdős. Let $\mathcal{F}$ be a family of d-sets, $d \geq 2$, from an underlying set X. We say that $\mathcal{F}$ is *2-colorable* if there exists a coloring of the elements of X with two colors such that in every set $A \in \mathcal{F}$, *both* colors occur.

It is clear that not every family $\mathcal{F}$ can be thus colored. If, for example, $\mathcal{F}$ is the family of all d-subsets of a $(2d-1)$-set, then by the pigeonhole principle, there must be a d-set of a single color. On the other hand, it is also clear that every subfamily of a two-colorable family is itself two-colorable. We would like to know the smallest number $m = m(d)$ for which there exists a family $\mathcal{F}$ with $|\mathcal{F}| = m$ that is not two-colorable. The example above shows that $m(d) \leq \binom{2d-1}{d}$. What sort of lower bound is there for $m(d)$?

We have $m(d) > 2^{d-1}$. That is, every family with at most 2^{d-1} d-sets is two-colorable.

Let $\mathcal{F}$ be given, with $|\mathcal{F}| \leq 2^{d-1}$. We color X randomly with two colors, whereby all colorings are equally probable. For $A \in \mathcal{F}$ let E_A be the event that the elements of A all have the same color. Since there are exactly two such colorings on A, we obtain

$$p(E_A) = \frac{2}{2^d} = \frac{1}{2^{d-1}}.$$

Thus with $|\mathcal{F}| \leq 2^{d-1}$ (where the events are not disjoint),

$$p\left(\bigcup_{A \in \mathcal{F}} E_A\right) < \sum_{A \in \mathcal{F}} p(E_A) = m\frac{1}{2^{d-1}} \leq 1.$$

Now $\bigcup_{A \in \mathcal{F}} E_A$ is the event that *some* set in $\mathcal{F}$ is of a single color, and we conclude from $p(\bigcup_{A \in \mathcal{F}} E_A) < 1$ that there *must* exist a two-coloring of S without single-colored sets, which is just what we wanted to show.

An upper bound for $m(d)$ of order of magnitude $d^2 2^d$ is known, where this time random sets and a fixed coloring are used. As for exact values, only the first two are known: $m(2) = 3$ and $m(3) = 7$.

Exercises for Chapter 1

1.1 Suppose that Dean B of County College determines that every student must enroll in exactly four courses in the history of mathematics from among the seven that are offered. The professors of the various courses specify the maximum enrollments in their courses as $51, 30, 30, 20, 25, 12, 18$. What can one conclude from this?

1.2 Suppose we are given n disjoint sets S_i. Let the first have a_1 elements, the second a_2, and so on. Show that the number of sets that contain at most one element from each S_i is equal to $(a_1+1)(a_2+1)\cdots(a_n+1)$. Apply this result to the following number-theoretic problem: Let $n = p_1^{a_1} p_2^{a_2} \cdots$ be the prime decomposition of n. Then n has exactly $t(n) = \prod(a_i + 1)$ divisors. Conclude that n is a square if and only if $t(n)$ is odd.

▷ **1.3** Let $N = \{1, 2, \ldots, 100\}$ and let A be a subset of N with $|A| = 55$. Show that A contains two numbers a and b such that $a - b = 9$. Does this hold as well for $|A| = 54$?

1.4 Number the twelve edges of a cube with the numbers 1 through 12 in such a way that the sum of the three edges meeting at each vertex is the same for each vertex.

▷ **1.5** In the parliament of country X there are 151 seats and three political parties. How many ways (i, j, k) are there of dividing up the seats such that no party has an absolute majority?

1.6 How many different words can be made from permutations of the letters in *ABRACADABRA*?

1.7 Show that $1! + 2! + \cdots + n!$ for $n > 3$ is never a square.

1.8 Show that for the binomial coefficients $\binom{n}{k}$ the following holds:

$$\binom{n}{0} < \binom{n}{1} < \cdots < \binom{n}{\lfloor n/2 \rfloor} = \binom{n}{\lceil n/2 \rceil} > \cdots > \binom{n}{n},$$

where for even n the two middle coefficients coincide.

▷ **1.9** Prove an analogous result for the Stirling numbers of the second kind. For each $n \geq 1$ there exists $M(n)$ such that $S_{n,0} < S_{n,1} < \cdots < S_{n,M(n)} > S_{n,M(n)+1} > \cdots > S_{n,n}$ or $S_{n,0} < S_{n,1} < \cdots < S_{n,M(n)-1} = S_{n,M(n)} > \cdots > S_{n,n}$, where $M(n) = M(n-1)$ or $M(n) = M(n-1) + 1$. The same result holds also for $s_{n,k}$. Hint: Use the recurrence relation for $S_{n,k}$ and finish the proof by induction.

▷ **1.10** Show that every natural number n possesses a unique representation $n = \sum_{k \geq 0} a_k k!$ with $0 \leq a_k \leq k$.

1.11 Derive the following recurrence relation for partition numbers $P_{n,k}$: $P_{n,1} = P_{n,n} = 1$ and $P_{n,k} = P_{n-k,1} + P_{n-k,2} + \cdots + P_{n-k,k}$.

1.12 The *Bell number* $\tilde{B}_n$ is the number of *all* set partitions of an n-set; that is, $\tilde{B}_n = \sum_{k=0}^{n} S_{n,k}$ with $\tilde{B}_0 = 1$. Show that

$$\tilde{B}_{n+1} = \sum_{k=0}^{n} \binom{n}{k} \tilde{B}_k \ .$$

▷ **1.13** Let $f_{n,k}$ be the number of k-subsets of $\{1, \ldots, n\}$ that contain no pairs of adjacent numbers. Show that

(a) $f_{n,k} = \binom{n-k+1}{k}$ (b) $\sum_k f_{n,k} = F_{n+2}$,

where F_n is the nth Fibonacci number (that is, $F_0 = 0$, $F_1 = 1$, $F_n = F_{n-1} + F_{n-2}$ $(n \geq 2)$).

1.14 Show that the sum of the binomial coefficients in Pascal's triangle along a diagonal from upper right to lower left is always the Fibonacci number F_{n+k+1}. Example: Starting with $n = 4$, $k = 3$ gives $4 + 10 + 6 + 1 = 21 = F_8$.

1.15 Prove that $\binom{n}{r}\binom{r}{k} = \binom{n}{k}\binom{n-k}{r-k}$ and conclude that $\sum_{k=0}^{m} \binom{n}{k}\binom{n-k}{m-k} = 2^m \binom{n}{m}$.

▷ **1.16** An ordinary deck of 52 playing cards is well shuffled. What is the probability that both the top and bottom cards are queens (on the assumption that all 52! permutations are equally probable)?

1.17 In a lottery, six numbers from the set $\{1, 2, \ldots, 49\}$ are selected. What is the probability that the chosen set of numbers contains two adjacent integers?

1.18 Suppose the random variable X can assume only the values 0 and 1. Show that $VX = EX \cdot E(1 - X)$.

▷ **1.19** Prove that in every set of $n + 1$ integers from the collection $\{1, 2, \ldots, 2n\}$ there is always a pair of integers that are relatively prime, and always a pair in which one of the pair is a divisor of the other. Does the result hold for n integers?

1.20 Construct a sequence of n^2 distinct numbers that contains neither a monotonically increasing subsequence nor a monotonically decreasing sequence of length $n + 1$.

▷ **1.21** The Euler φ-function is defined by

$$\varphi(n) = |\{k : 1 \leq k \leq n, \ k \text{ relatively prime to } n\}|.$$

Prove that

$$\sum_{d \mid n} \varphi(d) = n \, .$$

1.22 Each square of a 4×7 checkerboard is colored either white or black. Show that there is always a rectangle whose four corner squares have the same color. Does the result hold for a 4×6 checkerboard?

▷ **1.23** Let $N = \{1, 2, 3, \ldots, 3n\}$. Matilda removes n elements of N. Show that she can now remove a further n numbers in such a way that the remaining n appear in the order odd, even, odd, even, ...

1.24 Matilda selects n points on the circumference of a circle and colors each point red or blue. Show that there are at most $\lfloor \frac{3n-4}{2} \rfloor$ chords connecting points of different colors that do not intersect in the interior of the circle.

▷ **1.25** Consider an $n \times n$ checkerboard with rows and columns numbered from 1 to n. A set T of n squares will be called a *transversal* if none of its squares are in the same row or column. In other words, $T = \{(1, \pi_1), \ldots, (n, \pi_n)\}$, where $(\pi_1, \ldots, \pi_n)$ is a permutation of $\{1, \ldots, n\}$. Now let $n \geq 4$ be an even number and place a number in each of the n^2 squares in such a way that every number appears exactly twice. Show that there is always a transversal containing n distinct numbers. Hint: The checkerboard contains r pairs of squares (in different rows and columns) that contain the same number. Construct the $(n!) \times r$ incidence matrix (m_{ij}) with $m_{ij} = 1$ if the ith transversal contains the jth pair. Now count in two different ways.

1.26 We would like to consider how we might effectively list all $n!$ permutations of $\{1, \ldots, n\}$. The most usual method is the lexicographic ordering. We say that $\pi = (\pi_1, \ldots, \pi_n)$ is lexicographically smaller than $\sigma = (\sigma_1, \ldots, \sigma_n)$ if for the smallest i with $\pi_i \neq \sigma_i$, we have $\pi_i < \sigma_i$. For example, for $n = 3$ we obtain the ordering $123, 132, 213, 231, 312, 321$. Show that the following algorithm finds the successor permutation σ to the permutation $\pi = (\pi_1, \ldots, \pi_n)$:

1. Search for the largest index r with $\pi_r < \pi_{r+1}$. If no such r exists, then $\pi = (n, \ldots, 2, 1)$ is the last permutation.
2. Search for the index $s > r$ with $\pi_s > \pi_r > \pi_{s+1}$.
3. $\sigma = (\pi_1, \ldots \pi_{r-1}, \pi_s, \pi_n, \ldots, \pi_{s+1}, \pi_r, \pi_{s-1}, \ldots, \pi_{r+1})$ is the successor permutation.

1.27 Analogously to the previous exercise, we would like to list all 2^n subsets of an n-set. As usual, we represent the subsets by $(0, 1)$-words of length n. The following list is called a Gray code. Suppose $G(n) = \{G_1, \ldots, G_{2^n}\}$ is the list for n. Then $G(n+1) = \{0G_1, 0G_2, \ldots, 0G_{2^n}, 1G_{2^n}, 1G_{2^n-1}, \ldots, 1G_1\}$. Prove the following: (a) Every pair of neighboring $(0, 1)$-words in $G(n)$ differ in exactly one place. (b) Let $G(n, k)$ be the subsequence of $G(n)$ with exactly k 1's. Show that consecutive words in $G(n, k)$ differ in exactly two places.

1.28 There are $4n$ participants in a bridge tournament playing at n tables. Each player requires one other player as partner, and every pair of partners requires another pair of partners as their opponents. In how many ways can the selection of partners and opponents occur?

▷ **1.29** In how many ways can the numbers $1, \ldots, n$ be arranged in a row so that aside from the first element, each number k has $k - 1$ or $k + 1$ as one of its predecessors (not necessarily the immediate one)? Examples: 3 2 4 5 1 6 and 4 3 5 2 1 6.

1.30 In how many ways can the numbers $1, \ldots, n$ be arranged in a circle so that adjacent numbers always differ by 1 or 2?

1.31 For a permutation $a_1 a_2 \ldots a_n$ of $\{1, \ldots, n\}$, an *inversion* is a pair a_i, a_j with $i < j$ but $a_i > a_j$. Example: 1 4 3 5 2 has the inversions $4, 3; 4, 2; 3, 2; 5, 2$. Let $I_{n,k}$ be the number of n permutations with exactly k inversions. Prove the following:

a. $I_{n,0} = 1$.
b. $I_{n,k} = I_{n,\binom{n}{2}-k}$ $(k = 0, \ldots, \binom{n}{2})$.
c. $I_{n,k} = I_{n-1,k} + I_{n,k-1}$ for $k < n$. Is this true as well for $k = n$?
d. $\sum_{k=0}^{\binom{n}{2}} (-1)^k I_{n,k} = 0$, $n \geq 2$.

▷ **1.32** Let $a_1, a_2, \ldots, a_n$ be a permutation of $\{1, \ldots, n\}$. We let b_j denote the number of elements to the left of j that are greater than j (and thus form an inversion with j). The sequence $b_1, \ldots, b_n$ is called the *inversion table* of $a_1, \ldots, a_n$. Show that $0 \leq b_j \leq n - j$ $(j = 1, \ldots, n)$, and prove that conversely, every sequence $b_1, \ldots, b_n$ with $0 \leq b_j \leq n - j$ $(\forall j)$ is the inversion table for exactly one permutation.

1.33 We return to the lexicographic ordering of permutations introduced in Exercise 26. We assign the number 0 to the smallest permutation $(1, 2, \ldots, n)$. We assign 1 to the next permutation, and so on, giving the last permutation $(n, n - 1, \ldots, 1)$ the number $n! - 1$. The problem is to determine the number $\ell_n(\pi)$ assigned to a given permutation $\pi = (\pi_1, \ldots, \pi_n)$. Show that $\ell_1(1) = 0$, $\ell_n(\pi) = (\pi_1 - 1)(n - 1)! + \ell_{n-1}(\pi')$, where $\pi' = (\pi_1', \ldots, \pi_{n-1}')$ is derived from π by deleting π_1 and reducing all $\pi_j > \pi_1$ by 1. Example: $\ell_4(2314) = 3! + \ell_3(213) = 3! + 2! + \ell_2(12) = 8$.

1.34 The converse of the previous exercise. Let ℓ, $0 \leq \ell \leq n! - 1$ be given. Determine the permutation π for which $\ell_n(\pi) = \ell$. Hint: From Exercise 10, we can represent ℓ in the form $\ell = a_{n-1}(n-1)! + a_{n-2}(n-2)! + \cdots + a_1 1!$ with $0 \leq a_k \leq k$.

1.35 Prove the following recurrence formulas for the Stirling numbers:

(a) $s_{n+1,k+1} = \sum_i \binom{i}{k} s_{n,i}$,   (b) $S_{n+1,k+1} = \sum_i \binom{n}{i} S_{i,k}$.

▷ **1.36** The *Euler numbers* $A_{n,k}$ count the permutations π of $\{1, \ldots, n\}$ with precisely k increases, that is, k places i for which $\pi_i < \pi_{i+1}$. For example, for $n = 3$, we have $A_{3,0} = 1$, $A_{3,1} = 4$, $A_{3,2} = 1$. Prove the recurrence

$$A_{n,k} = (n - k)A_{n-1,k-1} + (k + 1)A_{n-1,k}$$

$(n > 0)$ with $A_{0,0} = 1$, $A_{0,k} = 0$ $(k > 0)$.

1.37 Many identities for binomial coefficients can be obtained by counting paths in lattices. Consider an $m \times n$ lattice such as the one shown in the figure for $m = 5$, $n = 3$. (Note that m and n are the numbers of edges.)

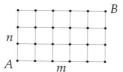

Show that the number of different paths from A to B that always move upward or to the right is equal to $\binom{m+n}{n}$. For example, for $m = 2$, $n = 2$ we have the $\binom{4}{2} = 6$ paths

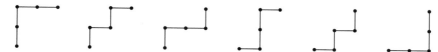

1.38 Two examples of the lattice method. (a) Classify the lattice paths according to a path's first meeting the right vertical, and derive equation (1.9) (see the figure on the left). (b) Prove the Vandermonde identity (1.13) by classifying the lattice paths according to their point of intersection with the diagonal shown in the right-hand figure.

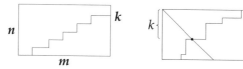

▷ **1.39** In how many ways can a king move from the lower left corner of a chess board to the upper right corner if it always moves up, to the right, or diagonally up and to the right? Hint: Let r equal the number of diagonal moves and sum over r.

1.40 Show that $r^{\underline{k}}(r - \frac{1}{2})^{\underline{k}} = \frac{(2r)^{\underline{2k}}}{2^{2k}}$ and conclude the validity of the formula $\binom{-\frac{1}{2}}{n} = \left(-\frac{1}{4}\right)^n \binom{2n}{n}$.

▷ **1.41** The following problem goes back to J. L. F. Bertrand (1822–1900). Two candidates A and B receive a and b votes in an election, with $a > b$. In how many ways can the ballots be arranged so that in counting, one vote after the other, A always has more votes than B. For example, for $a = 4$, $b = 2$, we have the following possibilities: $AAAABB$, $AAABAB$, $AAABBA$, $AABAAB$, $AABABA$. Show that the answer is $\frac{a-b}{a+b}\binom{a+b}{a}$. Hint: Draw a sequence as points (x, y), where y is the number of votes for A minus the number of votes for B when x votes have been counted. The desired sequences are then the paths from $(0, 0)$ to $(a + b, a - b)$ that after $(0, 0)$ never again touch the x-axis.

1.42 Show that $(1+\sqrt{3})^{2n+1} + (1-\sqrt{3})^{2n+1}$ represents a natural number for every $n \geq 0$. Hint: Use the binomial theorem. Since $0 < |1 - \sqrt{3}| < 1$, it must be that $-(1 - \sqrt{3})^{2n+1}$ is the fractional part of $(1 + \sqrt{3})^{2n+1}$. Conclude that the integer part of $(1 + \sqrt{3})^{2n+1}$ always contains 2^{n+1} as a factor.

▷ **1.43** Let $a_n = \frac{1}{\binom{n}{0}} + \frac{1}{\binom{n}{1}} + \cdots + \frac{1}{\binom{n}{n}}$. Show that $a_n = \frac{n+1}{2n}a_{n-1} + 1$ and determine $\lim_{n \to \infty} a_n$ (if the limit exists). Hint: Show that $a_n > 2 + \frac{2}{n}$ and $a_{n+1} < a_n$ for $n \geq 4$.

1.44 It is clear that n points on a line divide the line into $n+1$ parts.

a. Let L_n denote the maximum number of pieces into which the plane can be divided by n lines. Determine a recurrence for L_n and calculate L_n.

b. Let M_n denote the maximum number of three-dimensional pieces into which 3-space $\mathbb{R}^3$ can be divided by n planes. Determine a recurrence for M_n and calculate M_n.

c. Generalize to $\mathbb{R}^n$.

▷ **1.45** Pascal's triangle (shifted somewhat) yields an astonishing primality test. We number the rows as usual with $0, 1, 2, \ldots, n, \ldots$, and likewise the columns. In the nth row we write the $n+1$ binomial coefficients $\binom{n}{0}, \binom{n}{1}, \ldots, \binom{n}{n}$, but shifted into the columns with numbers $2n$ through $3n$ inclusive. Finally, we draw a circle around any of these $n+1$ numbers that are multiples of n. The first rows and columns look like this:

$n\backslash k$	0	1	2	3	4	5	6	7	8	9	10	11	12
0	1												
1			①	①									
2					1	②	1						
3							1	③	③	1			
4									1	④	6	④	1

Show that a number k is prime if and only if all the elements in the kth column are circled. Hint: For k even the problem is easy, and for k odd, first prove that the element in the nth row and kth column is $\binom{n}{k-2n}$.

▷ **1.46** Two dice have the same probability distribution. Show that the probability of throwing the two dice and obtaining the same number on each is always at least $\frac{1}{6}$.

1.47 The following important inequalities estimate the distance from X to the expectation EX. Prove Markov's inequality: Let X be a random variable that takes on only nonnegative values. Then $p(X \geq \alpha) \leq \frac{EX}{\alpha}$ for $\alpha \geq 0$. Conclude from this Chebyshev's inequality: $p(|X - EX| \geq \alpha) \leq \frac{VX}{\alpha^2}$ for a random variable X and $\alpha \geq 0$.

1.48 Estimate, with the help of the previous exercise, the probability that a permutation has $k+1$ fixed points (all permutations of equal probability).

▷ **1.49** A group of n hunters shoot simultaneously at r rabbits, with every shot hitting a rabbit. Every hunter hits each rabbit with equal probability, and if a rabbit is hit, it is killed. What is the expectation for the number of surviving rabbits? Show with the help of Markov's inequality that for $n \geq r(\log r + 5)$, the probability that no rabbit survives is greater than 0.99.

1.50 A random number generator chooses one of the numbers $1, 2, \ldots 9$, all with equal probability. Determine the probability that after n selections ($n > 1$), the product of these numbers is divisible by 10. Hint: Consider the random variables X_k, the number of times k is chosen, $1 \le k \le 9$.

▷ **1.51** A deck of playing cards with n cards contains three aces. The deck is shuffled, with all permutations equally probable. The cards are then dealt out one after the other until two aces have appeared. Show that the expectation for the number of dealt cards is $\frac{n+1}{2}$.

1.52 Let $(p_1, \ldots, p_6)$ and $(q_1, \ldots, q_6)$ be the probability distributions for two dice. Show that p_i, q_j can never be chosen such that the different possible sums $2, 3, \ldots, 12$ of the throws are equally probable (that is, equal to $\frac{1}{11}$).

▷ **1.53** Let x be a real number. Then among the numbers $x, 2x, 3x, \ldots, (n-1)x$ there is at least one that differs from an integer by at most $\frac{1}{n}$.

1.54 Prove the following general theorem of Ramsey: Let k and $\ell_1, \ldots, \ell_r$ be given. Then there is a least number $R(k; \ell_1, \ldots, \ell_r)$ such that the following holds: If N is an n-set with $n \ge R(k; \ell_1, \ldots, \ell_r)$ and if the k-subsets of N are colored in some way with the colors $1, \ldots, r$, then there is a color i such that in some ℓ_i-subset of N, all k-subsets are colored with i. Hint: Induction.

▷ **1.55** Show that if the Ramsey numbers $R(k-1, \ell)$ and $R(k, \ell-1)$ are both even, then $R(k, \ell) < R(k-1, \ell) + R(k, \ell-1)$. Calculate $R(3, 4)$.

1.56 For the two-coloring problem of families of sets discussed in Section 1.6, prove that $m(2) = 3$, $m(3) = 7$.

▷ **1.57** Prove that $R(k, k) \ge 2^{k/2}$. Hint: $R(2, 2) = 2$, $R(3, 3) = 6$. So assume that $k \ge 4$. Let $n < 2^{k/2}$. Altogether, there are $2^{\binom{n}{2}}$ patterns of acquaintance. Let A denote the event, $|A| = k$, that all the persons are mutually acquainted. Now use the probabilistic method of Section 1.6.

1.58 In country Z, every pair of cities is linked by exactly one of the three modes of transportation bus, train, air, where all three possibilities occur. No city is served by all three forms of transportation, and no three cities are linked by the same mode of transport. Determine the maximum number of cities in country Z. Hint: Consider the possible modes of transport from a fixed city.

▷ **1.59** Suppose that n different numbers (n is very large) are written on n scraps of paper and then the scraps are mixed together in a hat. We now pull one scrap of paper after another out of the hat. Our task is to determine the largest number. When the scrap on which we believe the largest number to be written has been drawn, we are to announce, "That is the largest number." It is not allowed to name a number that was previously drawn. Since we know nothing about the size of the numbers or their arrangement, the task seems hopeless. Nevertheless, there is an algorithm that names the correct number with probability greater than $\frac{1}{3}$. Hint: Let s numbers be drawn, and then choose the first number to be drawn that is larger than all the previous numbers drawn.

1.60 Let $a_1, a_2, a_3, \ldots$ be an infinite sequence of natural numbers. Then there exists either an infinite strictly monotonically increasing subsequence $a_{i_1} < a_{i_2} < a_{i_3} < \cdots$ (where $i_1 < i_2 < i_3 < \cdots$) or a strictly monotonically decreasing subsequence or an infinite constant subsequence $a_{j_1} = a_{j_2} = a_{j_3} = \cdots$.

Chapter 2

Summation

Many counting problems can be reduced to the evaluation of sums, and conversely, the result of an enumeration can often be represented as a sum. In this chapter we shall investigate some of the standard methods for calculating sums.

2.1. Direct Methods

We generally write a sum in the form $\sum_{k=0}^{n} a_k$ or $\sum_{0 \le k \le n} a_k$. The running index is frequently denoted by k. If we wish to sum the even numbers between 0 and 100, we can write

$$\sum_{\substack{k=0 \\ k \text{ even}}}^{100} k \qquad \text{or} \qquad \sum_{k=1}^{50} 2k.$$

The following notation is often more convenient: $\sum_{k=0}^{100} k \, [k \text{ even}]$. The expression in brackets, which is to be multiplied, means that

$$[k \text{ has property } E] = \begin{cases} 1 & \text{if } k \text{ satisfies property } E, \\ 0 & \text{otherwise.} \end{cases}$$

One of the most elementary (and useful) techniques is that of **index transformation**. Let $i \ge 0$. Then

$$\sum_{k=m}^{n} a_k = \sum_{k=m+i}^{n+i} a_{k-i} = \sum_{k=m-i}^{n-i} a_{k+i} \, .$$

Thus lowering the running index by i corresponds to raising the limits of summation by i, and conversely. As a further example, with the transformations $k \to n - k$ and $k \to m + k$ we obtain

$$\sum_{k=m}^{n} a_k = \sum_{k=0}^{n-m} a_{n-k} = \sum_{k=0}^{n-m} a_{m+k} .$$

Consider, for example, the *arithmetic sum* $S = 0 \cdot a + 1 \cdot a + \cdots + n \cdot a = \sum_{k=0}^{n} ka$. With the transformation $k \to n - k$ we see that $S = \sum_{k=0}^{n}(n-k)a$, and therefore $2S = \sum_{k=0}^{n} ka + \sum_{k=0}^{n}(n - k)a = \sum_{k=0}^{n} na = n \sum_{k=0}^{n} a = n(n+1)a$. That is, $S = \frac{n(n+1)}{2}a$.

Suppose we have a rectangular array of real numbers $a_i a_j$ for $i, j = 1, \ldots, n$. If we sum all the numbers, we obtain $S = \sum_{1 \le i,j \le n} a_i a_j = (\sum_{i=1}^{n} a_i)(\sum_{j=1}^{n} a_j) = (\sum_{k=1}^{n} a_k)^2$. Our task is now to sum all the products $a_i a_j$ below and including the main diagonal, that is, to determine $\underline{S} = \sum_{1 \le j \le i \le n} a_i a_j$. We first observe that the sum above and including the diagonal is $\overline{S} = \sum_{1 \le i \le j \le n} a_i a_j = \sum_{1 \le i \le j \le n} a_j a_i = \underline{S}$. From $S = \underline{S} + \overline{S} - \sum_{i=1}^{n} a_i^2 = 2\underline{S} - \sum_{k=1}^{n} a_k^2$ we at once obtain

$$\underline{S} = \frac{1}{2}\left(\left(\sum_{k=1}^{n} a_k\right)^2 + \sum_{k=1}^{n} a_k^2 \right) .$$

Which direct methods shall we try out first for calculating sums? The first to try is certainly **induction**. A simple example is the summation of the first n odd integers: $S_n = \sum_{k=1}^{n}(2k - 1)$. We begin with a table of small values:

n	1	2	3	4	5	6
S_n	1	4	9	16	25	36

This should suffice for conjecturing that the answer is $S_n = n^2$. For $n = 1$ we have $S_1 = 1^2 = 1$. From the assumption $S_n = n^2$ it follows that $S_{n+1} = S_n + (2n + 1) = n^2 + 2n + 1 = (n + 1)^2$, and the correctness of the conjecture follows by induction.

The disadvantage of induction as a technique is clear. We have to be able to "guess" the correct answer. Moreover, it frequently happens that the truth of the case $n + 1$ does not follow easily from the assumption of the case n. This second difficulty can often be obviated by use of a more refined variant of induction.

Let us examine the arithmetic–geometric inequality. Let $a_1, \ldots, a_n$ be a collection of nonnegative real numbers. Then for all $n \ge 1$, we have

(Pn) $\sqrt[n]{a_1 a_2 \cdots a_n} \le \dfrac{a_1 + \cdots + a_n}{n},$

or equivalently,

$$a_1 a_2 \cdots a_n \le \left(\frac{a_1 + \cdots + a_n}{n} \right)^n .$$

For $n = 1$ the assertion is clear, and for $n = 2$ we have $a_1 a_2 \le (\frac{a_1 + a_2}{2})^2 \iff 4 a_1 a_2 \le a_1^2 + 2 a_1 a_2 + a_2^2 \iff 0 \le a_1^2 - 2 a_1 a_2 + a_2^2 = (a_1 - a_2)^2$, and so assertion $(P2)$ is true as well. The transition from n to $n+1$ requires some effort, and so instead, we proceed in two steps:

(a) $(Pn) \implies (P(n-1))$,

(b) $(Pn) \wedge (P2) \implies P(2n)$.

The combination of these two steps yields the entire result (clear?). To prove (a), we set $b = \sum_{i=1}^{n-1} \frac{a_i}{n-1}$ and obtain

$$\left(\prod_{k=1}^{n-1} a_k \right) \sum_{k=1}^{n-1} \frac{a_k}{n-1} = \left(\prod_{k=1}^{n-1} a_k \right) b \overset{(Pn)}{\le} \left(\frac{\sum_{k=1}^{n-1} a_k + b}{n} \right)^n = \left(\frac{n \sum_{k=1}^{n-1} a_k}{n(n-1)} \right)^n$$

$$= \left(\frac{1}{n-1} \right)^n \left(\sum_{k=1}^{n-1} a_k \right)^n ,$$

and hence

$$\prod_{k=1}^{n-1} a_k \le \left(\frac{1}{n-1} \right)^{n-1} \left(\sum_{k=1}^{n-1} a_k \right)^{n-1} .$$

For (b) we have

$$\prod_{k=1}^{2n} a_k = \left(\prod_{k=1}^{n} a_k \right) \left(\prod_{k=n+1}^{2n} a_k \right) \overset{(Pn)}{\le} \left(\sum_{k=1}^{n} \frac{a_k}{n} \right)^n \left(\sum_{k=n+1}^{2n} \frac{a_k}{n} \right)^n$$

$$\overset{(P2)}{\le} \left(\frac{1}{2} \right)^{2n} \left(\sum_{k=1}^{2n} \frac{a_k}{n} \right)^{2n} = \left(\frac{1}{2n} \right)^{2n} \left(\sum_{k=1}^{2n} a_k \right)^{2n} ,$$

and we are done.

Another useful method consists in **isolating** the first and last terms of a sum. Let $S_n = \sum_{k=0}^{n} a_k$. Then with an index transformation we have

$$S_{n+1} = S_n + a_{n+1} = a_0 + \sum_{k=1}^{n+1} a_k = a_0 + \sum_{k=0}^{n} a_{k+1} .$$

The idea is to compare the last sum with S_n. Let us look at a couple of examples. First we consider the *geometric sum* $S_n = 1 + a^1 + a^2 + \cdots + a^n = \sum_{k=0}^{n} a^k$. Isolating the terms yields

$$S_{n+1} = S_n + a^{n+1} = 1 + \sum_{k=0}^{n} a^{k+1} = 1 + a \sum_{k=0}^{n} a^k = 1 + a S_n ,$$

and we obtain $S_n + a^{n+1} = 1 + aS_n$. That is, $S_n = \frac{a^{n+1}-1}{a-1}$ for $a \neq 1$. For $a = 1$ the result is of course $S_n = n + 1$. Now let us determine $S_n = \sum_{k=0}^{n} k2^k$. Our method gives us

$$S_{n+1} = S_n + (n+1)2^{n+1} = \sum_{k=0}^{n}(k+1)2^{k+1} = 2\sum_{k=0}^{n} k2^k + 2\sum_{k=0}^{n} 2^k$$

$$= 2S_n + 2^{n+2} - 2 ,$$

and hence

$$S_n = (n-1)2^{n+1} + 2 .$$

Once a formula has been proved, it is a good idea to verify it for small values. For $n = 4$ we obtain $S_4 = 2^1 + 2\cdot2^2 + 3\cdot2^3 + 4\cdot2^4 = 2 + 8 + 24 + 64 = 98$, and on the right-hand side, $3 \cdot 2^5 + 2 = 96 + 2 = 98$.

Now we would like to look at the second item mentioned at the beginning of the chapter: the representation of a counting function by a summation formula. The simplest form is as follows: Suppose the values T_n $(n \geq 0)$ that we seek are given by means of a recurrence formula:

$$T_0 = \alpha,$$
$$a_n T_n = b_n T_{n-1} + c_n \qquad (n \geq 1) .$$

In the formula we can express T_{n-1} in terms of T_{n-2}, T_{n-2} in terms of T_{n-3}, and so on, until we reach T_0. The result will be an expression in a_k, b_k, c_k, and α. The following approach simplifies the calculation considerably. We multiply both sides of the recurrence formula by a **summation factor** s_n that satisfies

(2.1) $$s_{n-1}a_{n-1} = s_n b_n .$$

With $S_n = s_n a_n T_n$ we thereby obtain

$$S_n = s_n(b_n T_{n-1} + c_n) = S_{n-1} + s_n c_n,$$

and thus

$$S_n = \sum_{k=1}^{n} s_k c_k + s_0 a_0 T_0,$$

and hence

(2.2) $$T_n = \frac{1}{s_n a_n}\left(\sum_{k=1}^{n} s_k c_k + s_0 a_0 T_0\right) .$$

How do we find the summation factors s_n? By iterating equation (2.1) we obtain

(2.3) $\quad s_n = \dfrac{a_{n-1}s_{n-1}}{b_n} = \dfrac{a_{n-1}a_{n-2}s_{n-2}}{b_n b_{n-1}} = \cdots = \dfrac{a_{n-1}a_{n-2}\cdots a_0}{b_n b_{n-1}\cdots b_1}, \quad s_0 = 1,$

or some suitable multiple. However, we must take care that all the a_i, b_j are nonzero.

As an example, we would like to calculate the number D_n of fixed-point-free permutations, the so-called *derangements*. We have $D_1 = 0$, $D_2 = 1$, and we set $D_0 = 1$. Let $n \geq 3$. We classify the fixed-point-free permutations π according to the image $\pi(1)$ of 1. Clearly, $\pi(1)$ can be any of the numbers $2, 3, \ldots, n$. Suppose $\pi(1) = i$. We now distinguish two cases: $\pi(i) = 1$ and $\pi(i) \neq 1$. In the first case, we have

$$\pi = \begin{pmatrix} 1 & \cdots & i & \cdots & n \\ i & \cdots & 1 & \cdots & \pi(n) \end{pmatrix}.$$

That is, the numbers $k \neq 1, i$ can be mapped in every possible fixed-point-free way, and we thereby obtain D_{n-2} permutations. In the second case, we have

$$\pi = \begin{pmatrix} 1 & \cdots & i & \cdots & n \\ i & \cdots & \pi(i) \neq 1 & \cdots & \pi(n) \end{pmatrix}.$$

We now replace i by 1 in the first row and remove the first place, thus obtaining a fixed-point-free permutation on $\{1, \ldots, n\} \setminus \{i\}$, and conversely, by again setting $1 \rightarrow i$, every such permutation yields a permutation of $\{1, \ldots, n\}$ with $\pi(i) \neq 1$. From the rule of equality it follows that in the second case the result is exactly D_{n-1} permutations. Since $\pi(1)$ can assume the $n - 1$ values $2, \ldots, n$, the summation rule yields the recurrence

(2.4)
$$D_n = (n - 1)(D_{n-1} + D_{n-2}),$$

and this recurrence holds also for $n = 2$, since we have set $D_0 = 1$. In order to be able to apply our technique of summation factors, we need a recurrence of order one. This presents no problem. From (2.4) we see that

$$D_n - nD_{n-1} = -(D_{n-1} - (n - 1)D_{n-2})$$
$$= D_{n-2} - (n - 2)D_{n-3}$$
$$\vdots$$
$$= (-1)^{n-1}(D_1 - D_0) = (-1)^n,$$

whence

(2.5)
$$D_n = nD_{n-1} + (-1)^n \qquad (n \geq 1),$$

and now we have obtained the desired form. With $a_n = 1$, $b_n = n$, $c_n = (-1)^n$ we obtain, in view of (2.3), the summation factor $s_n = \frac{1}{n!}$, and then from (2.2),

$$D_n = n!\left(\sum_{k=1}^{n} \frac{(-1)^k}{k!} + 1 \right) = n! \sum_{k=0}^{n} \frac{(-1)^k}{k!},$$

or

$$\frac{D_n}{n!} = \sum_{k=0}^{n} \frac{(-1)^k}{k!}.$$

From calculus, we know that $\sum_{k=0}^{n} \frac{(-1)^k}{k!}$ converges as $n \to \infty$ to e^{-1}. From this we can derive the following surprising result: If we choose a permutation at random, the probability of having chosen a fixed-point-free permutation, for large n, is approximately $e^{-1} \approx \frac{1}{2.71828} > \frac{1}{3}$. We offer the following amusing interpretation: If a gust of wind sweeps the manuscript pages of a book out of the author's hands and blows them about, thoroughly mixing them up, the probability that afterward not a single page is in its correct place is greater than $\frac{1}{3}$, a truly depressing realization.

2.2. The Calculus of Finite Differences

We may regard the sum $\sum_{k=a}^{b} g(k)$ as a discrete analogue of the definite integral $\int_a^b g(x)\,dx$. The fundamental theorem of differential calculus yields the following well-known method of evaluating such an integral. Let D denote the differential operator. Let f be an *antiderivative* of g, that is, $g = Df$. Then

(2.6)
$$\int_a^b g(x)\,dx = f(b) - f(a).$$

We would like to investigate whether in the discrete case we can find an analogous "differential operator," one that would make possible the calculation of sums as in (2.6).

In calculus, $Df(x)$ is approximated by quotients of the form $\frac{f(x+h)-f(x)}{h}$. In the discrete case, the best approximation is for $h = 1$, which is $f(x+1) - f(x)$.

For a function $f(x)$ we define the **translation operator** E^a with *step size a* by $E^a : f(x) \mapsto f(x+a)$, where we set $E = E^1$ and $I = E^0$. Thus I is the *identity*. We now define the two fundamental difference operators $\Delta = E - I$ and $\nabla = I - E^{-1}$, that is,

$$\Delta : f(x) \mapsto f(x+1) - f(x),$$
$$\nabla : f(x) \mapsto f(x) - f(x-1).$$

The operator Δ is called the **forward difference operator**, and ∇ the **backward difference operator**.

For example, we have $\Delta(x^3) = (x+1)^3 - x^3 = 3x^2 + 3x + 1$. That is, Δ maps the polynomial x^3 to the second-degree polynomial $3x^2 + 3x + 1$. In general, Δ reduces the degree of a polynomial by 1, since the highest powers cancel.

We can add operators in the usual way and multiply by a scalar. We also have an operator product, *composition*:

$$(P + Q)f = Pf + Qf\,,$$
$$(\alpha P)f = \alpha(Pf)\,,$$
$$(QP)f = Q(Pf)\,.$$

The same rules of calculation as for real numbers hold for these operators, with the exception of the existence of a multiplicative inverse. For example, let us calculate Δ^n. Since $\Delta = E - I$, we have $\Delta^n = (E - I)^n$, and by applying the binomial theorem to $(E-I)^n$, we obtain the important formula

$$\Delta^n f(x) = (E - I)^n f(x) = \sum_{k=0}^{n}(-1)^{n-k}\binom{n}{k}E^k f(x)$$

$$(2.7) \qquad = \sum_{k=0}^{n}(-1)^{n-k}\binom{n}{k}f(x + k)\,.$$

In particular, for $x = 0$, we obtain

$$(2.8) \qquad \Delta^n f(0) = \sum_{k=0}^{n}(-1)^{n-k}\binom{n}{k}f(k)\,.$$

Therefore, we can calculate $\Delta^n f(x)$ at the point $x = 0$ (or at any other point) *without* knowing the polynomial $\Delta^n f(x)$. As an example, let us consider $\Delta^3(x^4)$. Here we have $\Delta^3(x^4)_{x=0} = \sum_{k=0}^{3}(-1)^{3-k}\binom{3}{k}k^4 = -0 + 3 \cdot 1 - 3 \cdot 2^4 + 3^4 = 36$.

And now back to our main topic. An important rule of differentiation is $Dx^n = nx^{n-1}$ for $n \in \mathbb{Z}$. For the difference operators Δ and ∇ there are also sequences with these properties, namely the *falling* and *rising factorials* $x^{\underline{n}} = x(x-1)\cdots(x-n+1)$ and $x^{\overline{n}} = x(x+1)\cdots(x+n-1)$, which we met earlier, in Section 1.4.

We have $(x+1)^{\underline{n}} = (x+1)x^{\underline{n-1}}$, $x^{\underline{n}} = x^{\underline{n-1}}(x - n + 1)$ and therefore

$$(2.9) \quad \Delta x^{\underline{n}} = (x+1)^{\underline{n}} - x^{\underline{n}} = (x+1)x^{\underline{n-1}} - x^{\underline{n-1}}(x - n + 1) = nx^{\underline{n-1}}\,,$$

and analogously,

$$(2.10) \quad \nabla x^{\overline{n}} = x^{\overline{n}} - (x - 1)^{\overline{n}} = x^{\overline{n-1}}(x + n - 1) - (x - 1)x^{\overline{n-1}} = nx^{\overline{n-1}}\,.$$

We would like to extend (2.9) and (2.10) to arbitrary $n \in \mathbb{Z}$. How, then, should we define $x^{\underline{n}}$ and $x^{\overline{n}}$ for $n < 0$? For the quotients $x^{\underline{n}}/x^{\underline{n-1}}$, we obtain $x - n + 1$, thus, for example, $x^{\underline{3}}/x^{\underline{2}} = x - 2$, $x^{\underline{2}}/x^{\underline{1}} = x - 1$, $x^{\underline{1}}/x^{\underline{0}} = x$. For the next quotient we should obtain $x^{\underline{0}}/x^{\underline{-1}} = 1/x^{\underline{-1}} = x + 1$, and so we *define* $x^{\underline{-1}} = \frac{1}{x+1}$, and then $x^{\underline{-2}} = \frac{1}{(x+1)(x+2)}$, and so on. We proceed analogously for $x^{\overline{n}}$. We summarize with the following definitions:

$$x^{\underline{n}} = x(x-1)\cdots(x-n+1), \quad n \geq 0,$$

$$x^{\underline{-n}} = \frac{1}{(x+1)\cdots(x+n)}, \quad n > 0,$$

$$x^{\overline{n}} = x(x+1)\cdots(x+n-1), \quad n \geq 0,$$

$$x^{\overline{-n}} = \frac{1}{(x-1)\cdots(x-n)}, \quad n > 0.$$

Formulas (2.9) and (2.10) now hold for all $n \in \mathbb{Z}$. Let us check this for Δ:

$$\Delta x^{\underline{-n}} = (x+1)^{\underline{-n}} - x^{\underline{-n}} = \frac{1}{(x+2)\cdots(x+n+1)} - \frac{1}{(x+1)\cdots(x+n)}$$

$$= \frac{1}{(x+1)\cdots(x+n+1)}(x+1-x-n-1)$$

$$= (-n)\frac{1}{(x+1)\cdots(x+n+1)} = (-n)x^{\underline{-n-1}}.$$

In conclusion, for all $n \in \mathbb{Z}$ we have

$$\Delta x^{\underline{n}} = n\,x^{\underline{n-1}} \quad \text{and} \quad \nabla x^{\overline{n}} = n\,x^{\overline{n-1}}.$$

We will hereinafter concentrate on the operator Δ. Let us first recall the analogous method from calculus. To compute $\int_a^b g(x)\,dx$, we determine an antiderivative f, that is, $Df = g$, and we then obtain $\int_a^b g(x)\,dx = f(b) - f(a)$.

In the discrete case we proceed analogously: We call f a (discrete) **antiderivative** of g if $\Delta f = g$. We write $f = \sum g$ and call f an **indefinite sum**. Thus

$$\Delta f = g \iff f = \sum g.$$

The following result is the precise analogue of the fundamental theorem of differential calculus:

Theorem 2.1. *Let f be an antiderivative of g. Then*

$$\sum_{k=a}^{b} g(k) = f(b+1) - f(a).$$

Proof. Since $\Delta f = g$, we have $f(k+1) - f(k) = g(k)$ for all k, and we thus obtain

$$\sum_{k=a}^{b} g(k) = \sum_{k=a}^{b} (f(k+1) - f(k)) = f(b+1) - f(a),$$

since the $f(k)$ in successive terms $a < k \leq b$ add to zero. □

We thus have the following method at our disposal: To determine the sum $\sum_{k=a}^{b} g(k)$, we find an antiderivative $f = \sum g$, and obtain

$$\sum_{k=a}^{b} g(k) = \sum_{a}^{b+1} g(x) = f(x) \big|_{a}^{b+1} = f(b+1) - f(a) .$$

Caution: The limits of summation for f are a and $b+1$. Thus in order to use our method effectively, we need a list of antiderivatives. We already know one example:

$$\sum x^n = \frac{x^{n+1}}{n+1} \quad \text{for } n \neq -1 .$$

What is $\sum x^{-1}$? From $x^{-1} = \frac{1}{x+1} = f(x+1) - f(x)$ we immediately obtain $f(x) = 1 + \frac{1}{2} + \cdots + \frac{1}{x}$, that is, $f(x) = H_x$, our familiar harmonic number. In sum, we have

(2.11)
$$\sum x^n = \begin{cases} \frac{x^{n+1}}{n+1}, & n \neq -1, \\ H_x, & n = -1 . \end{cases}$$

Thus H_x is the discrete analogue of the logarithm, and that is why the harmonic numbers appear in many summation formulas. What, then, is the analogue of e^x? We seek a function $f(x)$ such that $f(x) = \Delta f(x) = f(x+1) - f(x)$, from which follows $f(x+1) = 2f(x)$, that is, $f(x) = 2^x$. Let us consider an arbitrary exponential function c^x $(c \neq 1)$. From $\Delta c^x = c^{x+1} - c^x = (c-1)c^x$ we conclude that

$$\sum c^x = \frac{c^x}{c-1} \quad (c \neq 1) .$$

We note that the operators Δ and $\sum$ are linear; that is, we always have $\Delta(\alpha f + \beta g) = \alpha \Delta f + \beta \Delta g$ and $\sum(\alpha f + \beta g) = \alpha \sum f + \beta \sum g$.

It is now time for us to apply our results. For example, if we wish to compute $\sum_{k=0}^{n} k^2$, we need an antiderivative of x^2. We don't know of one, but we have $x^2 = x(x-1) + x = x^{\underline{2}} + x^{\underline{1}}$ and thereby obtain

$$\sum_{k=0}^{n} k^2 = \sum_{0}^{n+1} x^2 = \sum_{0}^{n+1} x^{\underline{2}} + \sum_{0}^{n+1} x^{\underline{1}} = \frac{x^{\underline{3}}}{3} \bigg|_{0}^{n+1} + \frac{x^{\underline{2}}}{2} \bigg|_{0}^{n+1}$$

$$= \frac{(n+1)^{\underline{3}}}{3} + \frac{(n+1)^{\underline{2}}}{2} = \frac{(n+1)n(n-1)}{3} + \frac{(n+1)n}{2}$$

$$= \frac{n(n+\frac{1}{2})(n+1)}{3} .$$

It is clear how this method can be applied to arbitrary power sums $\sum_{k=0}^{n} k^m$. We know from Section 1.4 that $x^m = \sum_{k=0}^{m} S_{m,k} x^{\underline{k}}$. Therefore, for $m \geq 1$

we have

$$\sum_{k=0}^{n} k^m = \sum_{0}^{n+1} x^m = \sum_{0}^{n+1} \left(\sum_{k=0}^{m} S_{m,k} x^{\underline{k}} \right) = \sum_{k=0}^{m} S_{m,k} \sum_{0}^{n+1} x^{\underline{k}}$$

$$= \sum_{k=0}^{m} S_{m,k} \left. \frac{x^{\underline{k+1}}}{k+1} \right|_{0}^{n+1} = \sum_{k=0}^{m} \left. \frac{S_{m,k}}{k+1} x^{\underline{k+1}} \right|_{0}^{n+1}$$

$$= \sum_{k=0}^{m} \frac{S_{m,k}}{k+1} (n+1) n \cdots (n-k+1) \,.$$

We have thus reduced the power sum to elementary quantities, Stirling numbers, and falling factorials. In particular, we see that $\sum_{k=0}^{n} k^m$ is a polynomial in n of degree $m+1$ with leading coefficient $\frac{1}{m+1}$ and constant term 0 (since $S_{m,0} = 0$ for $m \geq 1$). In Exercises 2.37 and 3.45 we will look at these polynomials more closely.

There is also a rule for **partial summation**. From

$$\Delta\big(u(x)v(x)\big) = u(x+1)v(x+1) - u(x)v(x)$$
$$= u(x+1)v(x+1) - u(x)v(x+1) + u(x)v(x+1) - u(x)v(x)$$
$$= (\Delta u(x))v(x+1) + u(x)(\Delta v(x)),$$

it follows that

$$\sum u \Delta v = uv - \sum (Ev)\Delta u \,,$$

which is precisely the analogue of partial integration, aside from the additional translation E.

We can now compute the known sum $\sum_{k=0}^{n} k 2^k$ as follows: We set $u(x) = x$, $\Delta v(x) = 2^x$, and on account of $\sum 2^x = 2^x, \Delta x = 1$, obtain

$$\sum_{k=0}^{n} k 2^k = \sum_{0}^{n+1} x 2^x = \left. x 2^x \right|_{0}^{n+1} - \sum_{0}^{n+1} 2^{x+1} = (n+1)2^{n+1} - \left. 2^{x+1} \right|_{0}^{n+1}$$

$$= (n+1)2^{n+1} - 2^{n+2} + 2 = (n-1)2^{n+1} + 2 \,.$$

A further example: We would like to sum the first n harmonic numbers. From $u(x) = H_x$, $\Delta v(x) = 1 = x^{\underline{0}}$ and taking into account (2.11), we obtain

$$\sum_{k=1}^{n} H_k = \sum_{1}^{n+1} H_x x^{\underline{0}} = \left. H_x x \right|_{1}^{n+1} - \sum_{1}^{n+1} \frac{1}{x+1}(x+1) = \left. H_x x \right|_{1}^{n+1} - \left. x \right|_{1}^{n+1}$$

$$= (n+1)H_{n+1} - 1 - (n+1) + 1 = (n+1)(H_{n+1} - 1) \,.$$

Of course, we could derive this result as well by means of our direct methods from Section 2.1, but with considerably more effort. On the other hand, the calculus of finite differences follows completely automatically. Since this worked so well for $\sum H_k$, let us try a more complicated example.

What is $\sum_{k=1}^{n} \binom{k}{m} H_k$? From binomial recurrence we have $\binom{x+1}{m+1} = \binom{x}{m} + \binom{x}{m+1}$, and therefore $\Delta\binom{x}{m+1} = \binom{x}{m}$, or $\sum\binom{x}{m} = \binom{x}{m+1}$. Partial summation with $u(x) = H_x$ and $\Delta v(x) = \binom{x}{m}$ yields

$$\sum_{k=1}^{n}\binom{k}{m}H_k = \sum_{1}^{n+1}\binom{x}{m}H_x = \binom{x}{m+1}H_x \bigg|_1^{n+1} - \sum_{1}^{n+1}\frac{1}{x+1}\binom{x+1}{m+1}$$

$$= \binom{x}{m+1}H_x\bigg|_1^{n+1} - \frac{1}{m+1}\sum_{1}^{n+1}\binom{x}{m}$$

$$= \binom{x}{m+1}H_x\bigg|_1^{n+1} - \frac{1}{m+1}\binom{x}{m+1}\bigg|_1^{n+1}$$

$$= \binom{n+1}{m+1}\left(H_{n+1} - \frac{1}{m+1}\right), \quad m \geq 0,$$

since the lower limits cancel.

There is yet another formula from calculus that we can adapt to the discrete situation. Let $f(x)$ be a polynomial, $f(x) = \sum_{k=0}^{n} a_k x^k$. Then we know that for the coefficients a_k, we have $a_k = \frac{f^{(k)}(0)}{k!} = \frac{D^k f(0)}{k!}$, where $D^k f$ is the kth derivative of f. The sum $f(x) = \sum_{k=0}^{n} \frac{D^k f(0)}{k!} x^k$ is the well-known *Taylor series* of f (at the point 0). In the calculus of finite differences, Δ corresponds to the differential operator D, while $x^{\underline{k}}$ corresponds to x^k, and indeed, for a polynomial of degree n, we have

(2.12)
$$f(x) = \sum_{k=0}^{n} \frac{\Delta^k f(0)}{k!} x^{\underline{k}} = \sum_{k=0}^{n} \Delta^k f(0)\binom{x}{k}.$$

Equation (2.12) is called the **Newton representation** of f. To prove this, we observe that f can be represented uniquely in the form $f(x) = \sum_{k=0}^{n} b_k x^{\underline{k}}$. If f has degree 0, then this is obviously true: $f = a_0 = a_0 x^{\underline{0}}$. If now a_n is the leading coefficient of f, then the polynomial $g(x) = f(x) - a_n x^{\underline{n}}$ has degree $n-1$, and the result follows by induction. It therefore remains to show that $b_k = \frac{\Delta^k f(0)}{k!}$. We observe first that

$$\Delta^k x^{\underline{i}} = i(i-1)\cdots(i-k+1)x^{\underline{i-k}} = i^{\underline{k}}x^{\underline{i-k}}.$$

From $f(x) = \sum_{i=0}^{n} b_i x^{\underline{i}}$ it follows, by the linearity of Δ, that $\Delta^k f(x) = \sum_{i=0}^{n} b_i i^{\underline{k}} x^{\underline{i-k}}$. For $i < k$ we have $i^{\underline{k}} = 0$, and for $i > k$ we have that $x^{\underline{i-k}}$ is equal to zero at zero. We therefore obtain

$$\Delta^k f(0) = b_k k^{\underline{k}} = k! b_k, \quad \text{that is,} \quad b_k = \frac{\Delta^k f(0)}{k!}.$$

As an example, consider $f(x) = x^n$. In this case we know from (1.16) that $b_k = S_{n,k}$, and we therefore conclude that $k!\, S_{n,k} = (\Delta^k x^n)_{x=0}$. From (2.3) we conclude (with summation index i) that

$$k!\, S_{n,k} = (\Delta^k x^n)_{x=0} = \sum_{i=0}^{k} (-1)^{k-i} \binom{k}{i} i^n \,,$$

and we thereby obtain a summation formula for the Stirling numbers of the second kind:

$$(2.13) \qquad S_{n,k} = \frac{1}{k!} \sum_{i=0}^{k} (-1)^{k-i} \binom{k}{i} i^n \,.$$

2.3. Inversion

Let us look now at the formulas (2.8) and (2.12) from the previous section, where we set $x = n$ in (2.12):

$$\Delta^n f(0) = \sum_{k=0}^{n} (-1)^{n-k} \binom{n}{k} f(k),$$

$$f(n) = \sum_{k=0}^{n} \binom{n}{k} \Delta^k f(0) \,.$$

If we set $u_k = f(k)$, $v_k = \Delta^k f(0)$, then we see that the first formula expresses the quantity v_n in terms of $u_0, u_1, \ldots, u_n$, while the second expresses the number u_n in terms of $v_0, v_1, \ldots, v_n$. We are thus dealing with an **inversion formula**. We ask whether a general principle underlies these formulas. The first formula came from the equation

$$(2.14) \qquad \Delta^n = (E - I)^n,$$

that is, we have expressed Δ in terms of E. If we turn things around, we see that

$$(2.15) \qquad E^n = (\Delta + I)^n \,,$$

and of course this yields the second formula, since $E^n = (\Delta + I)^n = \sum_{k=0}^{n} \binom{n}{k} \Delta^k$ applied to f implies

$$f(x + n) = \sum_{k=0}^{n} \binom{n}{k} \Delta^k f(x) \,,$$

and so with $x = 0$, we have

$$f(n) = \sum_{k=0}^{n} \binom{n}{k} \Delta^k f(0) \,.$$

The crucial fact is thus the relationship between (2.14) and (2.15), and this is simply a twofold application of the binomial theorem. If we set $E = x$ and $\Delta = x - 1$, then (2.14) and (2.15) reduce to the formulas

$$(x-1)^n = \sum_{k=0}^{n} (-1)^{n-k} \binom{n}{k} x^k \quad \text{and} \quad x^n = \sum_{k=0}^{n} \binom{n}{k} (x-1)^k \,.$$

We can see now what the general principle is. A **basis sequence** $(p_0(x), p_1(x), \dots)$ is a sequence of polynomials of degree $p_n = n$. Therefore, $p_0(x)$ is a nonzero constant, $p_1(x)$ has degree 1, and so on. Our standard examples are the powers (x^n) and the falling and rising factorials $(x^{\underline{n}})$ and $(x^{\overline{n}})$. If $f(x)$ is an arbitrary polynomial of degree n, we can represent $f(x)$ uniquely as a linear combination of $p_k(x)$, $0 \le k \le n$. We have already carried out a proof for the falling factorials $x^{\underline{k}}$ in the previous section, and it can be applied word for word to any basis sequence. In the language of linear algebra, the polynomials $p_0(x), p_1(x), \dots, p_n(x)$ form a basis in the vector space of all polynomials of degree less than or equal to n.

Now let $(p_n(x))$ and $(q_n(x))$ be two basis sequences. Then we can express every $q_n(x)$ uniquely in terms of $p_0(x), \dots, p_n(x)$; and conversely, every $p_n(x)$ can be expressed in terms of $q_0(x), \dots, q_n(x)$. That is, there exist unique coefficients $a_{n,k}$ and $b_{n,k}$ with

$$(2.16) \qquad\qquad q_n(x) = \sum_{k=0}^{n} a_{n,k} p_k(x),$$

$$(2.17) \qquad\qquad p_n(x) = \sum_{k=0}^{n} b_{n,k} q_k(x) \,.$$

We call $a_{n,k}$, $b_{n,k}$ the **connection coefficients**, where we set $a_{n,k} = b_{n,k} = 0$ for $n < k$. The coefficients $(a_{n,k})$ and $(b_{n,k})$ form two (infinite) lower triangular matrices. The relations (2.16) and (2.17) can be expressed as matrix equations as follows: Let $A = (a_{i,j})$, $B = (b_{i,j})$, $0 \le i, j \le n$. Then

$$\sum_{k \ge 0} a_{n,k} b_{k,m} = [n = m] \,.$$

That is, the matrices A and B are inverses of each other: $A = B^{-1}$.

Theorem 2.2. *Let $(p_n(x))$ and $(q_n(x))$ be two basis sequences with connection coefficients $a_{n,k}$ and $b_{n,k}$. Then for two sequences of integers $u_0, u_1, u_2, \dots$ and $v_0, v_1, v_2, \dots$, we have*

$$v_n = \sum_{k=0}^{n} a_{n,k} u_k \ (\forall n) \iff u_n = \sum_{k=0}^{n} b_{n,k} v_k \ (\forall n) \,.$$

Proof. Since the matrices $A = (a_{i,j})$, $B = (b_{i,j})$, $0 \leq i, j \leq n$, are inverses, for two vectors $u = (u_0, \ldots, u_n)$ and $v = (v_0, \ldots, v_n)$, we have

$$v = Au \iff u = Bv.$$

$\square$

Thus every pair of basis sequences yields an inversion formula, provided that we can determine the connection coefficients. Let us return yet again to our example

$$x^n = \sum_{k=0}^{n} \binom{n}{k}(x-1)^k, \qquad (x-1)^n = \sum_{k=0}^{n}(-1)^{n-k}\binom{n}{k}x^k.$$

For two sequences $u_0, \ldots, u_n$ and $v_0, \ldots, v_n$ we then have, from Theorem 2.2,

$$(2.18) \qquad v_n = \sum_{k=0}^{n}\binom{n}{k}u_k \ (\forall n) \iff u_n = \sum_{k=0}^{n}(-1)^{n-k}\binom{n}{k}v_k \ (\forall n).$$

Formula (2.18) is called **binomial inversion**. By setting $u_n \to (-1)^n u_n$, we can express this in a symmetric form:

$$(2.19) \qquad v_n = \sum_{k=0}^{n}(-1)^k\binom{n}{k}u_k \ (\forall n) \iff u_n = \sum_{k=0}^{n}(-1)^k\binom{n}{k}v_k \ (\forall n).$$

The method of inversion can be thought of as follows: We would like to determine a counting function (that is, a sequence of coefficients). If we can express a *known* function in terms of the one to be determined on one side of the inversion formula, then the desired sequence will be expressed on the other side of the formula.

As an example, let us look once more at the derangement numbers D_n. Let $d(n, k)$ denote the number of permutations of length n with exactly k fixed points, so that $d(n, 0) = D_n$. Since we can choose the k fixed points in $\binom{n}{k}$ ways, we have

$$d(n, k) = \binom{n}{k}D_{n-k},$$

and therefore

$$(2.20) \qquad n! = \sum_{k=0}^{n}d(n,k) = \sum_{k=0}^{n}\binom{n}{k}D_{n-k} = \sum_{k=0}^{n}\binom{n}{k}D_k.$$

If we now apply binomial inversion (2.18) with $u_n = D_n$, $v_n = n!$, we obtain our old summation formula

$$D_n = \sum_{k=0}^{n}(-1)^{n-k}\binom{n}{k}k! = n!\sum_{k=0}^{n}\frac{(-1)^{n-k}}{(n-k)!} = n!\sum_{k=0}^{n}\frac{(-1)^k}{k!}.$$

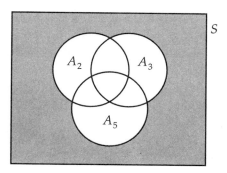

Figure 1. The shaded area represents the subset of the set S of positive integers less than or equal to 30 that are not multiples of 2, 3, or 5.

Let us look once again at the basis sequences (x^n) and $(x^{\underline{n}})$. From the relations

$$x^n = \sum_{k=0}^{n} S_{n,k} x^{\underline{k}}, \qquad x^{\underline{n}} = \sum_{k=0}^{n} (-1)^{n-k} s_{n,k} x^k$$

in (1.16) and (1.17), we obtain the **Stirling inversion** formula

$$v_n = \sum_{k=0}^{n} S_{n,k} u_k \ (\forall n) \iff u_n = \sum_{k=0}^{n} (-1)^{n-k} s_{n,k} v_k \ (\forall n),$$

and also, in particular, $\sum_{k\geq 0} S_{n,k}(-1)^{k-m} s_{k,m} = [n = m]$.

If we check this for $n = 7$, $m = 3$ with the help of our tables of Stirling numbers from Chapter 1, we obtain $\sum_{k\geq 0} S_{7,k}(-1)^{k-3} s_{k,3} = 301 - 350 \cdot 6 + 140 \cdot 35 - 21 \cdot 225 + 1624 = 0$.

2.4. Inclusion–Exclusion

Consider the following problem: How many integers are there between 1 and 30 that are relatively prime to 30? To be sure, we could simply write down all the integers from 1 to 30 and read off the list of those that are relatively prime to 30. But of course our real goal is to be able to answer the question for an arbitrary positive integer n, and the methods that we have developed thus far are of no use to us. Let us try the following approach. Since $30 = 2 \cdot 3 \cdot 5$ is the prime decomposition of 30, we are looking for all integers that are multiples neither of 2 nor of 3 nor of 5. So let us set $S = \{1, 2, \ldots, 30\}$ and let A_2 denote the set of multiples of 2 less than or equal to 30, and analogously define A_3 and A_5 for the corresponding sets of multiples of 3 and of 5. We are now looking for the number of elements in $S \setminus (A_2 \cup A_3 \cup A_5)$. This set appears as the shaded area in the set diagram of Figure 1.

Every element of S belongs to exactly one of the eight regions in the diagram. If we begin our calculation by writing $|S| - |A_2| - |A_3| - |A_5|$, then we have certainly removed all the elements of $A_2 \cup A_3 \cup A_5$ from the total, but we have done so twice for some of them, since an element of $A_2 \cap A_3$, for example, has been removed twice. Let us fix this problem by adding back these extra reductions. This gives us $|S| - |A_2| - |A_3| - |A_5| + |A_2 \cap A_3| + |A_2 \cap A_5| + |A_3 \cap A_5|$. We are making progress: All the elements of S not in one of the three sets A_i, $i = 1, 2, 3$, have been removed only once, with the exception, however, of those in $A_2 \cap A_3 \cap A_5$. These elements have been removed three times, but they have also been counted three times. So in fact, we haven't yet removed them effectively at all. If we now subtract their number from our total, we obtain the correct formula:

$$\left| S \setminus \left(\bigcup_{i=1}^{3} A_i \right) \right| = |S| - |A_2| - |A_3| - |A_5| + |A_2 \cap A_3| + |A_2 \cap A_5|$$
$$+ |A_3 \cap A_5| - |A_2 \cap A_3 \cap A_5|.$$

To answer our original question, we now need to determine $|A_i|$, $|A_i \cap A_j|$, $|A_2 \cap A_3 \cap A_5|$. But that is simple. Since A_2 denotes the multiples of 2, we have $|A_2| = \frac{30}{2} = 15$, and analogously, $|A_3| = 10$, $|A_5| = 6$. Continuing, we note that $A_2 \cap A_3$ contains the multiples of 6, and therefore $|A_2 \cap A_3| = \frac{30}{6} = 5$, and analogously, $|A_2 \cap A_5| = 3$, $|A_3 \cap A_5| = 2$, and finally, $|A_2 \cap A_3 \cap A_5| = 1$, since 30 is the only multiple of all of 2, 3, and 5 in S. We have solved our problem: The number of integers less than or equal to 30 and relatively prime to 30 is $30 - 15 - 10 - 6 + 5 + 3 + 2 - 1 = 8$. The integers in question are $1, 7, 11, 13, 17, 19, 23, 29$.

Thus to calculate $|S \setminus (A_2 \cup A_3 \cup A_5)|$, we exclude certain integers, then include those that were excluded too many times, then exclude those that have now been included too many times, and so on. We have thus developed an *inclusion–exclusion* method, which is valid for arbitrary sets S and subsets $B_1, B_2, \ldots, B_m$ of S.

Theorem 2.3. *Let $B_1, \ldots, B_m$ be subsets of a finite set S. Then*

$$\left| S \setminus \bigcup_{i=1}^{m} B_i \right| = |S| - \sum_{i=1}^{m} |B_i| + \sum_{1 \le i < j \le m} |B_i \cap B_j| - \cdots + (-1)^m |B_1 \cap \cdots \cap B_m|.$$

Proof. To prove this, we need only consider how often an element $x \in S$ is counted. If $x \in S \setminus \bigcup_{i=1}^{m} B_i$, then x is counted once on the right-hand side, namely, in $|S|$. If we choose $x \in \bigcup_{i=1}^{m} B_i$, or more precisely, x in $B_{i_1}, \ldots, B_{i_k}$, but not in the other B_j's, then x is first counted once (in $|S|$), then removed k times (in $|B_{i_1}|, \ldots, |B_{i_k}|$), then added back $\binom{k}{2}$ times (in $|B_{i_1} \cap B_{i_2}|, |B_{i_1} \cap B_{i_3}|, \ldots$), then removed $\binom{k}{3}$ times, and so on. Altogether,

x is counted exactly

$$1 - \binom{k}{1} + \binom{k}{2} - \binom{k}{3} + \cdots + (-1)^k \binom{k}{k}$$

times, and that sum is zero, as we know from (1.12). □

In applications, the formula is usually used in the following form, which is what is generally known as the inclusion–exclusion principle.

Theorem 2.4 (Inclusion–Exclusion Principle). *Let S be as set with n elements, and suppose $E_1, \ldots, E_m$ is a collection of properties that every element of S either does or does not possess. We let $N(E_{i_1} \ldots E_{i_k})$ denote the number of elements of S that satisfy properties $E_{i_1}, \ldots, E_{i_k}$ (and possibly others as well). Then the number $\overline{N}$ of elements that satisfy none of the properties is*

$$(2.21) \quad \overline{N} = n - \sum_{i=1}^{m} N(E_i) + \sum_{1 \le i < j \le m} N(E_i E_j) - \cdots + (-1)^m N(E_1 \ldots E_m).$$

Proof. To prove the theorem, we need only set $B_i = \{x \in S : x \text{ satisfies } E_i\}$. Then $N(E_{i_1} \ldots E_{i_k}) = |B_{i_1} \cap \cdots \cap B_{i_k}|$, $\overline{N} = |S \setminus \bigcup_{i=1}^{m} B_i|$, and (2.21) is simply the formula that we obtained earlier. □

In our original example, E_2 was the property "is divisible by 2," while E_3 was "is divisible by 3," and E_5 "is divisible by 5," and the problem consisted in determining the number $\overline{N}$. Exercise 2.43 determines the general result for arbitrary n.

In many problems, $N(E_{i_1} \ldots E_{i_k})$ depends only on the number k; that is, we have $N(E_{i_1} \ldots E_{i_k}) = N(E_{j_1} \ldots E_{j_k})$ for each pair of k-subsets of $\{E_1, \ldots, E_m\}$. We may therefore set $N_k = N(E_{i_1} \ldots E_{i_k})$ for arbitrary k-sets, and formula (2.21) takes the following simple form (with $N_0 = n$):

$$(2.22) \qquad \overline{N} = \sum_{k=0}^{m} (-1)^k \binom{m}{k} N_k \,.$$

The inclusion–exclusion principle is of irresistible simplicity, and consequently, it is applicable in a wide variety of situations. We illustrate this with a couple of additional examples.

First, we return to our old friends the derangement numbers D_n. It is clear how to proceed. We let S denote the set of n-permutations, and for $i = 1, \ldots, n$ we let E_i denote the property that i is a fixed point. If we select as fixed points $i_1, \ldots, i_k$, then the remaining points can be permuted arbitrarily. Therefore, we have $N(E_{i_1} \ldots E_{i_k}) = (n - k)!$ for every k-set

$\{i_1, \ldots, i_k\} \subseteq \{1, \ldots, n\}$, and by (2.22) we obtain

$$D_n = \sum_{k=0}^{n} (-1)^k \binom{n}{k} (n-k)! = n! \sum_{k=0}^{n} \frac{(-1)^k}{k!} .$$

In the previous section we proved the formula for D_n using binomial inversion, and in fact, one can show that the inclusion–exclusion principle represents an inversion over a suitable structure (see the cited literature for more on this topic).

As a somewhat more difficult example, we consider the n-set $\{a_1, \ldots, a_n\}$ and ask how many words of length $2n$ can be formed that contain each a_i exactly twice, with no pair of equal elements appearing next to each other. For $n = 2$, for example, we have only two words, namely, $a_1 a_2 a_1 a_2$ and $a_2 a_1 a_2 a_1$. We can interpret the problem by imagining a long dining table and asking in how many ways n married couples can be seated in such a way that no husband is sitting next to his wife.[1] Let S denote the set of all words of length $2n$ from the set $\{1, 2, \ldots, n\}$ in which the number i appears exactly twice. If E_i denotes the property that a word contains the number i twice in a row, then we are looking for $\overline{N}$. Let us consider a k-set $\{i_1, \ldots, i_k\}$. How many words contain all of $i_1, i_2, \ldots, i_k$ as neighboring pairs? We first ask in how many ways we can place the $2k$ numbers $i_1, i_1, \ldots, i_k, i_k$ as neighboring pairs. Consider the k initial placements of the k pairs. These are the k places between 1 and $2n - 1$ that differ by at least 2 (since the place immediately following is occupied by the second element). The number of these k choices from $\{1, 2, \ldots, 2n - 1\}$ was calculated in Exercise 1.13 as $\binom{2n-k}{k}$. We can now permute the k pairs in $k!$ ways, and insert the remaining $n - k$ pairs in $\frac{(2n-2k)!}{2^{n-k}}$ ways in the available places (the denominator is due to the double counting of the $n - k$ pairs). We thereby obtain $N_k = \binom{2n-k}{k} k! \frac{(2n-2k)!}{2^{n-k}}$ and therefore, from (2.22),

$$\overline{N} = \sum_{k=0}^{n} (-1)^k \binom{n}{k} \binom{2n-k}{k} k! \frac{(2n-2k)!}{2^{n-k}}.$$

The transformation $k \to n - k$ and simplification yields the final result

$$\overline{N} = \sum_{k=0}^{n} (-1)^{n-k} \binom{n}{k} \binom{n+k}{n-k} (n-k)! \frac{(2k)!}{2^k} = \sum_{k=0}^{n} (-1)^{n-k} \binom{n}{k} \frac{(n+k)!}{2^k}.$$

Let us return briefly to the topic of number partitions. How many partitions are there of the number 7 in which all summands are odd, and how many partitions are there in which the numbers are all distinct? We

---

[1] To keep the problem simple, we are assuming that each married couple consists of one man and one woman.

shall denote these numbers by $p_{\text{odd}}(7)$ and $p_{\text{distinct}}(7)$. Table 1 shows the partitions in question.

Odd	Distinct
7	7
$5 + 1 + 1$	$6 + 1$
$3 + 3 + 1$	$5 + 2$
$3 + 1 + 1 + 1 + 1$	$4 + 3$
$1 + 1 + 1 + 1 + 1 + 1 + 1$	$4 + 2 + 1$

Table 1. Partitions of 7 consisting of odd numbers and distinct numbers.

We have $p_{\text{odd}}(7) = p_{\text{distinct}}(7) = 5$. If we experiment with other small integers, we get the same result, that $p_{\text{odd}}(n) = p_{\text{distinct}}(n)$. Is this a general law? With inclusion–exclusion we can answer the question. Let $p(n)$ be the number of all partitions of n. We first calculate $p_{\text{odd}}(n)$. Let S denote the set of all partitions of n, and E_i the property that an even summand i appears. How many partitions contain 2? Clearly, $p(n-2)$, since we have merely to cross out the 2 in any partition. The result is now clear:

$$p_{\text{odd}}(n) = p(n)$$
$$- p(n-2) - p(n-4) - p(n-6) - \cdots$$
$$+ p(n-2-4) + p(n-2-6) + p(n-2-8) + \cdots$$
$$- p(n-2-4-6) - \cdots .$$

Now on to $p_{\text{distinct}}(n)$. Here we have that E_i is the property that i appears more than once as a summand. We then obtain

$$p_{\text{distinct}}(n) = p(n)$$
$$- p(n-1-1) - p(n-2-2) - p(n-3-3) - \cdots$$
$$+ p(n-1-1-2-2) + p(n-1-1-3-3) + \cdots ,$$

and we see that the two calculations agree line for line, and so indeed, $p_{\text{odd}}(n) = p_{\text{distinct}}(n)$.

The inclusion–exclusion principle is also used extensively in probability theory. Suppose we have n balls and r containers, $n \geq r$. We toss the n balls randomly into the containers. What is the probability that no container remains empty? Let Ω denote the set of all r^n outcomes, all equiprobable. We let A_i denote the event that container i remains empty. Then the inclusion–exclusion principle tells us that for the desired probability $\bar{p}$, we have

$$\bar{p} = 1 - \sum_{i=1}^{r} p(A_i) + \sum_{1 \le i < j \le r} p(A_i \cap A_j) \mp \cdots .$$

The probability for $A_{i_1} \cap \cdots \cap A_{i_k}$ is now clearly $p(A_{i_1} \cap \cdots \cap A_{i_k}) = \frac{(r-k)^n}{r^n}$ (favorable cases divided by total number of cases), and it follows that

$$\bar{p} = \sum_{k=0}^{r} (-1)^k \binom{r}{k} \frac{(r-k)^n}{r^n} = \frac{1}{r^n} \sum_{k=0}^{r} (-1)^{r-k} \binom{r}{k} k^n .$$

Since $\bar{p}$ is the number $r! S_{n,r}$ of surjective mappings (favorable) divided by the number r^n of all mappings (total number of possibilities), we have with the help of the inclusion–exclusion principle again proved the formula (2.13).

Exercises for Chapter 2

▷ **2.1** With the help of the summation factor, solve the recurrence $T_0 = 3$, $2T_n = nT_{n-1} + 3 \cdot n!$ $(n > 0)$.

2.2 Compute $\sum_{k=0}^{n} k x^k$ $(x \ne 1)$: (a) Using the method of isolation of terms. (b) Using the double sum $\sum_{1 \le j \le k \le n} x^k$. (c) By differentiating $\sum_{k=0}^{n} x^k$.

2.3 Calculate $\sum_k [1 \le j \le k \le n]$ as a function of j and n.

▷ **2.4** Write down the number 1 and subtract 1. Multiply the result by 2 and add 1. Multiply the result by 3 and subtract 1. Multiply by 4 and add 1, and so on. Finally, multiply the result by n and add $(-1)^n$. Show that the result is D_n. For example, $4(3(2(1-1)+1)-1)+1 = 9 = D_4$.

2.5 In an array with three rows and n columns there appear, in some order, n red, n white, and n green stones. Show that regardless of the arrangement, one can rearrange the stones in each row in such a way that the stones in each column are all of different colors.

▷ **2.6** Compute $\sum_{k=1}^{n-1} \frac{H_k}{(k+1)(k+2)}$ using partial summation.

2.7 Compute $\sum_{k=1}^{n} \frac{2k+1}{k(k+1)}$ in two different ways: (a) By partial fraction decomposition $\frac{1}{k(k+1)} = \frac{1}{k} - \frac{1}{k+1}$. (b) By partial summation.

2.8 Prove the following analogue of the binomial theorem:

$$(x+y)^{\underline{n}} = \sum_{k=0}^{n} \binom{n}{k} x^{\underline{k}} y^{\underline{n-k}}, \qquad (x+y)^{\overline{n}} = \sum_{k=0}^{n} \binom{n}{k} x^{\overline{k}} y^{\overline{n-k}} .$$

▷ **2.9** The Lah numbers (Ivo Lah, 1955) are defined by

$$L_{n,k} = (-1)^n \frac{n!}{k!} \binom{n-1}{k-1} \qquad (n, k \ge 0).$$

Prove that $(-x)^{\underline{n}} = \sum_{k=0}^{n} L_{n,k} x^{\underline{k}}$ and derive from this the Lah inversion formula:

$$v_n = \sum_{k=0}^{n} L_{n,k} u_k \ (\forall n) \iff u_n = \sum_{k=0}^{n} L_{n,k} v_k \ (\forall n).$$

2.10 How many sequences of letters can one make from the letters E, H, I, R, S, W that contain none of the subsequences WIR, IHR, SIE? For example, RSEWIH is acceptable, but not RSWIHE.

2.11 How many natural numbers less than or equal to one million are neither a square, a cube, or a fifth power?

2.12 Determine the smallest natural number with precisely 28 divisors.

▷ **2.13** We know that $\binom{n-m}{r-m}$ is equal to the number of r-subsets of an n-set that contain a fixed m-set M. Conclude from the inclusion–exclusion formula that $\binom{n-m}{r-m} = \sum_{k=0}^{m} (-1)^k \binom{m}{k} \binom{n-k}{r}$.

2.14 Let $A_n = \sum_{k=0}^{n} n^{\underline{k}}$ $(n \geq 0)$. Derive a recurrence for A_n and calculate A_n.

2.15 Suppose that at Discrete University, 60% of all professors play tennis, 65% play poker, and 50% play chess; 45% play two of these games. What is the largest possible percentage of the university's professorate that engage in all three activities?

2.16 How many integer solutions are there to the equation $x_1 + x_2 + x_3 + x_4 = 30$ satisfying: (a) $0 \leq x_i \leq 10$? (b) $-10 \leq x_i \leq 20$? (c) $0 \leq x_i$, $x_1 \leq 5$, $x_2 \leq 10$, $x_3 \leq 15$, $x_4 \leq 21$?

▷ **2.17** What is the probability that a hand of thirteen playing cards (from an ordinary 52-card deck) contains the following cards? (a) At least one card of each suit? (b) An ace, a king, a queen, and a jack?

▷ **2.18** The game "The Tower of Hanoi" was invented in 1883 by the French mathematician Edouard Lucas, at a time when Hanoi seemed a romantic and far-away place. Three spikes, labeled A, B, C, are stuck in the ground. On spike A are placed n disks of differing diameters arranged in order from smallest at the top to largest at the bottom. A move in the game consists in moving the top disk from one of the spikes onto another spike, with the proviso that no disk may be placed on top of a smaller disk. Develop a recurrence for each of the following numbers and calculate them: (a) T_n, the smallest number of moves to transfer all the disks on spike A to spike B. (b) S_n, as in the previous part, but with the restriction that only moves between A and C and between B and C are allowed. (c) R_n, the number of moves to transfer the disks from spike A to spike B when the tower consists of $2n$ disks, two of each size, where each pair of like-sized disks are indistinguishable (that is, it doesn't matter which lies above the other).

2.19 Calculate $\sum_{k=1}^{n}(-1)^k k$ and $\sum_{k=1}^{n}(-1)^k k^2$ using the method of isolation of terms.

▷ **2.20** The following problem is often called the Josephus problem (after the first-century Jewish general and historian Flavius Josephus, who took part in a Jewish revolt against the Romans): Legend has it that n Jewish rebels have been besieged by the Romans, and they have decided to commit suicide rather than surrender. They do so in the following manner: They form a circle, and those in the circle are numbered clockwise from 1 to n. Starting the count at 1, every second person is executed until only one person remains, who is supposed to kill himself. Determine the number $J(n)$ of the last person remaining. Example: If $n = 10$, the people will be eliminated in the following order: $2, 4, 6, 8, 10, 3, 7, 1, 9$. Thus $J(10) = 5$. Hint: Prove the recurrence $J(2n) = 2J(n) - 1$ $(n \geq 1)$, $J(2n+1) = 2J(n) + 1$ $(n \geq 1)$, $J(1) = 1$.

2.21 Let $J(n)$ be defined as in the previous exercise. Show that if $n = \sum_{k=0}^{m} a_k 2^k$, with $a_k \in \{0, 1\}$, then $J(n) = \sum_{k=1}^{m} a_{k-1} 2^k + a_m$.

▷ **2.22** Calculate $S = \sum_{1 \leq j < k \leq n} (a_k - a_j)(b_k - b_j)$, and derive from the result Chebyshev's inequality (after Pafnuty Chebyshev, 1821–1894):

$$\left(\sum_{k=1}^{n} a_k \right) \left(\sum_{k=1}^{n} b_k \right) \leq n \sum_{k=1}^{n} a_k b_k \quad \text{for} \quad a_1 \leq \cdots \leq a_n, \ b_1 \leq \cdots \leq b_n.$$

Hint: $2S = \sum_{1 \leq j, k \leq n} (a_k - a_j)(b_k - b_j)$.

2.23 The following five exercises are to be solved by induction. (a) On a line are given n segments, each pair of which intersect. Prove that there is a point that belongs to all n segments. (b) In the plane are placed n circular disks each three of which have points in common. Show that there is a point common to all the disks. (c) In three-dimensional space are placed n spheres in such a way that every four of them have points in common. Show that the intersection of all the spheres is nonempty. Hint: Let $s_1, \ldots, s_{n+1}$ denote the segments. Consider $s = s_n \cap s_{n+1}$, and then $s_1, \ldots, s_{n-1}, s$.

▷ **2.24** Suppose there are n farmhouses $H_1, \ldots, H_n$ and n wells $B_1, \ldots, B_n$, no three collinear. Each farmhouse is to be assigned a well, and a straight path is to be built from each farmhouse to its well. Show that there is a way of assigning wells to farmhouses such that no two of the n paths cross.

2.25 Let M be a point on the straight line g in the plane. Let $\overrightarrow{MP_1}, \ldots, \overrightarrow{MP_n}$ denote unit vectors, where the points P_i all lie on the same side of g. Show that if n is odd, then $\left| \overrightarrow{MP_1} + \overrightarrow{MP_2} + \cdots + \overrightarrow{MP_n} \right| \geq 1$, where $|\vec{a}|$ is the length of the vector $\vec{a}$. Hint: Draw a unit semicircle about M.

▷ **2.26** In the plane, $n \geq 3$ points are given. The maximum distance between two points is d. Show that this maximum distance d between points can occur at most n times. Hint: Think about what happens when at least three points stand at the maximum distance from a particular point.

2.27 There are $2n + 1$ weights, each of which weighs an integer number of grams. If any one of the weights is removed, the ones that remain can always be divided into two groups of n weights each such that each group has the same total weight. Prove that as a result of these conditions, all the weights must weigh the same amount.

2.28 Calculate $\Delta(c^x)$ and use the result to compute the sum $\sum_{k=1}^{n} \frac{(-2)^k}{k}$.

▷ **2.29** Let $a_0, a_1, \ldots, a_n \in \mathbb{C}$ be arbitrary. (a) Prove that $\sum_{k=0}^{n}(-1)^k \binom{n}{k}(a_0 + a_1 k + \cdots + a_n k^n) = (-1)^n n! a_n$. (b) From the above result derive the formula

$$\sum_{k=0}^{n} \binom{n}{k}\binom{n+k}{n}(-1)^k = (-1)^n.$$

Hint: In part (a) use $\Delta^n(f(x)) = \sum_{k=0}^{n} \binom{n}{k}(-1)^{n-k} f(x+k)$.

▷ **2.30** The number 15 can be represented in four ways as the sum of consecutive integers: $15 = 15$, $7 + 8$, $4 + 5 + 6$, $1 + 2 + 3 + 4 + 5$. Determine the number of such representations for arbitrary n. Hint: Consider $\sum_{k}^{\ell} x = \frac{1}{2}(\ell^2 - k^2) = \frac{1}{2}(\ell - k)(\ell + k - 1)$.

2.31 Verify the first terms of Simpson's formula $\int_{x}^{x+2} p(t)dt = 2(I + \Delta + \frac{\Delta^2}{6} - \frac{\Delta^4}{180} + \frac{\Delta^6}{180} \pm \cdots)p(x)$, where $p(x)$ is a polynomial.

2.32 Develop a calculus of finite differences with the operator ∇ instead of Δ.

▷ **2.33** Calculate $\sum_{k=0}^{n}(-1)^k / \binom{n}{k}$. Hint: Consider $\sum(-1)^x / \binom{n}{x}$.

2.34 Using partial summation, calculate $\sum_{k=1}^{n} \frac{H_k}{k}$.

2.35 Find the indefinite sum $\sum(-1)^k \binom{m}{k} H_k$. Hint: Partial summation.

2.36 Express $\sum_{k=1}^{n} H_k^2$ in terms of n and H_n.

▷ **2.37** Let $S_m(n) = \sum_{k=0}^{n-1} k^m$, $m \geq 1$. We know from Section 2.2 that $S_m(n)$ is a polynomial in n of degree $m + 1$ with leading coefficient $\frac{1}{m+1}$ and such that $S_m(0) = 0$. Prove the following: (a) $S_m(1 - n) = (-1)^{m+1} S_m(n)$. (b) $S_m(1) = 0$. (c) $S_m(\frac{1}{2}) = (-1)^{m+1} S_m(\frac{1}{2})$. (d) $S_m(\frac{1}{2}) = 0$ for even m. Hint: Set $\sum_{k=a}^{b-1} k^m = \sum_{a}^{b} x^m$ and use summation with suitable limits a and b.

2.38 Determine the connection coefficients of the bases $\{x^{\overline{n}}\}$ and $\{x^{\underline{n}}\}$.

▷ **2.39** Determine the numbers $a_n \in \mathbb{N}_0$ from the identity $n! = a_0 + a_1 n^{\underline{1}} + a_2 n^{\underline{2}} + \cdots$.

2.40 Show that $\sum_k \binom{n}{k} \frac{(-1)^k}{x+k} = \frac{1}{x} \binom{x+n}{n}^{-1}$ and derive a new formula by means of binomial inversion.

2.41 Prove the following formula, known as the Chebyshev inversion formula:

$$v_n = \sum_{k=0}^{\lfloor n/2 \rfloor} \binom{n}{k} u_{n-2k} (\forall n) \iff u_n = \sum_{k=0}^{\lfloor n/2 \rfloor} (-1)^k \frac{n}{n-k} \binom{n-k}{k} v_{n-2k} \, (\forall n).$$

Hint: Consider the polynomial $t_n(x) = \sum_{k=0}^{\lfloor n/2 \rfloor} (-1)^k \frac{n}{n-k} \binom{n-k}{k} x^{n-2k}$ and invert.

▷ **2.42** Prove that the number of integer partitions of n not all of whose summands are divisible by 3 is equal to the number of partitions in which no summand appears more than twice. Example: $4 = 4$, $2 + 2$, $2 + 1 + 1$, $1 + 1 + 1 + 1$ and $4 = 4$, $3 + 1$, $2 + 2$, $2 + 1 + 1$. Generalize the result from 3 to d.

2.43 Let $\varphi(n)$ denote the number of integers k, $1 \leq k \leq n$, relatively prime to n. Prove using inclusion–exclusion that $\varphi(n) = n(1 - \frac{1}{p_1}) \cdots (1 - \frac{1}{p_t})$, where p_i are the prime divisors of n.

2.44 Generalize the inclusion–exclusion principle as follows: Let $E_1, \ldots, E_m$ be properties of the elements of an n-set S. Show that the number of elements that satisfy exactly t properties is given by

$$\sum_{i_1 < \cdots < i_t} N(E_{i_1} \ldots E_{i_t}) - \binom{t+1}{t} \sum N(E_{i_1} \ldots E_{i_{t+1}}) + \cdots \pm \binom{m}{t} N(E_1 \ldots E_m).$$

▷ **2.45** Use the previous exercise to calculate the number of permutations of n items with exactly t fixed points.

Chapter 3

Generating Functions

We come now to the third, and by far richest in applications, method of enumeration theory. We view the desired counting coefficients $a_0, a_1, a_2, \ldots$ as coefficients of a power series $\sum_{n \geq 0} a_n z^n$. We can now calculate with this power series. That is, we operate with the set of coefficients *as a whole*. We shall see that this technique makes many seemingly intractable problems quite easy to solve.

3.1. Definitions and Examples

Suppose we are presented with a sequence $a_0, a_1, a_2, \ldots$. The **generating function** of (a_n) is the formal series $A(z) = \sum_{n \geq 0} a_n z^n$. Two remarks are in order: The variable name z indicates that we are working over the complex numbers $\mathbb{C}$, even though generally we shall be dealing with integers. By "formal," we mean that we are considering the powers z^n as nothing more than placeholders for the calculation, and the question of convergence will be completely ignored. Frequently, it is advantageous to leave the index range unrestricted. We then write $\sum_n a_n z^n$, with the understanding that $a_n = 0$ for $n < 0$. For the coefficients a_n of z^n we also set $a_n = [z^n]A(z)$.

We now proceed to calculate with our formal series. The *sum of* $\sum a_n z^n$ and $\sum b_n z^n$ is of course $\sum (a_n + b_n) z^n$, and a *multiple* $c \sum a_n z^n$ of a series is the series $\sum (ca_n) z^n$. We also have a *product*. If $A(z) = \sum a_n z^n$, $B(z) = \sum b_n z^n$, we set

$$(3.1) \qquad A(z)\, B(z) = \sum_{n \geq 0} \left(\sum_{k=0}^{n} a_k b_{n-k} \right) z^n.$$

The product (3.1) is called the **convolution** of $A(z)$ and $B(z)$. It is obtained simply by multiplying out the formal series. What, then, is the coefficient of z^n in the product $A(z) B(z)$? We must take products of the coefficient a_k of z^k in $A(z)$ with those of b_{n-k} of z^{n-k} in $B(z)$ and then sum all these products. We see that the series $A(z) = 0$ and $A(z) = 1$ are the additive and multiplicative identities in the operations of addition and multiplication of formal series.

Formal series are subject to the usual rules of arithmetic, with the exception that $A(z)$ does not necessarily possess a multiplicative inverse. However, the question of when there exists for a series $A(z)$ an inverse series $B(z)$ such that $A(z) B(z) = 1$ is easy to answer. Since we must have $a_0 b_0 = 1$, we see that $a_0 \neq 0$ is a necessary condition for $A(z)$ to possess an inverse. But in fact, this condition is also sufficient. Let $A(z) = \sum a_n z^n$ with $a_0 \neq 0$. For the desired inverse series $B(z) = \sum b_n z^n$, we must have $b_0 = a_0^{-1}$. Let us assume, then, that $b_0, b_1, \ldots, b_{n-1}$ have already been determined. It then follows from $\sum_{k=0}^n a_k b_{n-k} = 0$ that $b_n = -a_0^{-1} \sum_{k=1}^n a_k b_{n-k}$ is well-defined.

Let us take as an example the *geometric* series $\sum_{n \geq 0} z^n$. From (3.1) we immediately obtain $\left(\sum_{n \geq 0} z^n \right)(1 - z) = 1$, and so the series $1 - z$ is seen to be the inverse of $\sum z^n$, and we write $\sum_{n \geq 0} z^n = \frac{1}{1-z}$.

Here is a list of the most important generating functions:

(a) $\sum_{n \geq 0} z^n = \frac{1}{1-z}$,

(b) $\sum_{n \geq 0} (-1)^n z^n = \frac{1}{1+z}$,

(c) $\sum_{n \geq 0} z^{2n} = \frac{1}{1-z^2}$,

(d) $\sum_{n \geq 0} \binom{c}{n} z^n = (1 + z)^c$ $(c \in \mathbb{C})$,

(e) $\sum_{n \geq 0} \binom{c+n-1}{n} z^n = (1 - z)^{-c}$ $(c \in \mathbb{C})$,

(f) $\sum_{n \geq 0} \binom{m+n}{n} z^n = \frac{1}{(1-z)^{m+1}}$ $(m \in \mathbb{Z})$,

(g) $\sum_{n \geq 0} \frac{z^n}{n!} = e^z$,

(h) $\sum_{n \geq 1} \frac{(-1)^{n+1} z^n}{n} = \log (1 + z)$.

For $c \in \mathbb{N}_0$, formula (d) is the binomial theorem. The general case is proved in calculus. For (e), we use the negation formula (1.10). We have $\binom{c+n-1}{n} z^n = \binom{-c}{n}(-z)^n$, and the result follows from (d). Formula (f) follows from (e) with $m = c - 1$, and the last two expressions are well-known series developments.

We can already demonstrate an important application of the convolution product. Let $A(z) = \sum_{n \geq 0} a_n z^n$ be given. Then we have $\frac{A(z)}{1-z} = \sum_{n \geq 0} (\sum_{k=0}^{n} a_k) z^n$, since all coefficients of $\frac{1}{1-z} = \sum_{n \geq 0} z^n$ are equal to 1. For example, with $A(z) = \frac{1}{1-z}$ we obtain $\frac{1}{(1-z)^2} = \sum_{n \geq 0} (n+1) z^n$, or more generally,

$$\frac{1}{(1-cz)^2} = \frac{1}{1-cz} \sum_{n \geq 0} (cz)^n = \sum_{n \geq 0} (n+1)(cz)^n = \sum_{n \geq 0} (n+1) c^n z^n \,.$$

The index transformation can also be easily expressed. If $A(z) = \sum_n a_n z^n$, then we have $z^m A(z) = \sum_n a_n z^{n+m} = \sum_n a_{n-m} z^n$. That is, multiplication by z^m corresponds to a reduction in index by m. For example, from formula (f) we obtain the equation $\sum_n \binom{n}{m} z^n = \sum_n \binom{n}{n-m} z^n = z^m (1-z)^{-m-1}$.

3.2. Solving Recurrences

Generating functions provide us with important methods for solving arbitrary recurrences with constant coefficients.

As an example, let us consider the simplest of all binary recurrences:

$$F_0 = 0, \quad F_1 = 1, \quad F_n = F_{n-1} + F_{n-2} \quad (n \geq 2) \,.$$

In honor of their discoverer, Leonardo of Pisa (c. 1175–c. 1250), called Fibonacci (*filius Bonacci* = son of Bonaccio) by a nineteenth-century writer, the numbers F_n are called **Fibonacci numbers**. They appear in so many problems that there is an entire mathematical journal devoted to their study.

Here is a table of the first eleven Fibonacci numbers:

n	0	1	2	3	4	5	6	7	8	9	10
F_n	0	1	1	2	3	5	8	13	21	34	55

How can we calculate the nth Fibonacci number F_n? The following method is typical of all recurrences:

1. Express the recurrence in a single formula, including the initial conditions. As always, $F_n = 0$ for $n < 0$. We have $F_n = F_{n-1} + F_{n-2}$ for $n = 0$ as well, but for $n = 1$ we have $F_1 = 1$, but the right-hand side is zero. Therefore, the complete recurrence is

$$F_n = F_{n-1} + F_{n-2} + [n = 1] \,.$$

2. Now interpret the equation in part 1 above with the help of generating functions. We know already that lowering the index corresponds to multiplying by a power of z. We therefore obtain

$$F(z) = \sum F_n z^n = \sum F_{n-1} z^n + \sum F_{n-2} z^n + \sum [n = 1] z^n$$
$$= z F(z) + z^2 F(z) + z \,.$$

3. Solve the equation for $F(z)$. This is easy:

$$F(z) = \frac{z}{1 - z - z^2}.$$

4. Express the right-hand side as a formal series and thereby determine the coefficients.

This is the most difficult step. First, we write $1 - z - z^2$ in the form $1 - z - z^2 = (1 - \alpha z)(1 - \beta z)$, and obtain, by means of a partial fraction decomposition, a and b with

$$\frac{1}{(1 - \alpha z)(1 - \beta z)} = \frac{a}{1 - \alpha z} + \frac{b}{1 - \beta z}.$$

Our task is now complete, and we have

$$F(z) = z\left(\frac{a}{1 - \alpha z} + \frac{b}{1 - \beta z}\right) = z\left(a \sum \alpha^n z^n + b \sum \beta^n z^n\right)$$

$$= \sum_n \left(a\alpha^{n-1} + b\beta^{n-1}\right) z^n$$

and thereby

(3.2) $F_n = a\alpha^{n-1} + b\beta^{n-1}.$

In order to obtain a complete solution to (3.2), we must first determine α and β, and then a and b.

If we set $q(z) = 1 - z - z^2$, then $q^R(z) = z^2 - z - 1$ is called the *reflected polynomial*, and we assert that from $q^R(z) = (z - \alpha)(z - \beta)$ follows $q(z) = (1 - \alpha z)(1 - \beta z)$, that is, α and β are precisely the zeros of the reflected polynomial.

We would like to prove this in general. Let $q(z) = 1 + q_1 z + \cdots + q_d z^d$ be a polynomial over $\mathbb{C}$ of degree $d \geq 1$ with constant coefficient 1. The reflected polynomial $q^R(z)$ arises as a result of reflecting the powers z^i, and so we have $q^R(z) = z^d + q_1 z^{d-1} + \cdots + q_d$, with $q_d \neq 0$. Clearly, we have $q(z) = z^d q^R(\frac{1}{z})$. Now let $\alpha_1, \ldots, \alpha_d$ be the zeros of $q^R(z)$, that is, $q^R(z) = (z - \alpha_1) \cdots (z - \alpha_d)$. Such a representation is always possible over $\mathbb{C}$, where of course the α_i's do not need to be distinct. From $q(z) = z^d q^R(\frac{1}{z})$ we obtain

$$q(z) = z^d \left(\frac{1}{z} - \alpha_1\right) \cdots \left(\frac{1}{z} - \alpha_d\right) = (1 - \alpha_1 z) \cdots (1 - \alpha_d z),$$

as asserted. The determination of $\alpha_1, \ldots, \alpha_d$ (or α, β in our example of the Fibonacci numbers) is therefore nothing more than determining the zeros of $q^R(z)$.

For the Fibonacci numbers we now have

$$q^R(z) = z^2 - z - 1 = \left(z - \frac{1+\sqrt{5}}{2}\right)\left(z - \frac{1-\sqrt{5}}{2}\right),$$

$$q(z) = 1 - z - z^2 = \left(1 - \frac{1+\sqrt{5}}{2}z\right)\left(1 - \frac{1-\sqrt{5}}{2}z\right).$$

The usual notation for these zeros is $\phi = \frac{1+\sqrt{5}}{2}$, $\hat{\phi} = \frac{1-\sqrt{5}}{2}$ (one also sees $\tau, \hat{\tau}$). The number ϕ is called the *golden section*. It is one of the foundational numbers in all of mathematics, known already in antiquity. The name "golden section" comes from the following problem: Construct a rectangle (as in the picture below) with side lengths r and s, with $r \geq s$, such that if you remove a square of side length s from the rectangle, the remaining rectangle, now with side lengths s and $r - s$, has the same proportions as the original rectangle (that is, $\frac{r}{s} = \frac{s}{r-s}$)? This is illustrated in the following diagram:

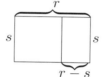

If $x = \frac{r}{s}$, then we should also have $x = \frac{r}{s} = \frac{s}{r-s} = \frac{1}{x-1}$. Thus the number x must satisfy the equation $x^2 - x - 1 = 0$, and so $x = \phi$, since $x \geq 1$ by assumption. From the equation $z^2 - z - 1 = (z - \phi)(z - \hat{\phi})$ we obtain the relations

$$\hat{\phi} = -\phi^{-1}, \quad \phi + \hat{\phi} = 1.$$

We come now to the second problem: the determination of a and b. We set a and b as unknown coefficients in the partial fraction decomposition:

$$\frac{1}{(1 - \phi z)(1 - \hat{\phi}z)} = \frac{a}{1 - \phi z} + \frac{b}{1 - \hat{\phi}z}.$$

Taking a common denominator, we obtain $(a + b) - (a\hat{\phi} + b\phi)z = 1$; that is, a and b must satisfy the system of equations

$$a + b = 1,$$

$$\hat{\phi}a + \phi b = 0,$$

which we solve to obtain $a = \frac{\phi}{\sqrt{5}}$, $b = -\frac{\hat{\phi}}{\sqrt{5}}$. Then from (3.2) we obtain

$$F_n = \frac{\phi}{\sqrt{5}}\phi^{n-1} - \frac{\hat{\phi}}{\sqrt{5}}\hat{\phi}^{n-1},$$

or equivalently,

$$(3.3) \qquad F_n = \frac{1}{\sqrt{5}} \left(\left(\frac{1 + \sqrt{5}}{2} \right)^n - \left(\frac{1 - \sqrt{5}}{2} \right)^n \right) .$$

Since $\left| \frac{1 - \sqrt{5}}{2} \right| < 1$, we see that F_n is the nearest integer to $\frac{1}{\sqrt{5}} \phi^n$. The following general theorem asserts that our steps 1 to 4 presented above always work.

Theorem 3.1. *Let $q_1, \ldots, q_d$ be a fixed sequence of complex numbers, $d \geq 1$, $q_d \neq 0$, $q(z) = 1 + q_1 z + \cdots + q_d z^d = (1 - \alpha_1 z)^{d_1} \cdots (1 - \alpha_k z)^{d_k}$, where the α_i are the distinct zeros of $q^R(z)$ with multiplicities d_i, $i = 1, \ldots, k$. For a counting function $f : \mathbb{N}_0 \to \mathbb{C}$ the following conditions are equivalent:*

(A1) *Recurrence of length d: For all $n \geq 0$, we have*

$$f(n + d) + q_1 f(n + d - 1) + \cdots + q_d f(n) = 0 .$$

(A2) *Generating function:*

$$F(z) = \sum_{n \geq 0} f(n) z^n = \frac{p(z)}{q(z)} ,$$

where $p(z)$ is a polynomial of degree less than d.

(A3) *Partial fraction decomposition:*

$$F(z) = \sum_{n \geq 0} f(n) z^n = \sum_{i=1}^{k} \frac{g_i(z)}{(1 - \alpha_i z)^{d_i}} ,$$

for polynomials $g_i(z)$ of degree less than d_i, $i = 1, \ldots, k$.

(A4) *Explicit representation:*

$$f(n) = \sum_{i=1}^{k} p_i(n) \alpha_i^n ,$$

where the $p_i(n)$ are polynomials in n of degree less than d_i, $i = 1, \ldots, k$.

Proof. We define the sets V_i by

$$V_i = \{ f : \mathbb{N}_0 \to \mathbb{C} : f \text{ satisfies } (Ai) \} , \quad i = 1, \ldots, 4 .$$

Each of these four sets is a vector space over $\mathbb{C}$, since sum and scalar product again satisfy the given condition. We see at once that each of these vector spaces has dimension d. In (A1) we may choose the initial values $f(0), \ldots, f(d - 1)$ arbitrarily; in (A2) the d coefficients $p_0, p_1, \ldots, p_{d-1}$ of $p(z)$; and in (A3), (A4) the d_i coefficients of $g_i(z)$ and $p_i(n)$ with $\sum_{i=1}^{k} d_i = d$. Therefore, if we can prove that $V_i \subseteq V_j$, it will follow that $V_i = V_j$.

Let $f \in V_2$. Then comparing coefficients for z^{d+n} in $q(z) \sum_{n \geq 0} f(n) z^n = p(z)$ immediately yields the recurrence (A1), that is, $f \in V_1$, and therefore $V_1 = V_2$.

Let $f \in V_3$. Taking a common denominator, we obtain

$$\sum_{n \geq 0} f(n) z^n = \frac{\sum_{i=1}^{k} g_i(z) \prod_{j \neq i} (1 - \alpha_j z)^{d_j}}{\prod_{i=1}^{k} (1 - \alpha_i z)^{d_i}} = \frac{p(z)}{q(z)}$$

with $\deg p(z) \leq \max(\deg g_i(z) + \sum_{j \neq i} d_j) < \sum_{i=1}^{k} d_i = d$, and we obtain $f \in V_2$, and thereby $V_1 = V_2 = V_3$.

Finally, we wish to show that $V_3 \subseteq V_4$. Let $f \in V_3$, $F(z) = \sum_{i=1}^{k} \frac{g_i(z)}{(1-\alpha_i z)^{d_i}}$. We consider a summand $\frac{g_i(z)}{(1-\alpha_i z)^{d_i}}$. From example (e) of the previous section we have

$$\frac{1}{(1 - \alpha_i z)^{d_i}} = \sum_{n \geq 0} \binom{d_i + n - 1}{n} \alpha_i^n z^n = \sum_{n \geq 0} \binom{d_i + n - 1}{d_i - 1} \alpha_i^n z^n .$$

Multiplication by $g_i(z) = g_0 + g_1 z + \cdots + g_{d_i-1} z^{d_i-1}$ represents a shift in index, and so we have

$$\frac{g_i(z)}{(1 - \alpha_i z)^{d_i}} = \sum_{n \geq 0} \left(\sum_{j=0}^{d_i-1} g_j \binom{d_i + n - j - 1}{d_i - 1} \alpha_i^{n-j} \right) z^n$$

$$= \sum_{n \geq 0} \left(\sum_{j=0}^{d_i-1} \alpha_i^{-j} g_j \binom{n + d_i - j - 1}{d_i - 1} \alpha_i^n \right) z^n.$$

We now write $p_i(n) = \sum_{j=0}^{d_i-1} \alpha_i^{-j} g_j \binom{n+d_i-j-1}{d_i-1}$. Then $p_i(n)$ is a polynomial in n of degree less than or equal to $d_i - 1$, and therefore we have $f \in V_4$, and the proof is complete. □

As an example, let us consider the recurrence

$$f(n + 2) - 6f(n + 1) + 9f(n) = 0$$

with initial values $f(0) = 0$, $f(1) = 1$. Here we have $q(z) = 1 - 6z + 9z^2 = (1 - 3z)^2$, and thus a solution of the form

$$f(n) = (a + bn)3^n .$$

From $0 = f(0) = a$ we obtain $a = 0$, and from $1 = f(1) = (a + b)3$ we have $b = \frac{1}{3}$. Thus the solution is $f(n) = n3^{n-1}$.

How do we calculate the polynomials $g_i(z)$ or $p_i(n)$? The situation is especially simple when the zeros $\alpha_1, \ldots, \alpha_d$ of $q^R(z)$ are all distinct, that is, when $d_i = 1$ for all i and $k = d$. In this case, the polynomials $g_i(z)$ are of degree 0, that is, $g_i(z) = a_i$ ($i = 1, \ldots, d$), and likewise, $p_i(n) = a_i$. From

$$\frac{p(z)}{q(z)} = \frac{\sum_{i=1}^{d} a_i \prod_{j \neq i} (1 - \alpha_j z)}{\prod_{i=1}^{d} (1 - \alpha_i z)}$$

we obtain $p(z) = \sum_{i=1}^{d} a_i \prod_{j \neq i} (1 - \alpha_j z)$. For $z = \frac{1}{\alpha_i}$ we therefore have

$$p\left(\frac{1}{\alpha_i}\right) = a_i \prod_{j \neq i} \left(1 - \frac{\alpha_j}{\alpha_i}\right),$$

since in a summand $a_h \prod_{j \neq h} (1 - \frac{\alpha_j}{\alpha_i})$ of $p(z)$ for $h \neq i$ the factor $1 - \frac{\alpha_i}{\alpha_i} = 0$ appears. We thus obtain the formula

$$(3.4) \qquad a_i = \frac{p(\frac{1}{\alpha_i})}{\prod_{j \neq i} (1 - \frac{\alpha_j}{\alpha_i})} \qquad (i = 1, \ldots, d).$$

The expression (3.4) can be further simplified. With $q(z) = \prod_{i=1}^{d} (1 - \alpha_i z)$ we have for the derivative $q'(z) = -\sum_{i=1}^{d} \prod_{j \neq i} (1 - \alpha_j z) \alpha_i$, and therefore $q'(\frac{1}{\alpha_i}) = -\prod_{j \neq i} (1 - \frac{\alpha_j}{\alpha_i}) \alpha_i$. Putting this into (3.4) yields

$$(3.5) \qquad a_i = \frac{-\alpha_i p(\frac{1}{\alpha_i})}{q'(\frac{1}{\alpha_i})} \qquad (i = 1, \ldots, d).$$

For example, we can shorten our calculation of the Fibonacci numbers, without traversing the path of partial fractions. Here we have $p(z) = z$, $q(z) = 1 - z - z^2$, $q'(z) = -1 - 2z$, $\alpha_1 = \phi$, $\alpha_2 = \hat{\phi}$, and therefore

$$a_i = \frac{-\alpha_i(\frac{1}{\alpha_i})}{-1 - \frac{2}{\alpha_i}} = \frac{\alpha_i}{\alpha_i + 2}.$$

We immediately compute $\frac{\phi}{\phi+2} = \frac{1}{\sqrt{5}}$, $\frac{\hat{\phi}}{\hat{\phi}+2} = -\frac{1}{\sqrt{5}}$ and again obtain (3.3).

Our method of generating functions also proves successful in solving simultaneous recurrences. Such a problem appeared in a 1980 mathematical competition: In expressing the number $(\sqrt{2} + \sqrt{3})^{1980}$ as a decimal fraction, what are the digits immediately to the right and to the left of the decimal point?

At first glance, the problem appears impossible. And in the current context, what could this problem possibly have to do with recurrence? Let us consider the general expression $(\sqrt{2} + \sqrt{3})^{2n}$. We obtain $(\sqrt{2} + \sqrt{3})^0 = 1$, $(\sqrt{2} + \sqrt{3})^2 = 5 + 2\sqrt{6}$, $(\sqrt{2} + \sqrt{3})^4 = (5 + 2\sqrt{6})^2 = 49 + 20\sqrt{6}$. Are all the expressions $(\sqrt{2} + \sqrt{3})^{2n}$ of the form $a_n + b_n\sqrt{6}$? By induction, the answer is clear:

$$(\sqrt{2}+\sqrt{3})^{2n} = (\sqrt{2}+\sqrt{3})^{2n-2}(\sqrt{2}+\sqrt{3})^2$$
$$= (a_{n-1}+b_{n-1}\sqrt{6})(5+2\sqrt{6})$$
$$= (5a_{n-1}+12b_{n-1}) + (2a_{n-1}+5b_{n-1})\sqrt{6}\,.$$

We thus obtain recurrences for the two sequences (a_n) and (b_n):

(3.6)
$$a_n = 5a_{n-1} + 12b_{n-1},$$
$$b_n = 2a_{n-1} + 5b_{n-1},$$

with initial values $a_0 = 1$, $b_0 = 0$.

Our solution of these recurrences proceeds in four steps:

Step 1.

$$a_n = 5a_{n-1} + 12b_{n-1} + [n=0] \qquad \text{(since } a_0 = 1\text{)},$$
$$b_n = 2a_{n-1} + 5b_{n-1}\,.$$

Step 2.   With $A(z) = \sum_{n\geq 0} a_n z^n$, $B(z) = \sum_{n\geq 0} b_n z^n$ we obtain

$$A(z) = 5zA(z) + 12zB(z) + 1,$$
$$B(z) = 2zA(z) + 5zB(z)\,.$$

Step 3.   We solve for $A(z)$. From the second equation we have $B(z) = \frac{2z\,A(z)}{1-5z}$, which placed into the first equation yields

$$A(z) = 5z\,A(z) + \frac{24z^2\,A(z)}{1-5z} + 1,$$

or

$$A(z) = \frac{1-5z}{1-10z+z^2}\,.$$

Step 4.   Here we have $q(z) = q^R(z)$, and we obtain

$$q(z) = \left(1 - \left(5+2\sqrt{6}\right)z\right)\left(1 - \left(5 - 2\sqrt{6}\right)z\right)\,.$$

Since the two zeros are distinct, we can use formula (3.5), and we easily obtain

(3.7)
$$a_n = \frac{1}{2}\left(\left(5+2\sqrt{6}\right)^n + \left(5-2\sqrt{6}\right)^n\right)$$

for the coefficients of $A(z)$. Very nice! We now know a_n (and b_n can be obtained analogously), but what does this tell us about the decimal digits of $(\sqrt{2}+\sqrt{3})^{2n} = a_n + b_n\sqrt{6}$ for $n = 990$?

First of all, we know that $(5+2\sqrt{6})^n = (\sqrt{2}+\sqrt{3})^{2n} = a_n + b_n\sqrt{6}$. That is, (3.7) yields $a_n = \frac{1}{2}(a_n + b_n\sqrt{6} + (5-2\sqrt{6})^n)$, or

(3.8)
$$a_n = b_n\sqrt{6} + (5-2\sqrt{6})^n\,.$$

Let us denote by $\{x\}$ the fractional part of a real number, that is, $x = \lfloor x \rfloor + \{x\}$, $0 \le \{x\} < 1$. Since a_n is an integer, it follows from (3.8) that $\{b_n\sqrt{6}\} + \{(5 - 2\sqrt{6})^n\} = 1$. We now have $(5 - 2\sqrt{6})^n \to 0$, since $5 - 2\sqrt{6} < 1$. That is, for n large, and certainly for $n = 990$, we have $(5 - 2\sqrt{6})^n = 0.00\ldots$ and therefore $\{b_{990}\sqrt{6}\} = 0.99\ldots$. The first decimal place of $a_{990} + b_{990}\sqrt{6}$ is therefore 9. Now let A be the units digit of a_{990}, and B that of $b_{990}\sqrt{6}$. That is, $a_{990} = \ldots A$ and $b_{990}\sqrt{6} = \ldots B.9\ldots$. From (3.8) we obtain $A \equiv B + 1 \pmod{10}$, and therefore the units digit of $a_{990} + b_{990}\sqrt{6}$ is equal to $A + B \equiv 2A - 1 \bmod 10$. Readers unfamiliar with modular arithmetic will have another look at this in Section 12.1. We now have only to determine A, and to that end we use the original recurrence (3.6). The first values of the units digit (modulo 10) are as follows:

n	a_n	b_n
0	1	0
1	5	2
2	9	0
3	5	8
4	1	0
5	5	2

The units digits of a_n thus repeat themselves periodically with period 4. In particular, $990 \equiv 2 \pmod 4$, that is, $A = 9$, and we obtain the result $2 \cdot 9 - 1 \equiv 7 \pmod{10}$, and so 7 is the last digit before the decimal point.

3.3. Generating Functions of Exponential Type

For many counting functions (a_n) it is advantageous to consider the function

$$\hat{A}(z) = \sum_{n \ge 0} \frac{a_n}{n!} z^n$$

instead of the usual generating functions. We call $\hat{A}(z)$ the **exponential generating function** of the sequence (a_n).

Multiplication of two exponential generating functions $\hat{A}(z) = \sum_{n \ge 0} \frac{a_n}{n!} z^n$, $\hat{B}(z) = \sum_{n \ge 0} \frac{b_n}{n!} z^n$ is easy. Let $\hat{C}(z) = \sum_{n \ge 0} \frac{c_n}{n!} z^n$, with $\hat{C}(z) = \hat{A}(z)\,\hat{B}(z)$. From the product formula (3.1) we obtain

$$\frac{c_n}{n!} = \sum_{k=0}^{n} \frac{a_k}{k!} \frac{b_{n-k}}{(n-k)!},$$

that is,

(3.9)
$$c_n = \sum_{k=0}^{n} \binom{n}{k} a_k b_{n-k}.$$

Because of the appearance of $\binom{n}{k}$, (3.9) is called **binomial convolution**. Suppose the counting functions (a_n), (b_n), and (c_n) are related by (3.9). Then we may conclude at once that $\hat{C}(z) = \hat{A}(z)\,\hat{B}(z)$.

Let us test this on a simple example. For the exponential function e^{az} we have $e^{az} = \sum_{n \geq 0} \frac{a^n}{n!} z^n$. That is, e^{az} is the exponential generating function of the geometric sequence $(a^0, a^1, a^2, \dots)$. From $e^{az} e^{bz} = e^{(a+b)z}$, we obtain immediately, with the help of (3.9),

$$(a+b)^n = \sum_{k=0}^{n} \binom{n}{k} a^k b^{n-k}\,.$$

That is, the binomial theorem is simply the binomial convolution of exponential functions.

Let us consider another example. In Section 3.1 we obtained $\sum_{n \geq 0} \binom{a}{n} z^n = (1+z)^a$. If we write the left-hand side as $\sum_{n \geq 0} \frac{a^{\underline{n}}}{n!} z^n$, we see that $(1+z)^a$ is the exponential generating function of the sequence $(a^{\underline{n}})$. Since

$$(1+z)^a (1+z)^b = (1+z)^{a+b},$$

we obtain, via binomial convolution,

$$(a+b)^{\underline{n}} = \sum_{k=0}^{n} \binom{n}{k} a^{\underline{k}} b^{\underline{n-k}},$$

or

$$\binom{a+b}{n} = \sum_{k=0}^{n} \binom{a}{k} \binom{b}{n-k},$$

the familiar Vandermonde identity (1.13).

What is the exponential generating function of the derangement numbers D_n? In Section 2.3 we obtained in (2.20) the relation $n! = \sum_{k=0}^{n} \binom{n}{k} D_k$. The sequence $(n!)$ is therefore the binomial convolution of the sequence (D_k) with the constant sequence $(1, 1, \dots)$, whose exponential generating function is of course e^z. From (3.9) it follows that $\hat{D}(z) = \sum_{n \geq 0} \frac{D_n}{n!} z^n$,

$$\hat{D}(z) e^z = \sum_{n \geq 0} \frac{n!}{n!} z^n = \sum_{n \geq 0} z^n = \frac{1}{1-z}\,,$$

and we conclude that $\hat{D}(z) = \frac{e^{-z}}{1-z}$. If we look at this as an equation between usual generating functions, then we know from Section 3.1 that the right-hand side sums the first n terms of e^{-z}. We thus again obtain our well-known formula

$$\frac{D_n}{n!} = \sum_{k=0}^{n} \frac{(-1)^k}{k!}\,.$$

In general, our binomial inversion formula (2.18) corresponds to the obvious equation

$$\hat{V}(z) = \hat{U}(z)e^z \iff \hat{U}(z) = \hat{V}(z)e^{-z},$$

with $\hat{U}(z) = \sum_{n \geq 0} \frac{u_n}{n!} z^n$ and $\hat{V}(z) = \sum_{n \geq 0} \frac{v_n}{n!} z^n$.

Here is another example, one that clarifies the concision of the method of generating functions. Let a_n be the number of mappings f from $\{1, \ldots, n\}$ to $\{1, \ldots, n\}$ with the property that if $j \in \text{im}(f)$, the image of f, then all $i < j$ are in $\text{im}(f)$, and $a_0 = 1$. For example, we obtain $a_2 = 3$ with the mappings $1 \to 1, 2 \to 1; 1 \to 1, 2 \to 2; 1 \to 2, 2 \to 1$. Assume that f maps exactly k elements to 1. Then these elements can be chosen in $\binom{n}{k}$ ways, and the remainder can be mapped in a_{n-k} ways to $\{2, \ldots, n\}$. We thereby obtain

$$a_n = \sum_{k=1}^{n} \binom{n}{k} a_{n-k}, \qquad \text{or} \qquad 2a_n = \sum_{k=0}^{n} \binom{n}{k} a_{n-k} + [n = 0].$$

For the exponential generating function this yields

$$2\hat{A}(z) = e^z \hat{A}(z) + 1,$$

that is,

$$\hat{A}(z) = \frac{1}{2 - e^z}.$$

By developing the right-hand side, we obtain

$$\frac{1}{2 - e^z} = \frac{1}{2} \cdot \frac{1}{1 - e^z/2} = \frac{1}{2} \sum_{k \geq 0} \left(\frac{e^z}{2}\right)^k = \sum_{k \geq 0} \frac{1}{2^{k+1}} \sum_{n \geq 0} \frac{k^n z^n}{n!},$$

and comparing coefficients for z^n yields

$$a_n = \sum_{k \geq 0} \frac{k^n}{2^{k+1}}.$$

We have therefore not only found a combinatorial interpretation for the series $\sum_{k \geq 0} \frac{k^n}{2^{k+1}}$. We know also that its value is a natural number, namely a_n. For $n = 2$, for example, we obtain $\sum_{k \geq 0} \frac{k^2}{2^{k+1}} = 3$, and for $n = 3$, we have $\sum_{k \geq 0} \frac{k^3}{2^{k+1}} = 13$.

Exercises for Chapter 3

3.1 Verify the identity $\sum_{k=0}^{n} \binom{n}{k}^2 = \binom{2n}{n}$ using generating functions.

▷ **3.2** A subset $A \subseteq \{1, \ldots, n\}$ is said to be *fat* if $k \geq |A|$ holds for all $k \in A$. For example, $\{3, 5, 6\}$ is fat, but not $\{2, 4, 5\}$. Let $f(n)$ be the number of fat subsets of $\{1, \ldots, n\}$, where $\varnothing$ is fat by definition. Prove the following: (a) $f(n) = F_{n+2}$ (Fibonacci number). (b) Derive from this that $F_{n+1} = \sum_{k=0}^{n} \binom{n-k}{k}$ and (c) $\sum_{k=0}^{n} \binom{n}{k} F_k = F_{2n}$.

3.3 The following exercises reveal some properties of the Fibonacci sequence. Prove the following: (a) $\sum_{k=0}^{n} F_k = F_{n+2} - 1$. (b) $\sum_{k=1}^{n} F_{2k-1} = F_{2n}$. (c) $\sum_{k=0}^{n} F_k^2 = F_n F_{n+1}$.

3.4 Consider the matrix $A = \begin{pmatrix} 1 & 1 \\ 1 & 0 \end{pmatrix}$. Prove that

$$A^n = \begin{pmatrix} F_{n+1} & F_n \\ F_n & F_{n-1} \end{pmatrix}$$

and derive from this that $F_{n+1}F_{n-1} - F_n^2 = (-1)^n$. Prove further that $F_{n+1}^2 - F_{n+1}F_n - F_n^2 = (-1)^n$ and show that conversely, from $|m^2 - mk - k^2| = 1$ for $m, k \in \mathbb{Z}$ one has $\{m, k\} = \{\pm F_{n+1}, \pm F_n\}$ for some $n \in \mathbb{N}_0$.

$\triangleright$ **3.5** The nth Lucas number is given by $L_n = F_{n-1} + F_{n+1}$. Show that $F_{2n} = F_n L_n$ and express L_n in terms of ϕ and $\hat{\phi}$.

3.6 Show that every natural number n possesses a unique representation $n = F_{m_1} + \cdots + F_{m_t}$ with $m_i \geq m_{i+1} + 2$, $m_t \geq 2$. Example: $33 = 21 + 8 + 3 + 1$.

3.7 Let A_n be the number of ways that a $2 \times n$ rectangle can be covered with 1×2 dominos. Determine a recurrence for A_n and calculate A_n.

3.8 Solve the general recurrence $a_0 = \alpha$, $a_1 = \beta$, $a_n = s a_{n-1} + t a_{n-2}$ $(n \geq 2)$.

$\triangleright$ **3.9** Let $f(n)$ be the number of words of length n in which the letters are all 0, 1, and 2 and in which two 0's in a row do not occur. For example, $f(1) = 3$, $f(2) = 8$, $f(3) = 22$. Calculate $f(n)$.

3.10 We considered the Bell number $\tilde{B}_n$ in Exercise 1.12. Determine the exponential generating function for the Bell numbers and prove that $\tilde{B}_n = \frac{1}{e} \sum_{k \geq 0} \frac{k^n}{k!}$.

3.11 Determine the generating function and exponential generating functions of $a_n = 2^n + 5^n$.

$\triangleright$ **3.12** Determine the exponential generating functions for the number a_k of k-permutations of n for fixed n and for the number b_n for fixed k.

3.13 Let $\hat{A}(z) = \sum_{n \geq 0} a_n \frac{z^n}{n!}$. Show that $(\hat{A}(z))' = \sum_{n \geq 0} a_{n+1} \frac{z^n}{n!}$, that is, a shift in the index corresponds to differentiation.

$\triangleright$ **3.14** Use the previous exercise to compute $\hat{S}_m(z) = \sum_{n \geq 0} S_{n,m} \frac{z^n}{n!}$, where the $S_{n,m}$ are Stirling numbers of the second kind. Conclude further that $\sum_{m \geq 0} \hat{S}_m(z) t^m = e^{t(e^z - 1)}$.

3.15 Let $A(z) = \sum a_n z^n$. Show that $\frac{1}{2}(A(z) + A(-z)) = \sum a_{2n} z^{2n}$ and $\frac{1}{2}(A(z) - A(-z)) = \sum a_{2n+1} z^{2n+1}$.

$\triangleright$ **3.16** We know that $F(z) = \frac{z}{1-z-z^2}$ is the generating function of the Fibonacci numbers. From this, determine the generating function $\sum_n F_{2n} z^n$ of the Fibonacci numbers with even index.

$\triangleright$ **3.17** Solve the recurrence $g_0 = 1$, $g_n = g_{n-1} + 2g_{n-2} + \cdots + ng_0$ by considering the generating function $G(z)$ and the previous exercise.

3.18 Determine the generating function of the harmonic numbers.

3.19 Use the previous exercise to determine $\sum_{k=0}^{n} H_k H_{n-k}$.

▷ **3.20** Every generating function $F(z) = \sum_{n \geq 0} a_n z^n$ with $a_0 = 1$ defines a polynomial sequence $(p_n(x))$ via $F(z)^x = \sum_{n \geq 0} p_n(x) z^n$, where $p_n(1) = a_n$ and $p_n(0) = [n = 0]$. Show that $p_n(x)$ is of degree n and prove the convolution formulas $\sum_{k=0}^{n} p_k(x) p_{n-k}(y) = p_n(x + y)$ and $(x + y) \sum_{k=0}^{n} k p_k(x) p_{n-k}(y) = n x p_n(x + y)$. Hint: For $n > 0$ one has that $[z^n] e^{x \log F(z)}$ is a polynomial of degree n in x that is a multiple of x.

3.21 Use the previous exercise to prove the following identies:
(a) $\sum_k \binom{tk+r}{k} \binom{tn-tk+s}{n-k} \frac{r}{tk+r} = \binom{tn+r+s}{n}$, (b) $\sum_k \binom{n}{k} (tk+r)^k (tn - tk + s)^{n-k} \frac{r}{tk+r} = (tn + r + s)^n$.

3.22 Let $p(n)$ be the number of integer partitions of n with $p(0) = 1$. Show that $p(n) = [z^n](\sum z^k)(\sum z^{2k})(\sum z^{3k}) \cdots = [z^n] \frac{1}{(1-z)(1-z^2)\cdots}$. Hint: $\sum z^k$ gives the number of 1-summands, $\sum z^{2k}$ the number of 2-summands, and so on.

3.23 Give a proof of $p_u(n) = p_v(n)$ from Section 2.4 using generating functions. Hint: The generating function of $p_u(n)$ is $P_u(z) = \frac{1}{(1-z)(1-z^3)(1-z^5)\cdots}$. Analogously, compute $P_v(z)$ and compare.

▷ **3.24** A probability generating function $P_X(z)$, or P-generating function for short, for a random variable $X : \Omega \longrightarrow \mathbb{N}_0$ is $P_X(z) = \sum p_n z^n$, where $p_n = p(X = n)$. Prove the following:
(a) $EX = P_X'(1)$, $P_X' =$ the derivative of P_X, (b) $VX = P_X''(1) + P_X'(1) - (P_X'(1))^2$, (c) $P_{X+Y}(z) = P_X(z) P_Y(z)$ for independent random variables X, Y.

3.25 The random variable X assumes the values $0, 1, \ldots, n-1$, each with probability $\frac{1}{n}$. Compute $P_X(z)$.

▷ **3.26** Let X be the number of heads that occur when a coin is tossed n times (each time with probability $\frac{1}{2}$). Calculate $P_X(z)$ and from that derive EX, VX.

3.27 Suppose that a coin (probability $\frac{1}{2}$) is tossed until n heads have appeared. Let X be the number of tosses, where we assume that tails appears at most s times in a row. Calculate $P_X(z)$, EX, and VX.

▷ **3.28** Let Ω be the set of permutations of $\{1, \ldots, n\}$ and X_n the random variable that associates with each permutation the number of inversions (see Exercise 1.31). Show that for the P-generating function $P_n(z)$ of X one has $P_n(z) = \prod_{i=1}^{n} \frac{1 + z + \cdots + z^i}{i}$. Conclude that $EX = \frac{n(n-1)}{4}$, $VX = \frac{n(2n+5)(n-1)}{72}$. Hint: Conclude from $I_{n,k} = I_{n-1,k} + I_{n-1,k-1} + \cdots + I_{n-1,k-n+1}$ the relationship $P_n(z) = \frac{1+z+\cdots+z^{n-1}}{n} P_{n-1}(z)$.

▷ **3.29** Consider a $3 \times n$ rectangle that we cover with 2×1 dominos. Calculate the number A_n of different arrangements (thus, for example, $A_0 = 1$, $A_1 = 0$, $A_2 = 3$). Hint: Let B_n be the number of arrangements in which the upper left corner remains uncovered, that is, $B_0 = 0$, $B_1 = 1$, $B_2 = 0$. Establish recurrences for $A(z)$, $B(z)$.

▷ **3.30** In how many ways can one build a $2 \times 2 \times n$ tower out of $2 \times 1 \times 1$ stones? Hint: Let a_n be the number sought, and b_n the number of ways in which there is a $2 \times 1 \times 1$ stone missing at the top. Solve $A(z)$, $B(z)$ simultaneously.

▷ **3.31** Let a_n be the number of decompositions of the natural number n into sums of powers of 2, where order is irrelevant. Example: $a_3 = 2$, since $3 = 2+1 = 1+1+1$; $a_4 = 4$, since $4 = 4 = 2+2 = 2+1+1 = 1+1+1+1$. Set $a_0 = 1$. Let $b_n = \sum_{k=0}^{n} a_k$. Calculate $A(z)$ and $B(z)$. What do you get for a_n, b_n?

3.32 Consider the sequence $a_n = (1 + \sqrt{2})^n + (1 - \sqrt{2})^n$. Show that a_n is always a natural number. For example, $a_0 = a_1 = 2$, $a_2 = 6$. Determine a recurrence for a_n and conclude that for $n \geq 1$, one has that $\lceil (1 + \sqrt{2})^n \rceil$ is even if and only if n is even.

3.33 Let F_n be the nth Fibonacci number. Prove that $F_{n+k} = F_k F_{n+1} + F_{k-1} F_n$ and in particular, that $F_{2n} = F_n F_{n+1} + F_{n-1} F_n$. F_{2n} is therefore a multiple of F_n. Show that in general, F_{kn} is a multiple of F_n.

▷ **3.34** Show that $2^{n-1} F_n = \sum_k \binom{n}{2k+1} 5^k$.

3.35 For $S \subseteq \mathbb{N}$ let $S + 1$ denote the set $\{x + 1 : x \in S\}$. How many subsets are there of $\{1, \ldots, n\}$ with the property $S \cup (S + 1) = \{1, \ldots, n\}$?

3.36 Determine the following recurrences: (a) The number of $(0, 1)$-words of length n with an even number of zeros. (b) The number of $(0, 1, 2)$-words of length n with an even number of zeros and an even number of ones. Establish the generating functions and determine the numbers of words.

3.37 You are given a deck of n cards, numbered 1 to n. Suppose the top card has number k. The top k cards are then placed in reverse order on top of the deck. Now the card on top has number ℓ, and the top ℓ cards have their order reversed. The process is continued until the card with number 1 is on top. For example,

$$32514 \longrightarrow 52314 \longrightarrow 41325 \longrightarrow 23145 \longrightarrow 32145 \longrightarrow 12345.$$

Show that number of steps in the process is at most $F_{n+1} - 1$, where F_n is the nth Fibonacci number. Hint: Show that you need at most $F_{n-1} - 1$ steps to bring the highest card to the top, and at most F_n steps thereafter.

▷ **3.38** Let $q \in \mathbb{N}$, $q > 1$. Consider the recurrence $f_{m,0} = 1$ $(m \geq 0)$, $f_{0,n} = 1$ $(n \geq 0)$, $f_{m,n} = f_{m,n-1} + (q - 1) f_{m-1,n-1}$ $(m, n \geq 1)$. Calculate $f_{m,n}$. Hint: Establish the generating function $F(x, y) = \sum_{m,n \geq 0} f_{m,n} x^m y^n$.

3.39 Prove the formula $\sum_{k=0}^{n} F_k F_{n-k} = \frac{1}{5}(2n F_{n+1} - (n + 1) F_n)$. Hint: Note that $\sum_{k=0}^{n} F_k F_{n-k} = [z^n] F(z)^2$. Express $F(z)^2$ in terms of ϕ and $\hat{\phi}$.

▷ **3.40** Establish the exponential generating function for g_n, defined as follows: $g_0 = 0$, $g_1 = 1$, $g_n = -2n g_{n-1} + \sum_k \binom{n}{k} g_k g_{n-k}$ $(n > 1)$.

3.41 Calculate $\sum_{0 < k < n} \frac{1}{k(n-k)}$ in two ways: (a) partial fraction decomposition; (b) generating functions.

▷ **3.42** Let $f(n)$ denote the number of cyclic permutations $(a_1, \ldots, a_n)$ of $\{1, \ldots, n\}$, such that a_i, a_{i+1} are never pairs of consecutive integers $1, 2$; $2, 3$; $\ldots$; $n, 1$. Examples: $f(1) = f(2) = 0$ and $f(3) = 1$, with $(1, 3, 2)$ the only possibility in the latter example. Show that $f(n) + f(n + 1) = D_n$ (derangement number) and establish the exponential generating function for $f(n)$.

3.43 Solve the recurrence $a_0 = a_1 = 1$, $a_n = a_{n-1} + (n - 1) a_{n-2}$ $(n \geq 2)$ using exponential generating functions.

3.44 The *Bernoulli numbers* B_n are defined by the recurrence $\sum_{k=0}^{n} \binom{n}{k} B_k = B_n + [n = 1]$, that is, $B_0 = 1$, $B_1 = -\frac{1}{2}$, $B_2 = \frac{1}{6}$, $B_3 = 0$. Show that for the exponential generating function $\hat{B}(z)$, one has $\hat{B}(z) = \frac{z}{e^z - 1}$.

▷ **3.45** Let $S_m(n) = \sum_{0 \le k < n} k^m$, the mth power sum. We know from Section 2.2 that $S_m(n)$ is a polynomial in n of degree $m + 1$ (see also Exercise 2.37). We would now like to determine the coefficients of this polynomial. (a) Let $\hat{S}(z, n) = \sum_{m \ge 0} S_m(n) \frac{z^m}{m!}$. Show that $\hat{S}(z, n) = \frac{e^{nz} - 1}{e^z - 1}$ and conclude, with the help of the previous exercise, that $\hat{S}(z, n) = \frac{e^{nz} - 1}{z} \hat{B}(z)$. (b) $B_m(x) = \sum_k \binom{m}{k} B_k x^{m-k}$ is called the mth *Bernoulli polynomial*. Show that for $\hat{B}(z, x) = \sum_{m \ge 0} B_m(x) \frac{z^m}{m!}$, one has $\hat{B}(z, x) = \frac{z e^{xz}}{e^z - 1}$. (c) Conclude that $\hat{S}(z, n) = \frac{1}{z}[\hat{B}(z, n) - \hat{B}(z, 0)]$. (d) By comparing coefficients, prove that for z^m in part (c), one has the formula $S_m(n) = \frac{1}{m+1} \sum_{k=0}^{m} \binom{m+1}{k} B_k n^{m+1-k}$. Hint for part (a): Write

$$\hat{S}(z, n) = \sum_{m \ge 0} \left(\sum_{0 \le k < n} k^m \right) \frac{z^m}{m!} = \sum_{0 \le k < n} \sum_{m \ge 0} \frac{(kz)^m}{m!}.$$

3.46 Let $T_{m,n} = \sum_{k=0}^{n-1} \binom{k}{m} \frac{1}{n-k}$, $m, n \ge 1$. Show that $T_{m,n} = \binom{n}{n-m}(H_n - H_m)$. Hint: $T_{m,n} = [z^n] \sum_{n \ge 0} \binom{n}{m} z^n \cdot \sum_{n \ge 1} \frac{z^n}{n}$ and $\sum_{n \ge 1} \frac{z^n}{n} = \log \frac{1}{1-z}$.

Chapter 4

Counting Patterns

Many counting problems are of a quite different type from those that we have thus far encountered. They are governed by symmetries on the underlying structure. In this chapter we shall see how such problems can be tackled.

4.1. Symmetries

The following examples should clarify what we are talking about. We shall solve these three problems in the course of the chapter.

A. Sylvane is wearing a string of n beads around her neck; each bead is colored either white or black. How many different bead patterns are there? First we must decide what is meant by "different": We shall consider two strings of beads identical if each is a rotation of the other.

For $n = 4$ we obtain six different strings:

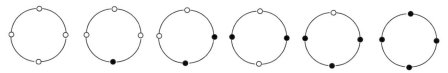

Let us now color the four beads such that W are white, S sepia, and R red. Now how many patterns are there? If we write down those patterns with $W \geq S \geq R$, we obtain

W	S	R	Patterns
4	0	0	1
3	1	0	1
2	2	0	2
2	1	1	3

Since we can permute the colors to obtain other patterns, we obtain altogether $3 \cdot 1 + 6 \cdot 1 + 3 \cdot 2 + 3 \cdot 3 = 24$ different strings of beads. Of course, what we would really like to know is the general result for n beads and r colors.

B. Consider a cube each of whose faces is colored white or sepia. How many different patterns are there? We consider two cubes identical if each is a rotation of the other. The picture below gives an example:

The following table can be checked most easily if you hold a cube in your hand:

Color	Type	Patterns
6	0	$2 \cdot 1$
5	1	$2 \cdot 1$
4	2	$2 \cdot 2$
3	3	$1 \cdot 2$
		10

Again we are interested in the general case, in which there are r colors available.

C. Now for a more challenging example. An alcohol consists of a hydroxyl (OH) group (univalent) and a certain number of carbon (C) atoms (tetravalent) saturated with hydrogen (H) atoms (univalent). How many alcohols are there with exactly n carbon atoms? For $n \leq 3$ one can easily write down all possibilities:

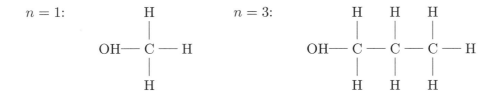

$n = 2$:

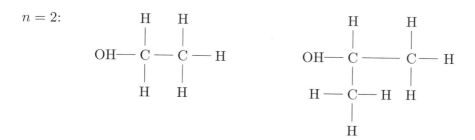

In all three examples we see that "equality" is determined by the symmetries of the structure. In example A, these are the n cyclic rotations. In example B it is the symmetries of the cube, while in example C it is the permutations of the three structures attached to OH—C. Let us examine these symmetries more closely. It turns out that they form a *group* G. We shall assume some familiarity with groups, and in particular with modular arithmetic. Sections 12.1 and 12.2 provide more precise information on this topic.

Let us consider the necklace problem with beads $N = \{1, 2, \ldots, n\}$. The symmetries G are the translations $g_i : a \mapsto a + i \pmod{n}$, $i = 0, 1, \ldots, n - 1$, and the product is composition, that is, $g_i g_j : a \mapsto (a + j) + i = a + (j + i) \pmod{n}$. Clearly, $g_i = g_1^i$, and we call $G = C_n$ the *cyclic group* of order n.

Now let us look at the cube. In geometry, one learns that every transformation that maps the cube to itself corresponds to a rotation about an axis. This axis can go through two opposite faces (more precisely, through the centers of the two faces), through opposite edges, or through diametrically opposed vertices. There are therefore three face axes, six edge axes, and four vertex axes. The figure shows that each face axis yields in addition to the identity three symmetries, where the upper and lower faces remain fixed.

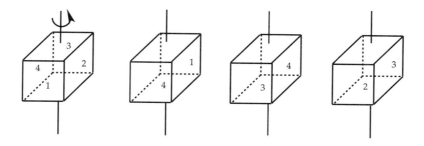

Altogether, the symmetry group of the cube contains twenty-four elements:

$$
\begin{array}{ll}
\text{Identity} & 1 \\
\text{Face axes} & 3 \cdot 3 \\
\text{Edge axes} & 6 \cdot 1 \\
\text{Vertex axes} & 4 \cdot 2 \\
\hline
& 24
\end{array}
$$

In Example C we obtain, as we have seen, the group S_3 of the six permutations on the three partial structures.

Permutations, which we considered in Section 1.3, will play a central role in what follows. The group of all permutations of N (with composition as multiplication) is called the *symmetric group* S_N. For $N = (1, \ldots, n\}$ we write S_n for short. Each subgroup $G \leq S_N$ is called a *permutation group* on N. For example, the group C_n from example A is the permutation group obtained from the n translations.

4.2. Statement of the Problem

We return to the necklace problem. Let $N = \{1, \ldots, n\}$ denote the set of beads, and $R = \{1, \ldots, r\}$ the colors. A colored string is then a mapping $f : N \longrightarrow R$, and we know that altogether, there are r^n mappings. For short, we write $R^N = \mathrm{map}(N, R)$.

Let $G = C_n$ be the cyclic group of order n. Two "equivalent" necklaces then clearly correspond to two mappings f and f' for which we have $f' = f \circ g$ for some $g \in G$ (we rotate by g and color with f).

We therefore call two mappings f, f' *equivalent*, in symbols $f' \sim f$, if there is some $g \in G$ such that $f' = f \circ g$. One may easily check that $\sim$ is an equivalence relation. For example, if $f' \sim f$ and $f'' \sim f'$ with $f' = f \circ g$ and $f'' = f' \circ h$, then we have $f'' = (f \circ g) \circ h = f \circ (gh)$, whence $f'' \sim f$. We shall call the equivalence classes of R^N under $\sim$ *patterns*, and our problem A then consists precisely in counting the number of patterns.

For example, if $G = S_N$, the full symmetric group on N, then we have the situation in which the elements of N are indistinguishable, and we obtain from the table at the end of Section 1.2 precisely $\binom{r+n-1}{n}$ patterns. To arrive at our actual problem, we require two generalizations. Consider our example B. The group G is the group of the twenty-four symmetries of the cube, and N is the set of faces of the cube. However, we could also color the edges ($|N| = 12$) or the vertices ($|N| = 8$). In each case, the group G *acts* on the associated set N. That is, each symmetry of the cube results in a *permutation* of the set N.

This idea leads to the following fundamental definition. Let G be a group. We say that G **acts via** τ on N if $\tau : G \longrightarrow S_N$, $g \mapsto \tau_g$, is a homomorphism of G into the symmetric group S_N. That is, we have

$\tau_{gh} = \tau_g \tau_h$ for all $g, h \in G$. In this general setting, we again set

$$f' \sim f \iff f' = f \circ \tau_g \text{ for } g \in G.$$

This again yields an equivalence relation on R^N, and we again call the equivalence classes **patterns**.

Thus in example B, τ_g specifies the actual permutation of the faces (or edges or vertices) resulting from the symmetry g of the cube.

For our second generalization, we could ask more specifically in example A how many necklace patterns contain exactly a_j beads of the color j. To clarify this idea, we associate with each color $j \in R$ a variable x_j and define the **weight** $w(f)$ of a mapping $f : N \longrightarrow R$ by

$$w(f) = \prod_{i \in N} x_{f(i)}.$$

If f, f' are equivalent mappings with $f' = f \circ \tau_g$, then we compute

$$w(f') = \prod_{i \in N} x_{f'(i)} = \prod_{i \in N} x_{(f \circ \tau_g)(i)} = \prod_{i \in N} x_{f(i)} = w(f),$$

since together with $i \in N$, we have that $\tau_g(i)$ runs through N as well. Therefore, equivalent mappings have the same weight, and we can unambiguously define the **weight** $w(M) = w(f)$ of a pattern M for $f \in M$.

For example, if in problem A we choose the variable W for white and S for sepia, then $W^2 S^2$ means that the necklace has two white beads and two sepia beads.

And so we finally arrive at the statement of the problem:

Problem. Given two sets N and R, with $|N| = n$ and $|R| = r$, together with a group G that acts on N and variables x_j ($j \in R$), let $\mathcal{M}$ denote the set of patterns. Calculate the **enumerator**

$$w(R^N; G) = \sum_{M \in \mathcal{M}} w(M).$$

In particular, if we set $x_j = 1$ for all $j \in R$, then each pattern has weight 1, and we obtain the *number* $|\mathcal{M}|$ of patterns.

Example A. Let $n = 4$, $r = 2$ with the variables W and S. Then as we have seen, the enumerator is

$$w(R^N; C_4) = W^4 + W^3 S + 2W^2 S^2 + W S^3 + S^4.$$

Example B. For the cube, we obtain

$$w(\text{faces}; G) = W^6 + W^5 S + 2W^4 S^2 + 2W^3 S^3 + 2W^2 S^4 + W S^5 + S^6.$$

Example C. Let x be the variable of the carbon atoms. Then we obtain

$$w(\text{alcohol}; S_3) = 1 + x + x^2 + 2x^3 + 3x^4 + 7x^5 + \cdots.$$

In this case we have before us a generating function, which we shall determine more precisely.

4.3. Patterns and the Cycle Indicator

We are now going to take a closer look at the patterns of R^N under the action τ of the group G. Let $f \in R^N$. Then the pattern $M(f)$ containing f consists precisely of the f' such that $f' = f \circ \tau_g$ for some $g \in G$. If we set $\tau_g f := f \circ \tau_g$, then we have $M(f) = \{\tau_g f : g \in G\}$.

We would like to deal with this situation in full generality. Let G be a group that acts on the set X via τ. We set, as usual, $y \sim x$ if $y = \tau_g x$ for some $g \in G$. Again $\sim$ is an equivalence relation, whose classes we call *patterns*. The pattern $M(x)$ that contains x is $M(x) = \{\tau_g x : g \in G\}$. The following concepts lead to one of the most fruitful theorems in combinatorics, which in particular yields the number $|\mathcal{M}|$ of patterns.

The **fixed group** G_x of $x \in X$ is $G_x = \{g \in G : \tau_g x = x\}$. Clearly, G_x is a subgroup of G. The **fixed-point set** of $g \in G$ is $X_g = \{x \in X : \tau_g x = x\}$. Counting the pairs $\{(x, g) : \tau_g x = x\}$ twice yields

$$(4.1) \qquad \sum_{x \in X} |G_x| = \sum_{g \in G} |X_g|,$$

and this leads to the following theorem, which is generally known as Burnside's lemma or the Burnside–Frobenius lemma.

Theorem 4.1. *Let the group G act on X. Then*

$$|\mathcal{M}| = \frac{1}{|G|} \sum_{g \in G} |X_g|.$$

Proof. We have $M(x) = \{\tau_g x : g \in G\}$. We thus obtain

$$\tau_g x = \tau_h x \iff \tau_{g^{-1}h} x = x \iff g^{-1}h \in G_x \iff h = ga, \ a \in G_x.$$

That is, for each $g \in G$, precisely $|G_x|$ group elements h (namely, $h = ga$ for $a \in G_x$) yield the *same* element $\tau_g x \in M(x)$. We therefore have

$$|M(x)| = \frac{|G|}{|G_x|} \text{ for all } x \in X.$$

In particular, for $y \in M(x)$ (that is, $M(y) = M(x)$) we obtain $|G_x| = |G_y|$, and thus

$$|G| = |M(x)||G_y| = \sum_{y \in M(x)} |G_y|.$$

Summation over the patterns, together with (4.1), yields

$$|\mathcal{M}||G| = \sum_{\text{patterns } M(x)} \sum_{y \in M(x)} |G_y| = \sum_{y \in X} |G_y| = \sum_{g \in G} |X_g|,$$

and we are done. □

The meaning of the lemma is clear. In many examples, it is easy to calculate the fixed-point set X_g, from which one then has the number $|\mathcal{M}|$ of patterns.

Let us test Burnside's lemma on our example A with $n = 4$, $r = 2$. In this case we have $X = R^N$. The identity $g_0 : a \mapsto a$ leaves all mappings unchanged, and so we have $|X_{g_0}| = 16$. If f is in X_{g_1} or X_{g_3}, then all colors must be the same (clear?), and so only necklaces of a single color come into question: $|X_{g_1}| = |X_{g_2}| = 2$. Finally, $f \in X_{g_2}$ requires that 1 and 3 have the same color, and likewise 2 and 4. We obtain $|X_{g_2}| = 4$ and altogether, from the lemma, our old result

$$|\mathcal{M}| = \frac{1}{4}(16 + 2 + 4 + 2) = 6.$$

Let us return to our initial problem, $X = R^N$ with $\tau_g f = f \circ \tau_g$. To calculate $|\mathcal{M}|$, we must determine, according to Theorem 4.1, the fixed-point set

$$R^N_g = \{f \in R^N : f \circ \tau_g = f\}.$$

Now the decomposition of τ_g into cycles comes into play. We introduced this concept in Section 1.3. The permutation τ_g decomposes N into the cycles

$$(a, \tau_g a, \tau_g^2 a, \dots)(b, \tau_g b, \tau_g^2 b, \dots)\dots.$$

From $f = f \circ \tau_g$ it follows that $f(a) = f(\tau_g a) = f(\tau_g^2 a) = \cdots$, and likewise for the other cycles. In other words, the mapping f is in the fixed set R^N_g precisely when f is *constant* on all cycles of τ_g.

We need one more concept, and then we can prove our main theorem. In Section 1.3 we defined the *type* of a permutation π on an n-set N by

$$t(\pi) = 1^{b_1(\pi)} 2^{b_2(\pi)} \dots n^{b_n(\pi)},$$

where $b_i(\pi)$ denotes the number of cycles of length i. Of course, we have $\sum_{i=1}^{n} i b_i = n$. Now let, as usual, G be a group that acts on R^N. Then the **cycle indicator** of G is the polynomial

$$(4.2) \qquad Z(G; z_1, \dots, z_n) = \frac{1}{|G|} \sum_{g \in G} z_1^{b_1(\tau_g)} z_2^{b_2(\tau_g)} \dots z_n^{b_n(\tau_g)},$$

in the variables $z_1, z_2, \dots, z_n$, where the $b_i(\tau_g)$ are as defined above.

As an illustration let us consider example B. The individual symmetries yield

Identity	z_1^6,
Face axes	$3(z_1^2 z_2^2 + 2z_1^2 z_4)$,
Edge axes	$6z_2^3$,
Vertex axes	$8z_3^2$.

For example, the eight rotations not equal to the identity around the vertex axes each yield two three-cycles on the faces, so that altogether, a contribution of $8z_3^2$ results. The cycle indicator is therefore

$$Z(G; z_1, \ldots, z_6) = \frac{1}{24} \left(z_1^6 + 3z_1^2 z_2^2 + 6z_1^2 z_4 + 6z_2^3 + 8z_3^2 \right).$$

4.4. Pólya's Theorem

The following classical result of counting theory traces the calculation of the enumerator back to the cycle indicator.

Theorem 4.2 (Pólya). *Let N and R be sets, $|N| = n$, $|R| = r$, G a group acting on N, and x_j, $j \in R$, variables. Then for the pattern enumerator we have*

$$w(R^N; G) = \sum_{M \in \mathcal{M}} w(M) = Z\left(G; \sum_{j \in R} x_j, \sum_{j \in R} x_j^2, \ldots, \sum_{j \in R} x_j^n \right).$$

That is, we obtain $w(R^N; G)$ by making the replacement $z_k \mapsto \sum_{j \in R} x_j^k$ in the cycle indicator.

Proof. Let M be a pattern of R^N under G. If we apply G to M, then we of course obtain only one pattern, namely M itself. From Theorem 4.1, we thus obtain

(4.3) $$1 = \frac{1}{|G|} \sum_{g \in G} |M_g|,$$

where $M_g = \{f \in M : f \circ \tau_g = f\}$. We know that all $f \in M$ have the same weight $w(f) = w(M)$. Multiplication of (4.3) by $w(M)$ therefore yields

$$w(M) = \frac{1}{|G|} \sum_{g \in G} |M_g| w(f) = \frac{1}{|G|} \sum_{g \in G} \sum_{\substack{f \in M \\ f \circ \tau_g = f}} w(f).$$

If we now sum over all patterns M, we obtain

(4.4) $$w(R^N; G) = \frac{1}{|G|} \sum_{g \in G} \sum_{\substack{f \in R^N \\ f \circ \tau_g = f}} w(f).$$

Let us examine one of the summands

$$\sum_{\substack{f \in R^N \\ f \circ \tau_g = f}} w(f)$$

of the inner sum. We know that $f \circ \tau_g = f$ holds precisely when f is constant on all cycles of τ_g. Let $t(\tau_g) = 1^{b_1} 2^{b_2} \ldots n^{b_n}$ be of type of τ_g, that is,

$$\tau_g = \underbrace{(\cdot)(\cdot) \cdots (\cdot)}_{b_1} \underbrace{(\cdot\cdot) \cdots (\cdot\cdot)}_{b_2} \cdots .$$

Then the images of the cycles under the mapping f are the elements

$$a_{11}, a_{12}, \ldots, a_{1b_1}; \ a_{21}, \ldots, a_{2b_2}; \ \ldots,$$

and therefore

$$w(f) = (x_{a_{11}} x_{a_{12}} \cdots x_{a_{1b_1}})(x_{a_{21}}^2 \cdots x_{a_{2b_2}}^2) \cdots .$$

Since now *all* mappings that are constant on the cycles appear, we obtain, by multiplying out,

$$\sum_{\substack{f \in R^N \\ f \circ \tau_g = f}} w(f) = \left(\sum_{j \in R} x_j\right)^{b_1} \left(\sum_{j \in R} x_j^2\right)^{b_2} \cdots \left(\sum_{j \in R} x_j^n\right)^{b_n}.$$

But this expression corresponds precisely to the summands of g in the cycle indicator, and with (4.4) we obtain

$$w(R^N; G) = \frac{1}{|G|} \sum_{g \in G} \left[\left(\sum_{j \in R} x_j\right)^{b_1(\tau_g)} \cdots \left(\sum_{j \in R} x_j^n\right)^{b_n(\tau_g)} \right]$$

$$= Z\left(G; \sum_{j \in R} x_j, \sum_{j \in R} x_j^2, \ldots, \sum_{j \in R} x_j^n\right),$$

which completes the proof. □

Let us set $x_j = 1$ for all $j \in R$. Then $\sum_{j \in R} x_j^k = r$ for all k, and we have the following corollary.

Corollary 4.3. *Under the hypotheses of Theorem 4.2,*

$$|\mathcal{M}| = Z(G; r, \ r, \ldots, \ r).$$

We can immediately write down yet another corollary. We consider the case $r = 2$ and interpret the patterns as patterns of colors, with the colors white and sepia. If we set $x_{\text{weiß}} = x$ and $x_{\text{schwarz}} = 1$, we obtain the following theorem:

Corollary 4.4. *Under the hypotheses of Theorem 4.2 with $r = 2$, we have*

$$\sum_{k=0}^{n} a_k x^k = Z(G; 1 + x, \ 1 + x^2, \ldots, 1 + x^n),$$

where a_k is the number of patterns in which the color white occurs exactly k times.

We are finally in a position to solve our problems. According to Pólya's theorem, we should determine the cycle indicator.

Example A. Let $N = \{1, \ldots, n\}$ and let $g_i : a \mapsto a + i \,(\mathrm{mod}\, n)$ be the translation by i positions. By symmetry, it is clear that g_i decomposes the set N into cycles of the same length d, for some d. Therefore the type is $t(g_i) = d^{n/d}$, and in particular, d is a divisor of n. In g_i each position a is translated by i, that is, d is the smallest integer such that we again arrive at a. Thus d is the smallest integer such that $n \mid di$. It follows, then, that $\frac{n}{\gcd(n,i)} \mid d$ and thus $d = \frac{n}{\gcd(n,i)}$, since with this d we of course have $n \mid di$. For example, if $n = 6$, $i = 4$, then we obtain the cycle decomposition $(1, 5, 3)(2, 6, 4)$, in agreement with $d = \frac{6}{\gcd(6,4)} = 3$. In the solution to Exercise 1.21 it is shown that there are precisely $\varphi(d)$ integers i such that $\gcd(i, n) = \frac{n}{d}$, where $\varphi(d)$ is Euler's φ-function. Altogether, we have obtained the appealingly simple formula

$$Z(C_n; z_1, \ldots, z_n) = \frac{1}{n} \sum_{d \mid n} \varphi(d) z_d^{n/d}.$$

Corollary 4.3 now takes care of the rest. The number of necklaces with n beads and r colors is

$$|\mathcal{M}_n| = \frac{1}{n} \sum_{d \mid n} \varphi(d) r^{n/d}.$$

For $n = 4$ we have

$$|\mathcal{M}_4| = \frac{1}{4}(r^4 + r^2 + 2r) = \frac{1}{4}r(r + 1)(r^2 - r + 2).$$

Thus for example, there are 70 necklaces with four colors, and 616 necklaces with seven colors.

Example B. Here we have already calculated the cycle indicator in Section 4.3. If we color the faces of the cube with r colors, then we obtain

$$|\mathcal{M}| = \frac{1}{24}(r^6 + 3r^4 + 12r^3 + 8r^2) = \frac{1}{24}r^2(r + 1)(r^3 - r^2 + 4r + 8),$$

and for example, for $r = 2$ and $r = 3$, we have $|\mathcal{M}| = 10$ and $|\mathcal{M}| = 57$.

Example C. We would like to determine the generating function $T(x) = \sum_{n \geq 0} t_n x^n$, where t_n is the number of alcohols with n carbon atoms. For $n \geq 1$ we call the carbon atom that abuts the OH the *root*. There are three bonds emanating from the root, and these can be permuted arbitrarily:

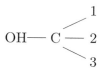

We set $N = \{1, 2, 3\}$. At each position i of N there hangs an alcohol A_i, where we regard the root as the OH radical of A_i. Let R be the set of alcohols A, where we associate with each A the weight x^n, where n is the number of carbon atoms. An alcohol then corresponds to a mapping $f : N \longrightarrow R$, and the various alcohols correspond precisely to the patterns of R^N under the symmetric group S_3. The cycle indicator of S_3 is easy to calculate:

$$Z(S_3; z_1, z_2, z_3) = \frac{1}{6}(z_1^3 + 3z_1 z_2 + 2z_3).$$

Furthermore, we have $\sum_{A \in R} x_A^k = T(x^k)$. If we now consider the root as well, then from Pólya's theorem we obtain the functional equation

$$T(x) = 1 + \frac{x}{6}\left[T(x)^3 + 3T(x)T(x^2) + 2T(x^3)\right].$$

By comparing coefficients, we obtain, with $t_0 = 1$,

$$t_n = \frac{1}{6}\left[\sum_{i+j+k=n-1} t_i t_j t_k + 3\sum_{i+2j=n-1} t_i t_j + 2t_{\frac{n-1}{3}}\right] \quad (n \geq 1).$$

The first few values are as follows:

n	0	1	2	3	4	5	6	7	8
t_n	1	1	1	2	4	8	17	39	89

To illustrate, we consider $n = 5$, where only the OH and C atoms are shown:

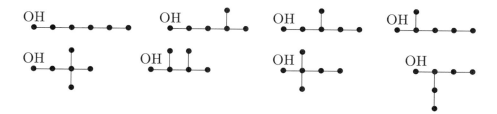

Finally, we shall consider a chess problem that nicely illustrates the power of the theory. In how many ways can we place n rooks on an $n \times n$ chessboard in such a way that none of the rooks is attacking any of the others (rooks attack along the horizontal and vertical lines of the board)? A moment's thought reveals that these positions correspond exactly to the $n!$ permutations π of the squares (i, π_i), $i = 1, \ldots, n$. But how many *patterns* are there? Here the symmetry group consists of four rotations of the board as well as reflections. This group is called the *dihedral group* D_4; it contains eight elements. For $n \leq 4$ we can immediately write down the patterns:

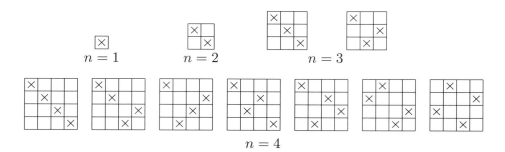

And what about for arbitrary n? No problem, thanks to Burnside's lemma! Let N^2 denote the set of all squares (i, j), and g_i the clockwise rotation through $k \cdot 90°$ for $k = 0, 1, 2, 3$. Together with the reflection h through the vertical central axis, we thus obtain all eight symmetries.

To apply the lemma, we must think about where a particular square (i, j) ends up under the action of a symmetry. The reader may check that the following table provides this information:

$$
\begin{array}{lll}
(i,j) & \xrightarrow{g_0} (i,\ j) \qquad\qquad & (i,j) \xrightarrow{hg_0} (i,\ n+1-j) \\[2mm]
 & \xrightarrow{g_1} (j,\ n+1-i) & \xrightarrow{hg_1} (j,\ i) \\[2mm]
(4.5) & \xrightarrow{g_2} (n+1-i,\ n+1-j) & \xrightarrow{hg_2} (n+1-i,\ j) \\[2mm]
 & \xrightarrow{g_3} (n+1-j,\ i), & \xrightarrow{hg_3} (n+1-j,\ n+1-i).
\end{array}
$$

In particular, we see that a permutation (i, π_i) is taken to a permutation. Therefore, the group D_4 acts on the set X of permutations. Now we must determine the fixed-point set X_g as well as its size, and then we will be done. We consider g_1. The other cases are solved just as easily. Let $\{(i, \pi_i) : i \in N\} \in X_{g_1}$. If $i \xrightarrow{\pi} j$, then from the table, we have $j \xrightarrow{\pi} n + 1 - i$. If we apply π to $n + 1 - i$, we obtain $n + 1 - i \xrightarrow{\pi} n + 1 - j$, and, on a further iteration, $n + 1 - j \xrightarrow{\pi} i$. Thus π decomposes into cycles of the form $(i,\ j,\ n + 1 - i,\ n + 1 - j)$. But watch out: it could happen that a cycle closes earlier.

If $i \xrightarrow{\pi} i$ is a fixed point, then we have $i \longrightarrow i \longrightarrow n + 1 - i = i$, and therefore $i = \frac{n+1}{2}$. This can happen only for an odd n, and it is immediately clear that the central square $(\frac{n+1}{2}, \frac{n+1}{2})$ remains fixed. Therefore, let $i \xrightarrow{\pi} j \neq i$. If we have a 2-cycle (i, j), we obtain $i \longrightarrow j \longrightarrow n + 1 - i = i \longrightarrow j = n + 1 - j$, and thus $i = j = \frac{n+1}{2}$, which is impossible. Thus there are no 2-cycles, and analogously, one sees that there are no 3-cycles. Therefore, if n is odd, $\pi \in X_{g_1}$ consists of a fixed point and $\frac{n-1}{4}$ 4-cycles, and if n is even, then π consists of $\frac{n}{4}$ 4-cycles. We conclude that n must be of the form $n = 4t + 1$ or $n = 4t$.

We see that in each 4-cycle $(i, j, n+1-i, n+1-j)$ there are two integers less than or equal to $\frac{n}{2}+1$ and two that are greater than or equal to $\frac{n}{2}+1$. We then have, according to the choice of $\{i, j, n+1-i, n+1-j\}$, two possible cycles. For example, for $n = 5$ we obtain the 4-cycles $(1, 2, 5, 4)$ and $(1, 4, 5, 2)$, which correspond to the arrangements $(1, 2), (2, 5), (3, 3), (4, 1), (5, 4)$ and $(1, 4), (2, 1), (3, 3), (4, 5), (5, 2)$:

Both positions are invariant under g_1, as of course they must be.

We still must count the number of ways in which the 4-sets can be chosen. But that is easy. First, to go with 1, we can choose one of the other integers $j \leq \frac{n}{2}$, and then the two that remain, n and $n + 1 - j$, are determined. We now choose the next pair, and so on. Altogether, we obtain

$$\frac{n-3}{2} \cdot \frac{n-7}{2} \cdots 1 \quad \text{for} \quad n = 4t + 1$$

and

$$\frac{n-2}{2} \cdot \frac{n-6}{2} \cdots 1 \quad \text{for} \quad n = 4t$$

possibilities. If we now consider that each 4-set yields a 4-cycle in two ways, we obtain

$$(4.6) \qquad |X_{g_1}| = \begin{cases} (n-3)(n-7)\cdots 2, & n = 4t + 1, \\ (n-2)(n-6)\cdots 2, & n = 4t, \end{cases}$$

and by symmetry, $|X_{g_3}| = |X_{g_1}|$. For g_2 we calculate, for $n \geq 2$,

$$(4.7) \qquad |X_{g_2}| = \begin{cases} (n-1)(n-3)\cdots 2, & n \text{ odd}, \\ n(n-2)\cdots 2, & n \text{ even}, \end{cases}$$

and of course we have $|X_{g_0}| = n!$.

Finally, from the table (4.5), for $n \geq 2$, we obtain

(4.8) $$|X_{hg_0}| = |X_{hg_2}| = 0 \,,$$

while the table shows for hg_1 (and by symmetry for hg_3) that $i \xrightarrow{\pi} j$ implies $j \xrightarrow{\pi} i$. Therefore, X_{hg_1} consists of all permutations (called *involutions*) whose cycles are of length 1 or 2. Let i_n denote the number of involutions.

Let a_1 denote the expression on the right-hand side of (4.6), and a_2 that of (4.7). Then Burnside's lemma gives us the result for the number $|\mathcal{M}_n|$ of rook placements for $n \geq 2$:

$$|\mathcal{M}_n| = \begin{cases} \frac{1}{8}(n! + 2a_1 + a_2 + 2i_n), & n = 4t,\ 4t + 1, \\ \frac{1}{8}(n! + a_2 + 2i_n), & n = 4t + 2,\ 4t + 3 \,. \end{cases}$$

We now need only a formula for i_n. In Exercise 4.20 we shall derive the recurrence relation $i_{n+1} = i_n + n i_{n-1}$ ($i_0 = 1$). The first values are thus

n	0	1	2	3	4	5	6	7	8
i_n	1	1	2	4	10	26	76	232	764

For $n = 4$ we calculate $|\mathcal{M}_4| = \frac{1}{8}(24 + 4 + 8 + 20) = 7$ as we have seen, and we also have $|\mathcal{M}_5| = 23$, $|\mathcal{M}_6| = 115$, $|\mathcal{M}_7| = 694$, $|\mathcal{M}_8| = 5282$.

Exercises for Chapter 4

▷ **4.1** The Zemblan flag is divided into n horizontal stripes. How many black–white patterns are there with exactly k white stripes, where two different flags are considered the same if one is identical to the other when flipped upside down. Check your result for $n = 5$, $k = 3$ and $n = 6$, $k = 4$.

4.2 Let G be the symmetry group of the cube. Determine the cycle indicator when N is (a) the set of edges or (b) the set of vertices. How many patterns are there with two or three colors in each case?

4.3 How many patterns do we obtain if we color the twelve edges of a cube with twelve different colors?

▷ **4.4** Consider the cycle indicator of the symmetric group S_n (without computing it) and conclude that $\sum_{k=0}^{n} s_{n,k} x^k = x^{\overline{n}}$, where $s_{n,k}$ is the Stirling number of the first kind.

4.5 How many face patterns (edge patterns, vertex patterns) of the cube are there in which black and white appear the same number of times?

4.6 A necklace can be removed, flipped around, and put on again backward. The symmetry group is then the dihedral group D_n, with $|D_n| = 2n$, $n \geq 3$. Determine the cycle indicator of D_n. Hint: The cyclic group C_n is a subgroup.

4.7 Let $n = 6$, $r = 2$. Show by considering C_6 and D_6 that there are exactly two colored necklaces that are different without flipping them around, but with flipping, they are the same. What are they?

▷ **4.8** Determine the symmetry group of the tetrahedron:

Show without calculation that whether we color the edges or the faces with r colors, we obtain the same number of patterns.

4.9 In analogy to the alcohol molecules with OH, consider molecules on which D (trivalent) and H (univalent) are bonded. Determine the generating function $\sum_{n \geq 0} d_n x^n$, where d_n is the number of D-atoms.

▷ **4.10** How many patterns of lattice paths from $(0,0)$ to (n,n) as in Exercise 1.37 are there if we consider paths identical if one can be manipulated (by rotating and flipping) in such a way as to cover the other. Example:

Hint: Each path corresponds to a sequence of n zeros and ones. Think about what group acts on such sequences.

▷ **4.11** Let $N = \{1, \ldots, n\}$. Determine the probability that in a random permutation of N, the number 1 is contained in a cycle of length k (all permutations are equally probable). What is the expectation for the length of the cycle that contains 1? What is the expectation for the number of cycles?

4.12 Determine the symmetry group of the octahedron and calculate the cycle indicator for the faces.

4.13 Show that there are exactly three vertex patterns of the octahedron with three vertices painted red, two blue, and one green. Draw them.

4.14 The Bates Motel is undergoing renovations. The owners plan on replacing the room keys with square 3×3 cards each with two holes punched in them, as in the figure below:

If the two sides of the card are indistinguishable, but there is a distinguished edge to be inserted into the lock, how many rooms can the motel accommodate with this system? Hint: Consider what group acts on the nine squares and apply Burnside's lemma.

4.15 Let $x_1, \ldots, x_r$ be variables. A polynomial $f(x_1, \ldots, x_r)$ is called *symmetric* if $f(x_{\pi(1)}, \ldots, x_{\pi(r)}) = f(x_1, \ldots, x_r)$ holds for every permutation $\pi \in S_r$. Important examples are the *elementary symmetric* functions

$$a_k(x_1, \ldots, x_r) = \sum_{i_1 < \cdots < i_k} x_{i_1} \cdots x_{i_k},$$

for example, $a_1(x_1, \ldots, x_r) = x_1 + \cdots + x_r$, and the *power functions*

$$s_k(x_1, \ldots, x_r) = \sum_{j=1}^{r} x_j^k.$$

Prove using Pólya's theorem the following formula of Waring:

$$a_n(x_1, \ldots, x_r) = Z(S_n; s_1, -s_2, s_3, \ldots, (-1)^{n-1} s_n).$$

▷ **4.16** Prove a result analogous to Pólya's theorem for the set of injective mappings $\mathrm{Inj}(N, R) : w(\mathrm{Inj}(N, R); G) = \frac{n!}{|G|} a_n(x_1, \ldots, x_r)$ with $a_n(x_1, \ldots, x_r)$ as in the previous exercise. What is the result for $x_1 = \cdots = x_r = 1$ and $G = \{\,\mathrm{id}\,\}$ or for $G = S_n$?

▷ **4.17** Determine the number of nonisomorphic multiplication tables for multiplication of two symbols $0, 1$. There are $2^4 = 16$ possible tables. Two multiplications $x \cdot y$ and $x \circ y$ are said to be *isomorphic* if there is a bijection φ such that $\varphi(x \cdot y) = \varphi(x) \circ \varphi(y)$. Hint: Use Burnside's lemma. Now determine the number for three symbols.

4.18 Let N be an n-set, and $G \le S_N$. We call two subsets A, B *G-equivalent* if $B = \{g(a) : a \in A\}$ for some $g \in G$. Let m_k be the number of patterns of k-sets. Show that $\sum_{k=0}^{n} m_k x^k = Z(G; 1 + x, 1 + x^2, \ldots, 1 + x^n)$. Conclude that $|\mathcal{M}| = \frac{1}{|G|} \sum_{g \in G} 2^{b(g)}$, where $b(g)$ is the number of cycles of g.

▷ **4.19** From the previous exercise, derive the following familiar formulas:

(a) $\sum_{k=0}^{n} \binom{n}{k} x^k = (1 + x)^n$,

(b) $(n+1)! = \sum_{g \in G} 2^{b(g)} = \sum_{k=0}^{n} s_{n,k} 2^k$,

(c) $\sum_{d|n} \varphi(d) = n$.

4.20 Prove that the number of fixed-point-free involutions is given by the formula $(n-1)(n-3)(n-5) \cdots$. For i_n the number of all involutions of $\{1, \ldots, n\}$ prove the recurrence relation $i_{n+1} = i_n + n i_{n-1}$ ($i_0 = 1$) and conclude that for the exponential generating function, $\sum_{n \ge 0} i_n \frac{z^n}{n!} = e^{z + z^2/2}$. From this calculate i_n.

▷ **4.21** Let $T_n = Z(S_n; z_1, \ldots, z_n)$. Show that $\sum_{n \ge 0} T_n y^n = \exp\left(\sum_{k \ge 1} z_k \frac{y^k}{k}\right)$ and derive once again the formula for the exponential generating function for involutions. What function does one obtain for permutations that have only three-cycles, or more generally, h-cycles?

4.22 Generalize the previous exercise as follows: Let $F(x) = \sum_{n \ge 0} a_n x^n$ be a generating function. Then

$$\sum_{n \ge 0} Z(S_n; F(x), F(x^2), \ldots, F(x^n)) y^n = \exp\left(\sum_{k \ge 1} F(x^k) \frac{y^k}{k}\right).$$

What is the result for $F(x) = 1$?

▷ **4.23** Let $N = \{1, 2, \ldots, n\}$, $R = \{W, S\}$, and let G be a group that acts on N. We call a pattern M *self-dual* if $f \in M$ implies $h \circ f \in M$, where h is the involution $W \longleftrightarrow S$. In other words, M is invariant under exchange of colors. Let $\overline{m}$ denote the number of self-dual patterns. Show that $\overline{m} = Z(G; 0, 2, 0, 2, \ldots)$. Hint: Use $h \circ f \sim f \Longleftrightarrow h \circ f = f \circ \tau_g$ for a $g \in G$.

4.24 Using the previous exercise, determine the number of self-dual patterns for the necklace problem. For a power of two, that is, $n = 2^s$, there is a particularly simple formula. What is it? Write down the four patterns for $n = 8$.

4.25 Complete the example of nonattacking rook patterns, verifying (4.5) and computing the sizes of the other fixed-point sets.

Chapter 5

Asymptotic Analysis

In the previous sections we have learned about a number of methods for working with a counting function $f(n)$: summation, recurrence relations, generating functions. But what are we to do when none of these methods works? In such cases, we shall try to estimate $f(n)$ from above and from below in order to obtain at least an idea of the order of magnitude of $f(n)$.

5.1. The Growth of Functions

It is well known from calculus that n, n^2, and c^n $(c > 1)$ all increase without bound as n approaches infinity and furthermore, that n^2 grows "faster" than n, and c^n "faster" than n^2. It is this idea of different rates of growth that we are now going to examine.

We consider functions $f(n)$ defined on the natural numbers. We are interested in how fast the values $|f(n)|$ grow; in particular, we are unconcerned whether the values of $f(n)$ are positive or negative. We want to know only how large they are in magnitude. We say that $g(n)$ **grows faster than** $f(n)$, in symbols $f(n) \prec g(n)$, if for every $\varepsilon > 0$ there exists $n_0(\varepsilon)$ such that

(5.1) $$|f(n)| \leq \varepsilon |g(n)| \quad \text{for all } n \geq n_0(\varepsilon).$$

In particular, $g(n) = 0$ always implies $f(n) = 0$ for $n \geq n_0(\varepsilon)$. If $g(n)$ has only finitely many zero values (which is almost always the case), then we have

$$f(n) \prec g(n) \iff \lim_{n \to \infty} \left| \frac{f(n)}{g(n)} \right| = 0 \, ,$$

where we leave undefined the quotients for the finitely many values $g(n) = 0$. We note first that the relation $\prec$ is *transitive*; that is, from $f(n) \prec g(n)$, $g(n) \prec h(n)$, it always follows that $f(n) \prec h(n)$.

For example, we have $n \prec n^2 \prec n^3$, or in general, $n^a \prec n^b$ for every pair of real numbers $a < b$. From calculus we know some further examples: $\log n \prec n$, and so $\log \log n \prec \log n$, and in general $\log n \prec n^\varepsilon$ for every $\varepsilon > 0$. If we apply the exponential function with base $c > 1$, then we obtain in our hierarchy

$$c \prec \log n \prec n^\varepsilon \prec c^n \prec n^n,$$

and so on. In short, the logarithmic function grows more slowly than every power, and every power more slowly than every exponential function.

Here is a question: Where does the factorial function $n!$ fit into this scheme? Certainly $n!$ cannot grow faster than n^n, and it is just as clear that $n!$ grows at least as fast as 2^n, since $n! \geq 2^n$ for $n \geq 4$. More precisely, we see that $n! \geq m^n$ for every natural number m, and thus we also have $n! \geq c^n$ for every $c > 1$ after some number n.

Our starting point was the notion of "faster" growth. Now we are going to discuss the idea of "at most" or "at least" as fast. In other words, we would like to estimate a function $f(n)$ asymptotically (that is, starting with some $n \geq n_0$) above and below. We shall employ the standard and suggestive notation ("O" is for "order")

(5.2) $O(g(n)) = \{ f(n) :$ there exists a constant $C > 0$

such that $|f(n)| \leq C |g(n)|$ for $n \geq n_0 \}$.

Thus $O(g(n))$ can be thought of as the set of all functions $f(n)$ that can be estimated asymptotically from above by $g(n)$. If $f(n)$ belongs to $O(g(n))$, we write $f(n) = O(g(n))$. This appears to be an odd notation, since it seems that one should write $f(n) \in O(g(n))$. However, tradition holds sway here, and moreover, the notation is helpful in computations, as we shall see. But we must never forget that $f(n) = O(g(n))$ is to be read from left to right (read "$f(n)$ is of order at most $g(n)$"). That is, it is a "one-sided" equation. If we failed to observe this caveat, we would find ourselves in the absurd situation of inferring from $n = O(n^2)$ and $n^2 = O(n^2)$ the false statement $n = n^2$.

Let us consider as an example the polynomial $p(n) = 2n^3 - n^2 + 6n + 100$. We obtain $p(n) = O(n^3)$, since

$$|p(n)| \leq 2|n^3| + |n^2| + 6|n| + 100 \leq 2|n^3| + |n^3| + 6|n^3| + 100 \leq 10|n^3|$$

for $n \geq 5$.

We have two observations to make on the use of the O notation. First, we are interested only in large values of n (think big). Second, $f(n) = O(g(n))$ has nothing to say about how fast $f(n)$ actually grows. In our example, we could as well have written $p(n) = O(n^4)$ or $p(n) = O(n^5)$. In general, we shall look for the best possible estimate.

For estimates from below, we have an analogous symbol:

(5.3) $\Omega\big(g(n)\big) = \{f(n) : \text{there exists a constant } C > 0,$

$$\text{such that } |f(n)| \geq C|g(n)| \text{ for } n \geq n_0\}.$$

We again write $f(n) = \Omega\big(g(n)\big)$ if $f(n)$ belongs to $\Omega\big(g(n)\big)$. Clearly, we have $f(n) = O\big(g(n)\big) \iff g(n) = \Omega\big(f(n)\big)$. It is apparent that O and Ω are transitive, that is, from $f(n) = O\big(g(n)\big)$, $g(n) = O\big(h(n)\big)$ it follows that $f(n) = O\big(h(n)\big)$, and analogously for Ω. In sum,

$$O\big(O(f(n))\big) = O\big(f(n)\big), \quad \Omega\big(\Omega(f(n))\big) = \Omega\big(f(n)\big).$$

Now we would like to introduce a symbol that encompasses both O and Ω:

(5.4)
$\Theta(g(n)) = \{f(n) : \text{there exist constants } C_1 > 0, \, C_2 > 0,$

$$\text{such that } C_1|g(n)| \leq |f(n)| \leq C_2|g(n)| \text{ for } n \geq n_0\}.$$

If we again set $f(n) = \Theta(g(n))$ if $f(n)$ belongs to $\Theta\big(g(n)\big)$, then we have $f(n) = \Theta\big(g(n)\big) \iff f(n) = O\big(g(n)\big)$ and $f(n) = \Omega\big(g(n)\big)$. The symbol Θ can then be interpreted to mean growth "at the same rate as," and we sometimes write

$$f(n) \asymp g(n) \iff f(n) = \Theta(g(n)) \iff g(n) = \Theta(f(n)).$$

A stronger version of $\asymp$ is obtained from

(5.5) $f(n) \sim g(n) \iff \lim_{n\to\infty} \left|\frac{f(n)}{g(n)}\right| = 1.$

If $f(n) \sim g(n)$, we say that $f(n)$ and $g(n)$ are **asymptotically equal**.

It is clear that $\asymp$ and $\sim$ are equivalence relations that are compatible with $\prec$, O and Ω. This means, for example, that from $f \prec g$ and $f \asymp f'$ we may conclude that $f' \prec g$ and that from $f = O(g)$ and $f \sim f'$ it follows that $f' = O(g)$.

Example 5.1. We can analyze polynomials without further ado. If $p(n)$ is a polynomial of degree d, then $\lim_{n\to\infty} \left|\frac{p(n)}{n^k}\right| = 0$ for $k > d$, and $p(n) \asymp n^d$, and we conclude for polynomials p and q that

(i) $p(n) \prec q(n) \iff \deg p(n) < \deg q(n)$;

(ii) $p(n) \asymp q(n) \iff \deg p(n) = \deg q(n)$;

(iii) $p(n) \sim q(n) \iff \deg p(n) = \deg q(n)$ and the leading coefficients are equal in absolute value.

Let us return to $n!$. Since $\log n! = \sum_{k=1}^{n} \log k$, it follows from constructing upper and lower sums for $\log x$ that

$$(5.6) \qquad \log(n-1)! < \int_1^n \log x \, dx = n \log n - n + 1 < \log n! \, ,$$

from which we may conclude that

$$e \left(\frac{n}{e} \right)^n < n! < \frac{(n+1)^{n+1}}{e^n} = (n+1) \left(\frac{n}{e} \right)^n \left(1 + \frac{1}{n} \right)^n < (n+1)e \left(\frac{n}{e} \right)^n .$$

The bounds agree up to a factor of $n+1$. The precise result is given by **Stirling's formula**, which states that $\lim_{n \to \infty} \frac{n!}{n^{n+\frac{1}{2}} e^{-n}} = \sqrt{2\pi}$, and thus

$$(5.7) \qquad\qquad n! \sim \sqrt{2\pi n} \left(\frac{n}{e} \right)^n .$$

From (5.7) we obtain $c^n \prec n! \prec n^n$, as one can immediately check. Let us look again at (5.6). The two inequalities yield

$$\log n! \sim n \log n .$$

We may obtain more precise information about the growth of $\log n!$ by taking the logarithm of (5.7):

$$\lim_{n \to \infty} \left(\log n! - \left(n \log n - n + \frac{\log n}{2} + \log \sqrt{2\pi} \right) \right) = 0 .$$

The number $\sigma = \log \sqrt{2\pi} \approx 0.919$ is called *Stirling's constant*, and the following formula, *Stirling's approximation*:

$$(5.8) \qquad\qquad \log n! = n \log n - n + \frac{\log n}{2} + \sigma + R(n),$$

with $R(n) \to 0$.

We see, then, that the growth of $\log n!$ can be determined even more precisely. For example, we also have

$$\log n! \sim n \log n - n + \frac{\log n}{2},$$

or in other words,

$$\log n! = n \log n - n + O(\log n) .$$

A propos the logarithm, we have used the *natural* logarithm $\log n$ in Stirling's approximation. In many algorithms, one does calculations to base 2, in which case the *binary logarithm* $\log_2 n$ comes into play. Yet another logarithm is the base-10 logarithm $\log_{10} n$ that one learns about in secondary school. For our O notation, it makes no difference which logarithm we use. Since any pair of logarithms differ by a positive factor, that is, $\log_b n = \frac{\log_a n}{\log_a b}$ $(a, b > 1)$, the expressions $O(\log n)$, $O(\log_2 n)$, and in general

$O(\log_a n)$ all have the same meaning. Since we shall be dealing mostly with the binary logarithm, we will represent it thus: $\lg n = \log_2 n$.

We are going to introduce yet one more symbol. We write

(5.9) $o(g(n)) = \{f(n) : \text{for every } \varepsilon > 0 \text{ there exists } n_0(\varepsilon)$

$\text{with } |f(n)| \le \varepsilon |g(n)| \text{ for } n \ge n_0(\varepsilon)\}.$

With usual notation, $f(n) = o(g(n))$ if $f(n)$ belongs to $o(g(n))$ means $f(n) = o(g(n))$, that is, $f(n) \prec g(n)$.

The notation $f(n) = o(1)$ means, therefore, that $f(n) \to 0$, while $f(n) = O(1)$ means that $f(n)$ remains bounded. For example, we can write (5.8) in the form

$$\log n! = n \log n - n + \frac{\log n}{2} + \sigma + o(1),$$

while if we take upper and lower sums of $\frac{1}{x}$, it follows that

$$\frac{1}{2} + \cdots + \frac{1}{n} < \log n < 1 + \frac{1}{2} + \cdots + \frac{1}{n-1},$$

that is,

$$H_n = \log n + O(1).$$

A further advantage of the O notation is that we can hide "superfluous details" in the O summand. For example, if we obtain in a calculation $\sqrt{n+a}$ for some constant a, we can immediately set $\sqrt{n+a} = \sqrt{n} + O(1)$, since $\sqrt{n+a} - \sqrt{n} \le |a|$, and we may now continue the calculation with the more convenient expression $\sqrt{n} + O(1)$.

5.2. Order of Magnitude of Recurrence Relations

Recall the Fibonacci recurrence relation: From $F_n = F_{n-1} + F_{n-2}$ and the initial values $F_0 = 0$, $F_1 = 1$ we were able to conclude from the methods of Section 3.2 that $F_n = \frac{1}{\sqrt{5}}(\phi^n - \hat{\phi}^n)$. Since $|\hat{\phi}| < 1$, we see that F_n is asymptotically equal to $\frac{1}{\sqrt{5}}\phi^n$, or, in our new notation,

$$F_n \sim \frac{1}{\sqrt{5}} \phi^n , \quad \phi = \frac{1 + \sqrt{5}}{2} ,$$

or, less precisely,

$$F_n \asymp \phi^n .$$

The Fibonacci numbers thus grow exponentially fast, and indeed at the rate of the exponential function with base approximately $\phi \approx 1.61$.

Does the rate of growth change if we maintain the same recurrence formula but change the initial values? For example, for $F_0 = -2$, $F_1 = 3$ we obtain the Fibonacci sequence $-2, 3, 1, 4, 5, 9, 14, \ldots$, while for $F_0 = 1$, $F_1 = -1$ we obtain the sequence $1, -1, 0, -1, -1, -2, -3, -5, \ldots$. Our four steps from Section 3.2 remain unchanged, and we again have $F(z) = \frac{c+dz}{1-z-z^2}$ with certain constants c, d determined by the initial conditions. With the help of Theorem 3.1 we obtain

$$F_n = a\phi^n + b\hat{\phi}^n$$

and thus again $F_n \asymp \phi^n$, or $F_n \sim a\phi^n$, except when $a = 0$. This certainly occurs if $F_0 = F_1 = 0$, and in that case, we obtain the zero sequence, as well as for one additional case (what is it?).

After a certain amount of manipulation, an explicit solution of a recurrence relation will always reveal the correct order of growth. A quite different question is whether we can determine the rate of growth of a counting function that is given by a recurrence relation for which we do not know the solution. This question takes on particular significance when we are unable to solve a recurrence relation explicitly.

For recurrences of fixed length and constant coefficients this represents nothing new. From formula (A4) in Theorem 3.1 we know that the largest root of $q^R(z)$ in absolute value dominates the remainder. In what follows we shall be interested in recurrence relations with coefficients some or all of which depend on n.

The simplest case $a_n T_n = b_n T_{n-1} + c_n$ was dealt with in Section 2.1. The method given there works because the elements T_n, T_{n-1} have sequential indices. In many, if not most, problems one employs a "divide and conquer" approach, which leads to completely different recurrence relations.

Let us consider an example. There are n participants in a tennis tournament. In the first round, every player who loses his or her match is eliminated. The winners are paired off for the second round, and the losers of these matches are again eliminated. The process is continued until a single winner remains. How many rounds $T(n)$ are required to determine a winner? To make things simple, let us assume that n is a power of 2, that is, $n = 2^k$. Our recurrence relation then looks as follows:

$$(5.10) \qquad T(n) = T(n/2) + 1 , \quad T(1) = 0 ,$$

since after the first round, $\frac{n}{2}$ players remain. Now, this recurrence relation is simple enough,

$$T(2^k) = T(2^{k-1}) + 1 = T(2^{k-2}) + 2 = \cdots = T(1) + k = k ,$$

and we conclude that $T(n) = \lg n, n = 2^k$.

The case $n \neq 2^k$ is now easily disposed of. In the first round, $\lfloor \frac{n}{2} \rfloor$ are eliminated, and $\lceil \frac{n}{2} \rceil$ remain. The general recurrence is then

$$(5.11) \qquad T(n) = T\left(\left\lceil \frac{n}{2} \right\rceil\right) + 1 , \quad T(1) = 0 ,$$

and we hypothesize that $T(n) = \lceil \lg n \rceil$. Note that $\lceil \lg n \rceil$ is the power k of 2 such that $2^{k-1} < n \leq 2^k$. In particular, we then have $2^{k-1} < n+1 \leq 2^k$, and therefore $\lceil \lg n \rceil = \lceil \lg(n+1) \rceil$, and hence also $\lceil \lg \frac{n}{2} \rceil = \lceil \lg \frac{n+1}{2} \rceil = \lceil \lg \lceil \frac{n}{2} \rceil \rceil$. If we go with our hypothesis $T(n) = \lceil \lg n \rceil$ in (5.11), we obtain

$$T(n) = \left\lceil \lg \left\lceil \frac{n}{2} \right\rceil \right\rceil + 1 = \left\lceil \lg \frac{n}{2} \right\rceil + 1 = \lceil \lg n - \lg 2 \rceil + 1 = \lceil \lg n \rceil \ .$$

Let us consider another example:

(5.12) $$T(n) = T(n/2) + n \ , \quad T(1) = 0 \ ,$$

again for $n = 2^k$. Iteration yields

$$T(2^k) = T(2^{k-1}) + 2^k = T(2^{k-2}) + 2^k + 2^{k-1} = \cdots = T(1) + 2^k + \cdots + 2^1$$
$$= 2^{k+1} - 2 \ ,$$

and hence $T(n) = 2n - 2 = \Theta(n)$, and for arbitrary n, the difference between $\lceil \frac{n}{2} \rceil$ and $\frac{n}{2}$ in $\Theta(n)$ vanishes.

Let us try something a bit more complex:

(5.13) $$T(n) = 3T\left(\left\lfloor \frac{n}{2} \right\rfloor\right) + n \ , \quad T(1) = 1 \ .$$

Iteration with $n = 2^k$ yields

$$T(2^k) = 3T(2^{k-1}) + 2^k = 3(3T(2^{k-2}) + 2^{k-1}) + 2^k$$
$$= 3^2 T(2^{k-2}) + 3 \cdot 2^{k-1} + 2^k$$
$$= \cdots = 3^k 2^0 + 3^{k-1} 2 + 3^{k-2} 2^2 + \cdots + 3^0 2^k \ .$$

By induction or using convolution of $\sum 3^n z^n$ and $\sum 2^n z^n$ we calculate that the right-hand side is equal to $3^{k+1} - 2^{k+1}$. Let us set $3 = 2^{\lg 3}$. We then obtain $3^{k+1} = 3 \cdot 3^k = 3 \cdot (2^{\lg 3})^k = 3n^{\lg 3}$, and hence

$$T(n) = 3n^{\lg 3} - 2n = \Theta(n^{\lg 3}) \quad \text{for } n = 2^k \ .$$

We again may expect that the rounding $\lfloor \frac{n}{2} \rfloor$ in $\Theta(n^{\lg 3})$ disappears.

Let us now consider the general recurrence relation

$$T(n) = a\,T(n/b) + f(n) \ , \quad T(1) = c \ ,$$

where we interpret $\frac{n}{b}$ as $\lfloor \frac{n}{b} \rfloor$ or $\lceil \frac{n}{b} \rceil$. The result clearly depends on a, b and $f(n)$. The following theorem describes the growth of $T(n)$.

Theorem 5.2. *Let $a \geq 1$, $b > 1$, and $T(n) = a\,T(n/b) + f(n)$.*
(a) *If $f(n) = O(n^{\log_b a - \varepsilon})$ for some $\varepsilon > 0$, then $T(n) = \Theta(n^{\log_b a})$.*
(b) *If $f(n) = \Theta(n^{\log_b a})$, then $T(n) = \Theta(n^{\log_b a} \lg n)$.*
(c) *If $f(n) = \Omega(n^{\log_b a + \varepsilon})$ for some $\varepsilon > 0$, and $a\,f(\frac{n}{b}) \leq c\,f(n)$ for some $c < 1$ and $n \geq n_0$, then $T(n) = \Theta(f(n))$.*

We are not going to prove this theorem. Rather, we will demonstrate its power by means of several examples. In essence, the theorem shows how the growth of $T(n)$ depends on the order of magnitude of $f(n)$. In case (c), $f(n)$ dominates, while in case (a), the summand $a\,T(n/b)$ dominates.

Example 5.3. Let us first consider $f(n) = C$, a constant. We have the following cases to consider:

$$
\begin{array}{lll}
a = 1, & T(n) = \Theta(\lg n), & \text{case (b),} \\
a = b > 1, & T(n) = \Theta(n), & \text{case (a),} \\
a \neq b,\ a, b > 1, & T(n) = \Theta(n^{\log_b a}), & \text{case (a).}
\end{array}
$$

Or for $f(n) = \Theta(n)$:

$$
\begin{array}{lll}
1 \leq a < b, & T(n) = \Theta(n), & \text{case (c),} \\
1 \leq a = b, & T(n) = \Theta(n \lg n), & \text{case (b),} \\
1 < b < a, & T(n) = \Theta(n^{\log_b a}), & \text{case (a).}
\end{array}
$$

Example 5.4. Now let us test the theorem on the example

$$T(n) = 4T(n/2) + n^2 \ .$$

Since $\log_2 4 = 2$, we have $f(n) = n^2 = n^{\log_2 4}$, that is, item (b) of the theorem comes into play, and we obtain $T(n) = \Theta(n^2 \lg n)$. Once we have determined the correct order of magnitude of $T(n)$, we can also of course try to prove the result inductively:

$$
T(n) = 4T\left(\frac{n}{2}\right) + n^2 = 4\Theta\left(\left(\frac{n}{2}\right)^2 \lg \frac{n}{2}\right) + n^2
$$
$$
= \Theta\left(n^2 \left(\lg n - 1\right)\right) + n^2 = \Theta\left(n^2 \lg n - n^2 + n^2\right) = \Theta\left(n^2 \lg n\right) \ .
$$

5.3. Running Times of Algorithms

An algorithm is by definition a step-by-step procedure for solving a particular problem. Everyone knows some algorithms and has used some of them since childhood. To add n numbers $a_1, a_2, \ldots, a_n$, we may first form the sum $a_1 + a_2$, then add in a_3, then a_4, and so on, until we have obtained the result $S = \sum_{i=1}^n a_i$. With this method we require $n - 1$ additions. Is there an algorithm for solving the same problem that uses fewer additions? When we think about algorithms, we often consider not just any old method for solving a problem, but one that satisfies some optimality condition, which often means finding the most efficient algorithm.

The field of combinatorics has long considered counting problems and the existence of particular configurations. We have learned some of its methods. But it is only with the added notion of "optimal configuration" that combinatorics has grown into the field that we today call discrete mathematics. It is certainly not by chance that the first general algorithms for difficult problems appeared along with the first fast computers. The simplex algorithm for solving large systems of inequalities is a prime example. We shall return to this algorithm later.

A particularly nice illustration of the three aspects counting, existence, and optimality is the famous *traveling salesman problem*: A traveling salesman wishes to visit n cities one after the other, visiting each city once and returning to his starting city. To travel between cities S_i and S_j entails some cost $c_{ij} \geq 0$, which may be in the form of dollars or perhaps distance. The salesman wishes to plan his route so as to minimize the total cost of visiting all n cities. To state this as a counting problem is trivial: There are $n!$ possible trips, or if we fix the starting city, $(n-1)!$. Among these $n!$ trips there is of course one (or perhaps several) that is the least expensive. So the question of existence is answered trivially. But how does the salesman determine which trip is the cheapest? For a small number of cities brute force gives the answer, but such a simple calculation of the cost of each route quickly becomes impossible on account of the exponential growth of $n!$. For a mere $n = 12$ cities the salesman must consider $11! = 39{,}916{,}800$ trips.

We see, then, that the question of optimality is intimately bound up with the use of computers. If an algorithm for solving a problem has a computation time that is exponential in the size of the input, we say that the algorithm has *exponential running time*. If every possible algorithm has exponential running time, then we consider the problem to be algorithmically unsolvable. Whether that is the case for the traveling salesman problem is an open question, and we shall see in Section 8.5 that this is *the* central problem in the theory of algorithmic computation.

Let us summarize the most important concepts about algorithms:

Structure of the Problem: We are given an *input*, and the algorithm produces an *output* by means of a sequence of instructional steps. In our problem of the traveling salesman, the input is the cost matrix (c_{ij}), $1 \leq i, j \leq n$, and the algorithm returns a permutation $i_1, i_2, \ldots, i_n$ such that $c_{i_1 i_2} + c_{i_2 i_3} + \cdots + c_{i_{n-1} i_n} + c_{i_n i_1}$ is minimal among all permutations.

Correctness of the Algorithm: We say that an algorithm is *correct* if the desired output is returned by the algorithm for every valid input. A proof of correctness, or as one also says, a *validation* of the algorithm, can be quite difficult. The algorithm might halt before delivering its output, produce an incorrect output, or go on forever without ever returning a result.

Design of Algorithms: An algorithm can be described in words, with a computer program, or be programmed directly in hardware. The data structures used by the algorithm also play an important role. We can input the data as a list, as a matrix, or recursively. The choice of the "correct" data structure is of central importance, and we will learn about different kinds of data structures in our discussion of these issues.

Analysis of Algorithms: Once correctness has been proven, one would like to know how "fast" the algorithm actually is. The *running time* of an algorithm is equivalent to the total number of individual computational steps required for execution of the algorithm as a function of the size of the input.

A couple of examples should make clear what all of this means.

Example 5.5. We would like an algorithm to evaluate a polynomial $p(x) = a_n x^n + a_{n-1} x^{n-1} + \cdots + a_0$ at the value c. How many multiplications do we need? The size of the input is conveniently taken to be n, the degree of the polynomial. Each time we perform a multiplication, we add 1 to our running total. We assume that all other operations, such as addition, are free.

We could proceed as follows: We compute the successive powers $c, c^2 = c \cdot c, c^3 = c^2 \cdot c, \ldots, c^n = c^{n-1} \cdot c$, which requires $n - 1$ multiplications, and then we use n additional multiplications to compute $p(c) = a_n c^n + a_{n-1} c^{n-1} + \cdots + a_1 c + a_0$. The running time is therefore $2n - 1$. But one can do better, for example using Horner's method:

$$p(c) = c(\cdots (c(c(c a_n + a_{n-1}) + a_{n-2}) + a_{n-3}) \cdots) + a_0 .$$

Here we need only n multiplications (on the left by c), and it can be shown that this is the best result possible. That is, there is no algorithm that can solve the problem with fewer than n multiplications.

Example 5.6. Our second example concerns sorting a set of numbers. The input is a set of n numbers $a_1, a_2, \ldots, a_n$, all different. The output is to be the same set of numbers, but sorted as $a'_1 < a'_2 < \cdots < a'_n$. A calculational step consists in comparing two numbers $a_i : a_j$, where we obtain the answer $a_i < a_j$ or $a_i > a_j$. The number of possible outputs is thus $n!$, and as we shall see later, every sorting algorithm requires at least $\lg n! = \Theta(n \lg n)$ comparisons.

How, then, shall we go about sorting the numbers? Let us employ the principle of divide and conquer. We divide the n numbers into two groups, as close to equal in size as possible: $n = \lfloor \frac{n}{2} \rfloor + \lceil \frac{n}{2} \rceil$. Let us not be bothered for now by rounding up and down, and take $n = \frac{n}{2} + \frac{n}{2}$. We now sort each of the two groups, and obtain $b_1 < b_2 < \cdots < b_{n/2}$ and $c_1 < c_2 < \cdots < c_{n/2}$. We now "unite" the two lists step by step. Clearly, the smallest number of all is either b_1 or c_1. Therefore, with a single comparison we can determine the smallest of the numbers. Let us assume, without loss of generality, that it is b_1. Removing this number to the output list being created by this process, we now have the two lists $b_2 < \cdots < b_n$ and $c_1 < \cdots < c_n$. The next-smallest number now must be b_2 or c_1, and a comparison tells us which it is. We add it to the output list, and consider the lists that remain. In this way, we see that the ordered list appears after at most n further comparisons (more precisely, $n-1$; why?). For the running time $T(n)$ of this "merge-sort" method we obtain the recurrence relation

$$T(n) = 2T(n/2) + n, \quad T(1) = 0,$$

and it follows from the previous section that

$$T(n) = \Theta(n \lg n).$$

Our algorithm is therefore asymptotically optimal. We shall return to sorting algorithms in Chapter 9, where we examine them in considerable detail.

Exercises for Chapter 5

5.1 Prove the validity of the following equations:

(a) $O(f(n))O(g(n)) = O(f(n)g(n))$,

(b) $O(f(n)g(n)) = f(n)O(g(n))$,

(c) $O(f(n)) + O(g(n)) = O(|f(n)| + |g(n)|)$.

▷ **5.2** Is the following assertion correct? From $f_1(n) \prec g_1(n)$, $f_2(n) \prec g_2(n)$ it follows that $f_1(n) + f_2(n) \prec g_1(n) + g_2(n)$.

5.3 Let $f(n) = n^2$ (n even) and $f(n) = 2n$ (n odd). Show that $f(n) = O(n^2)$, but not $f(n) = o(n^2)$ and not $n^2 = O(f(n))$.

▷ **5.4** Show that for $f \prec h$ there is always g such that $f \prec g \prec h$.

5.5 In each case below, find a function $g(n) \neq \Theta(f(n))$ of the form $g \colon \mathbb{N} \longrightarrow \mathbb{R}$ such that $f(n) = O(g(n))$ holds for the following function f:
(a) $f(n) = \binom{n}{2}$, (b) $f(n) = \frac{5n^3+1}{n+3}$, (c) $f(n) = \frac{n^2 3^n}{2^n}$, (d) $f(n) = n!$.

▷ **5.6** What is wrong with the following argument? Let $T(n) = 2T(\lfloor \frac{n}{2} \rfloor) + n$, $T(1) = 0$. We assume $T(\lfloor \frac{n}{2} \rfloor) = O(\lfloor \frac{n}{2} \rfloor)$ inductively with $T(\frac{n}{2}) \leq c\frac{n}{2}$. It follows that $T(n) \leq 2c\lfloor \frac{n}{2} \rfloor + n \leq (c+1)n = O(n)$.

5.7 Let $T(n) = 2T(\lfloor \sqrt{n} \rfloor) + \lg n$. Using the substitution $n = 2^m$, show that $T(n) = O(\lg n \lg \lg n)$.

5.8 Determine the order of magnitude of $T(n)$ in the following cases: (a) $T(n) = 3T(n-1)+n^2 2^n$, (b) $T(n) = 3T(n-1)+\frac{n+1}{n+2}3^n$, (c) $T(n) = 2T(n-1)+\frac{1+n^2}{3+n^2}2^{n-1}$, (d) $T(n) = 2T(n/3) + n\sqrt{n}$.

5.9 Suppose we have an algorithm that for an input of length n, executes n steps, where the ith step requires i^2 operations. Show that the running time of the algorithm is $O(n^3)$.

▷ **5.10** An *addition chain* for n is a sequence $1 = a_1, a_2, \ldots, a_m = n$ such that for every k, we have $a_k = a_i + a_j$ for some $i, j < k$. Example: $n = 19$, $a_1 = 1$, $a_2 = 2$, $a_3 = 4$, $a_4 = 8 = 4 + 4$, $a_5 = 9 = 8 + 1$, $a_6 = 17 = 9 + 8$, $a_7 = 19 = 17 + 2$. Let $\ell(n)$ be the minimal length of an addition chain for n. Show that $\lg n \leq \ell(n) \leq 2 \lg n$. Are there integers n for which $\ell(n) = \lg n$?

5.11 Show that every permutation $a_1 a_2 \ldots a_n$ can be brought via successive exchanges of neighboring elements into the form $12 \ldots n$. Example: $3124 \longrightarrow 3214 \longrightarrow 2314 \longrightarrow 2134 \longrightarrow 1234$. What is the minimal number of exchanges?

▷ **5.12** Verify carefully whether the following equation is correct: $\sum_{k=0}^{n}(k^2 + O(k)) = \frac{n^3}{3} + O(n^2)$. Note that this is a set comparison (from left to right).

5.13 Show that $O(x + y)^2 = O(x^2) + O(y^2)$ for real numbers x, y.

5.14 Prove that $(1 - \frac{1}{n})^n = e^{-1}(1 - \frac{1}{2n} - \frac{5}{24n^2} + O(\frac{1}{n^3}))$. Hint: Write $(1 - \frac{1}{n})^n = e^{n \log(1 - \frac{1}{n})}$.

▷ **5.15** Let $a_n = O(f(n))$ and $b_n = O(f(n))$.
Prove or disprove that the convolution $\sum_{k=0}^{n} a_k b_{n-k}$ is $O(f(n))$ for each of the following functions: (a) $f(n) = n^{-a}$, $a > 1$, (b) $f(n) = a^{-n}$, $a > 1$. Hint: In (a) we have $\sum_{k \geq 0} |f(k)| < \infty$, from which the validity follows.

5.16 We know that $n! \prec n^n$. Show that conversely, $n^n < (n!)^2$ for $n \geq 3$. Is it also true that $n^n \prec (n!)^2$?

5.17 Solve the recurrence relation $T(n) = 2T(\lfloor \frac{n}{2} \rfloor) + n^2$, $T(1) = 0$.

▷ **5.18** Show that $T(n) = T(\frac{n}{4}) + T(\frac{3n}{4}) + n$ has the solution $T(n) = O(n \lg n)$ by developing the right-hand side. That is, write $n = \frac{n}{4} + \frac{3n}{4} = (\frac{n}{16} + \frac{3n}{16}) + (\frac{3n}{16} + \frac{9n}{16})$, and so on.

5.19 Use the method of the previous problem to solve $T(n) = T(an) + T((1 - a)n) + n$, $0 < a < 1$.

▷ **5.20** Suppose Algorithm A has a running time given by the recurrence relation $T(n) = 7T(\frac{n}{2}) + n^2$. Algorithm A' is specified by the recurrence relation $S(n) = \alpha S(\frac{n}{4}) + n^2$. What is the largest α for which A' is faster than A, that is, for which running time $(A') \prec$ running time (A)?

5.21 Give best possible upper and lower bounds for the following recurrence relations: (a) $T(n) = 2T(\frac{n}{2}) + n^3$, (b) $T(n) = 3T(\frac{n}{4}) + \sqrt{n}$, (c) $T(n) = T(\sqrt{n}) + 10$.

5.22 Estimate the rate of growth of the largest binomial coefficient $f(n) = \binom{n}{n/2}$. Hint: Stirling's formula.

▷ **5.23** We shall meet the following recurrence relation in sorting problems: $T(n) = \frac{2}{n} \sum_{k=0}^{n-1} T(k) + an + b$, $T(0) = 0$ with $a > 0$. Show that $n \prec T(n) \prec n^2$. Hint: Try out $T(n) = cn$ and $T(n) = cn^2$.

5.24 Show that for $T(n)$ in the previous exercise, $T(n) = cn \lg n + o(n \lg n)$ holds with $c = a \log 4$. Hint: Try out $T(n) = cn \lg n$.

▷ **5.25** The Euclidean algorithm for computing the greatest common divisor of two integers operates by successive divisions with remainder. Example: If we are given the numbers 154 and 56, we then have $154 = 2 \cdot 56 + 42$, $56 = 1 \cdot 42 + 14$, $42 = 3 \cdot 14$, and thus $14 = \gcd(154, 56)$. Suppose $a > b$ are two integers. Show that from $b < F_{n+1}$ (Fibonacci number) it follows that the number of computational steps in computing $\gcd(a, b)$ is at most $n - 1$, and conclude from this that the number of steps is $O(\log b)$. Show further that the number of steps is at most five times the number of digits in b (given in decimal representation).

▷ **5.26** A waiter has n pizzas of n different sizes on a platter. Before he serves them, he wishes to arrange them in an aesthetically pleasing order, with the smallest on top, then the next-smallest, and so on, with the largest pizza at the bottom. An operation consists in picking up the top k pizzas as a group and flipping them over all at once. How many such moves are required to achieve the desired result? In terms of permutations, an operation consists in a flip of the first k elements. Example: $3241 \longrightarrow 2341 \longrightarrow 4321 \longrightarrow 1234$. For a permutation π let $\ell(\pi)$ be the minimal number of flips, and $\ell(n) = \max_\pi \ell(\pi)$. Show that $n \le \ell(n) \le 2n - 3$. Hint: For the upper bound, it is clear that by flipping n to the top and then to the end, it takes at most two flips to bring n into its proper position. Now forget n and do the same thing with $n - 1$, and so on. So you need at most $2(n - 2)$ flips from the nth to the third, and then one more to put the second into position.

5.27 Here is a problem related to the previous exercise. We are given a permutation π of $\{1, \ldots, n\}$, and we would like to bring it into the form $1, 2, \ldots, n$. In each step we can place the first element at an arbitrary position. The number of steps in an optimal algorithm is $\ell(\pi)$ with $\ell(n) = \max \ell(\pi)$. Calculate $\ell(n)$.

5.28 One can determine whether the number $n \ge 3$ is prime in the following way: Test each of the integers $2, 3, \ldots, \lfloor \sqrt{n} \rfloor$ as to whether it is a factor of n. If any of these integers is a factor, then n is not a prime, while if none of them is, then n is prime. Why can we stop the test at $\lfloor \sqrt{n} \rfloor$? How many steps does this algorithm require at most?

▷ **5.29** In a variant of the Tower of Hanoi problem (see Exercise 2.18) we are given four spikes A, B, C, D and the problem is to move a tower of n disks under the usual conditions from A to D. Let W_n be the minimal number of moves. Show that

$$W_{\binom{n+1}{2}} \le 2W_{\binom{n}{2}} + T_n \,,$$

where T_n is the minimal number for three spikes. Determine a function $f(n)$ with $W_{\binom{n+1}{2}} \le f(n)$. Hint: Consider

$$U_n = \frac{W_{\binom{n+1}{2}} - 1}{2^n}.$$

5.30 We begin with the numbers $a_1 = 1$ and $a_2 = 2$. At each step, $a_\ell = a_i + a_j + a_k$, for some choice of (not necessarily distinct) indices $i, j, k < \ell$, that is, we may add together any three numbers whose indices satisfy the given condition. Can every positive integer n be attained via such a three-member addition chain? If yes, estimate the length $m(n)$ from above and below. Generalize to k-member chains with the initial values $a_1 = 1$, $a_2 = 2$, $\ldots$, $a_{k-1} = k - 1$.

5.31 To compute the product of two $n \times n$ matrices A, B, we use the usual method requiring n^3 multiplications, namely n for each of the n^2 inner products of rows in A and columns in B. In particular, eight multiplications are required in the case $n = 2$. The following interesting method of Strassen requires only seven multiplications. Let $A = \begin{pmatrix} a & b \\ c & d \end{pmatrix}$, $B = \begin{pmatrix} \alpha & \beta \\ \gamma & \delta \end{pmatrix}$. Show that each element of AB is a sum of terms $\pm m_i$, where $m_1 = (a+d)(\alpha+\delta)$, $m_2 = (c+d)\alpha$, $m_3 = a(\beta - \delta)$, $m_4 = d(\gamma - \alpha)$, $m_5 = (a+b)\delta$, $m_6 = (a-c)(\alpha+\beta)$, $m_7 = (b-d)(\gamma+\delta)$. How many additions and subtractions are required in the usual method, and how many in Strassen's? Find a method that requires $O(n^{\lg 7})$ multiplications for the product of two $n \times n$ matrices. Hint: Split the matrices in half and use a recurrence starting with $n = 2$.

$\triangleright$ **5.32** Let N be a natural number that we would like to write in binary representation $a_k a_{k-1} \ldots a_0$. Show how one can use the Euclidean algorithm to estimate the running time $f(n)$, where n is the number of digits of N in decimal representation. Hint: Divide N successively by 2.

5.33 From Exercise 5.11 it follows that every permutation π can be brought by exchange of neighboring elements into any other permutation σ. What is the minimal number of exchanges that will suffice for every pair π and σ?

Bibliography for Part 1

Ideas and methods for counting constitute the classical theme of combinatorics. A more extensive collection of summation and inversion formulas can be found in the books of Riordan and Knuth. The subject of sums and differences is handled in great detail in Graham–Knuth–Patashnik. Some topics, such as the generalization of the difference calculus to arbitrary posets (e.g., Möbius inversion), have been omitted from this book entirely. The reader wishing to learn something about this may consult the book by Aigner. A deeper introduction to the theory of generating functions is offered in the books of Aigner, Stanley, and Wilf. Very nice introductions to probability theory are to be found in the book of Billingsley and in the classic by Feller. One should also include the book of Alon and Spencer, which contains an outstanding presentation of methods of particular use in discrete mathematics. Those wishing to learn more about asymptotic analysis are directed again to Graham, Knuth, and Patashnik, and for a more advanced treatment, to the book of Greene and Knuth. A nice selection of applications can be found in Montroll. And finally, the book of Matoušek and Nešetril and that of Lovász are highly recommended, in particular the latter, which is a virtual treasure trove of new problems (in addition to hints and solutions).

M. AIGNER: *Combinatorial Theory.* Springer-Verlag.

N. ALON AND J. SPENCER: *The Probabilistic Method.* Wiley Publications.

P. BILLINGSLEY: *Probability and Measure.* Wiley.

W. FELLER: *Probability Theory and its Applications.* Wiley Publications.

R. GRAHAM, D. KNUTH, AND O. PATASHNIK: *Concrete Mathematics.* Addison-Wesley.

D. GREENE AND D. KNUTH: *Mathematics for the Analysis of Algorithms.* Birkhäuser.

D. KNUTH: *The Art of Computer Programming I, Fundamental Algorithms.* Addison-Wesley.

L. LOVÁSZ: *Combinatorial Problems and Exercises.* North-Holland.

J. MATOUŠEK AND J. NEŠETRIL: *Invitation to Discrete Mathematics.* Oxford.

E. MONTROLL: *Applied Combinatorial Mathematics* (Beckenbach, ed.). Wiley.

J. RIORDAN: *Combinatorial Identities.* J. Wiley & Sons.

R. STANLEY: *Enumerative Combinatorics I, II.* Cambridge University Press.

H. WILF: *generatingfunctionology.* Academic Press.

Part 2

Graphs and Algorithms

In the first part of this book we learned about various methods for counting finite sets. Now we would like to turn our attention to the sets themselves. The sets that occur in a problem usually have a prescribed structure, which we make use of in investigating the problem and counting the sets. Or else we bring the sets into a particular form, such as a form suitable for input into a computer. The creation of good *data structures* is one of the most important tasks of information science.

Some structures are so familiar that we use them without much ado. For example, *Lists* $a_1 < a_2 < \cdots < a_n$, in which the elements are ordered linearly, or *matrices* (a_{ij}), in which the elements are arranged in rows and columns. Or more generally, we consider *schemes* $(a_{i_1 i_2 \ldots i_k})$ with k indices, or triangular schemes such as Pascal's triangle, which we looked at in the first part.

The simplest structure on a set is generated by a *binary relation*. Two elements are either related or not, and such relations are naturally expressed as **graphs**. That is, graphs are nothing but sets on which a binary relation has been defined.

For example, the natural numbers possess the "less-than" relation $<$, by which $1 < 2 < 3 < 4 < \cdots$, and the divisibility relation $|$, by which $2 \mid 4$, $3 \mid 12$, $5 \nmid 11$. The divisibility relation, like the less-than relation, is *transitive*, and it therefore produces an *order* on the set $\mathbb{N}$ of natural numbers. In Chapter 12 we shall discuss calculation with residue classes on $\mathbb{N}$, which is an example of an *equivalence relation*.

Graphs constitute the fundamental data structure of discrete mathematics, and the reason for this is that in practically every situation one can define a reasonable binary relation, which can then be modeled using graphs. We have already mentioned the $<$ relation and the divisibility relation. Another example is the inclusion relation $A \subseteq B$ for sets, from which we obtain the *set diagram* or *Boolean algebra* of all subsets of a fixed set S. In an algorithm we can consider two program steps to be related if they can occur in sequence when the program is executed. In this way we obtain a *flow chart*. Even extramathematical questions can be usefully modeled with graphs. For example, to study the social structure of a group, we can analyze the *dependency relations* among individuals in order to analyze the group's *hierarchical* structure. *Transit systems* can be studied by placing two locations in binary relation if they are connected by a road.

The most important application of graphs is the optimization of the types of situations just described. For example, we would like not only to describe a transit system, but to determine the optimum traffic flow. Practically the entire field of combinatorial optimization and the construction of good **algorithms** uses graphs as the underlying data structure. We shall go into all of this in considerable detail, particularly in Chapters 8–10. We shall first assemble, in the opening two chapters of this part, the necessary concepts.

Chapter 6

Graphs

6.1. Definitions and Examples

A **graph** $G = (V, E)$ consists of a (finite) **vertex set** V and a set $E \subseteq \binom{V}{2}$ of pairs $\{u, v\}$, $u \neq v$, of vertices, called **edges**.

The terms vertex and edge refer to the usual pictorial representation of a graph. For example, let $V = \{1, 2, 3, 4, 5\}$, $E = \{\{1, 2\}, \{1, 3\}, \{2, 3\}, \{2, 4\}, \{2, 5\}, \{3, 5\}\}$. We then may draw the graph G as the following diagram:

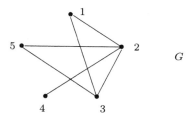

This also explains the name "graph": the edge–vertex system is reminiscent of the usual graphical representation of a function. Here as well, such a representation is of course only an aid in helping us to picture the abstract structure $G = (V, E)$.

Every set V with a binary relation can be interpreted as a graph by setting $\{u, v\} \in E$ if u and v are in the given relation. For example, consider the divisor relation: Let $V = \{1, 2, \ldots, 8\}$ and $\{i, j\} \in E$ if i, j are related by $i \mid j$ or $j \mid i$. Here is the associated graph:

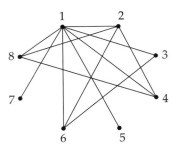

In the future we will omit the braces in our notation for edges and simply write $uv \in E$.

Sometimes we will allow **loops** uu, as well as **multiple edges** between vertices u, v. In such cases we speak of a **multigraph**. For example,

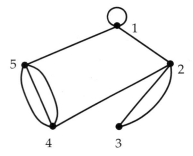

is a multigraph with a loop at 1, a double edge $\{2,3\}$, and a triple edge $\{4,5\}$. Since the terminology is not uniform among various authors, we stress again that in our definition, a graph *cannot* possess loops or multiple edges.

Some graphs have attained such an importance that they have their own names:

1.   If $|V| = n$ and $E = \binom{V}{2}$, we speak of a **complete graph** K_n:

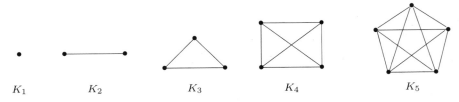

For complete graphs K_n we therefore have $|V| = n$, $|E| = \binom{n}{2}$.

2.   We call a graph $G = (S + T, E)$ **bipartite** if V consists of two disjoint sets S and T and every edge has one vertex in S and the other in T. If *all* possible edges between S and T are present, we then speak of a **complete**

bipartite graph $K_{S,T}$, or $K_{m,n}$ if $|S| = m$, $|T| = n$. Here is an example:

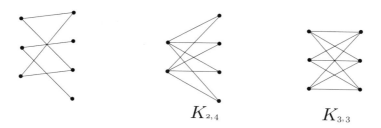

$$K_{2,4} \qquad K_{3,3}$$

For $K_{m,n}$ we thus have $|V| = m + n$, $|E| = mn$.

Bipartite graphs are an excellent tool for dealing with assignment problems. Thus if S is a set of people and T a set of jobs, we set $uv \in E$ with $u \in S$, $v \in T$ if person u is a suitable candidate for job v. The resulting bipartite graph thus models the person–job assignment problem, to which we shall return later in greater detail.

3. As an obvious generalization, we have the **complete k-partite graphs** $K_{n_1,\ldots,n_k}$ with $V = V_1 + \cdots + V_k$, $|V_i| = n_i$ $(i = 1,\ldots,k)$, and $E = \{uv : u \in V_i, v \in V_j, i \neq j\}$. We obtain $|V| = \sum_{i=1}^{k} n_i$, $|E| = \sum_{i<j} n_i n_j$. It is clear how arbitrary k-partite graphs are defined.

4. A **hypercube** Q_n has as its vertex set all $0, 1$ sequences of length n, and thus $|V| = 2^n$. We set $uv \in E$ if the sequences u and v differ at precisely one location. The graph Q_3 in the following figure justifies the name:

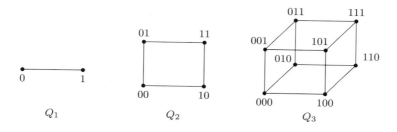

$$Q_1 \qquad Q_2 \qquad Q_3$$

For the edge number we obtain $|E(Q_1)| = 1$, $|E(Q_2)| = 4$, $|E(Q_3)| = 12$, and in general, $|E(Q_n)| = n2^{n-1}$. (Proof?)

If we interpret as usual the $0, 1$ sequences as characteristic vectors of subsets of a fixed n-set S, then Q_n has as its vertices the 2^n subsets, and we have $AB \in E$ if $A \subseteq B$, and B contains exactly one element more than A (or conversely). In this interpretation, Q_n is then precisely the diagram of the Boolean algebra $\mathcal{B}(S)$. For $S = \{1, 2, 3, 4\}$ we obtain

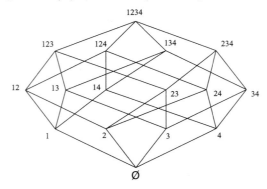

5.   If we think of a system of roads interconnecting a number of nodal points as a graph, then of primary interest is the question how one gets from one node to another. A **path** P_n in a graph consists of a sequence $u_1, u_2, \ldots, u_n$ of distinct vertices with $u_i u_{i+1} \in E$ for all i. The *length* of the path is the number $n - 1$ of edges $u_i u_{i+1}$:

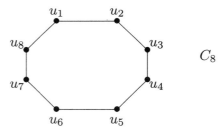

A **circuit** C_n is a sequence of distinct vertices $u_1, u_2, \ldots, u_n$ with $u_i u_{i+1} \in E$ ($i = 1, \ldots, n - 1$) and $u_n u_1 \in E$. The *length* of C_n is the number n of vertices and edges:

6.   A well-known graph that will become familiar to us as well is the **Petersen graph**:

As with every mathematical structure, we have a notion of isomorphism, one that tells us in this case when two graphs have the same structure. Two graphs $G = (V, E)$ and $G' = (V', E')$ are said to be **isomorphic**, written $G \cong G'$, if there is a bijection $\varphi : V \to V'$ such that $uv \in E \Leftrightarrow \varphi(u)\varphi(v) \in E'$. For example, two complete graphs on n vertices are isomorphic, which is why we chose a general symbol K_n, and the same holds for complete k-partite graphs. Moreover, we have, for example, $K_{2,2} \cong C_4 \cong Q_2$.

What exactly do isomorphic graphs have in common? For example, are the following two graphs isomorphic?

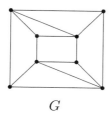
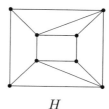

$$G \qquad\qquad H$$

Both graphs have eight vertices and fourteen edges, but is that enough to make $G \cong H$? Before we answer this question, we need to assemble a couple of concepts.

If $uv \in E$, we call u and v **neighboring** or **adjacent**. If $u \in V$, $k \in E$, with $u \in k$, we say that u and k are **incident** and that u is an **end vertex** of k. Likewise, we call two edges k and ℓ **incident** if they have a common end vertex, that is, if $k \cap \ell \neq \varnothing$. We denote the set of **neighbors** of $u \in V$ by $N(u)$ and call $d(u) = |N(u)|$ the **degree** of u. The vertex u is said to be **isolated** if $d(u) = 0$. For the size of V and E we generally use the letters $n = |V|$, $q = |E|$, and we call n the **order** and q the **size** of the graph G.

For every graph we can construct the *degree sequence $d_1 \geq d_2 \geq \cdots \geq d_n$*, with $d_i = d(u_i)$, $i = 1, \ldots, n$, and it is clear that isomorphic graphs have the same degree sequence. Moreover, an isomorphism maps every vertex to a vertex of the same degree.

Let us return to our example. Both G and H have four vertices of degree 4 and four vertices of degree 3, and therefore the degree sequences $4, 4, 4, 4, 3, 3, 3, 3$ are identical. Are G and H therefore isomorphic? No. In G, none of the vertices of degree 3 are neighboring, which is not the case in H, and it follows that $G \not\cong H$.

The degree of a graph yields a simple but useful theorem:

Theorem 6.1. *Let $G = (V, E)$ be a graph. Then*

$$\sum_{u \in V} d(u) = 2|E|.$$

Proof. We count the pairs (u, k), $u \in V$, $k \in E$ with $u \in k$, in two ways. Summing over the vertices, we obtain $\sum_{u \in V} d(u)$, and over the edges, $2|E|$, since each edge has two end vertices. $\qquad\square$

Corollary 6.2. *Every graph has an even number of vertices of odd degree.*

Proof. Let V_0 and V_1 denote respectively the sets of vertices of even and of odd degree. From Theorem 6.1 we have $2|E| = \sum_{u \in V_0} d(u) + \sum_{u \in V_1} d(u)$. Since the left-hand side and the first summand on the right-hand side are even numbers, it follows that $\sum_{u \in V_1} d(u)$ must be even, and thus the number $|V_1|$ of summands must be even. $\qquad\square$

We call a graph r-**regular**, if $d(u) = r$ for every degree. For example, K_n is $(n-1)$-regular, C_n is 2-regular, and Q_n is n-regular. Theorem 6.1 then yields for r-regular graphs the relation $r|V| = 2|E|$. We have for example $|E(Q_n)| = n2^{n-1}$.

6.2. Representation of Graphs

The pictorial representation of a graph is helpful as a visualization tool, but it is clearly not what we need for representing a graph in a computer program. For such applications, what is a suitable data structure for $G = (V, E)$? We could specify all the pairs $uv \in E$, or perhaps for each $u \in V$ we could give the list of neighbors. Two additional representations are more attractive for a number of applications: the **adjacency matrix** A and the **incidence matrix** B. We number the vertices $u_1, \ldots, u_n$ and the edges $k_1, \ldots, k_q$. The adjacency matrix is the $n \times n$ matrix $A = (a_{ij})$ with

$$a_{ij} = \begin{cases} 1, & \text{if } u_i u_j \in E, \\ 0, & \text{otherwise.} \end{cases}$$

Thus A is a symmetric matrix with zeros along the main diagonal whose row and column sums are equal to the degrees.

The incidence matrix $B = (b_{ij})$ is the $n \times q$ matrix $B = (b_{ij})$ with

$$b_{ij} = \begin{cases} 1 & \text{if } u_i \in k_j, \\ 0 & \text{otherwise.} \end{cases}$$

Counting the ones in B by rows and by columns corresponds exactly to Theorem 6.1. Let B^T be the transpose of the matrix B. For the symmetric $n \times n$ matrix $M = BB^T$, $M = (m_{ij})$, we then have

$$m_{ij} = \begin{cases} d(u_i) & \text{if } i = j, \\ a_{ij} & \text{if } i \neq j, \end{cases}$$

that is,

$$M = \begin{pmatrix} d(u_1) & & 0 \\ & \ddots & \\ 0 & & d(u_n) \end{pmatrix} + A.$$

For bipartite graphs $G = (S + T, E)$ on the defining vertex sets S and T we have yet another useful representation. Let $S = \{u_1, \ldots, u_m\}$, $T = \{v_1, \ldots, v_n\}$. Then $D = (d_{ij})$ is given by

$$d_{ij} = \begin{cases} 1 & \text{if } u_i v_j \in E, \\ 0 & \text{otherwise.} \end{cases}$$

As an example, we have

$$\longrightarrow \quad D = \begin{pmatrix} 1 & 1 & 0 & 1 \\ 1 & 0 & 0 & 0 \\ 0 & 0 & 1 & 1 \end{pmatrix}.$$

The double counting in D (by rows and by columns) thus corresponds to the equation $\sum_{u \in S} d(u) = \sum_{v \in T} d(v)$.

Apart from their convenience, how are these matrices useful as computer input? Let us consider the ℓth power A^ℓ of the adjacency matrix with the usual matrix product. We assert that the entry $A^\ell(i, j)$ is equal to the number of walks along the edges from u_i to u_j of length ℓ. Here, edges and vertices can be traversed multiple times, but ℓ edges altogether, with start vertex u_i and end vertex u_j. For $\ell = 1$ this is precisely the definition of A. We then continue by induction, and we have

$$A^\ell(i, j) = \sum_{k=1}^{n} A^{\ell-1}(i, k) A(k, j).$$

If we classify these routes by the last vertex u_k before u_j, then the right-hand side is our old friend the summation rule. In the next chapter we shall consider applications of the incidence matrix.

Let us look again at the adjacency matrix. Every numbering f of V yields an adjacency matrix A_f. Are there particulary "good" numberings?

Example 6.3. Let us consider the following two numberings:

$$\xrightarrow{f} \quad \begin{pmatrix} 0 & 0 & 1 & 1 & 0 \\ 0 & 0 & 1 & 1 & 1 \\ 1 & 1 & 0 & 0 & 0 \\ 1 & 1 & 0 & 0 & 1 \\ 0 & 1 & 0 & 1 & 0 \end{pmatrix} = A_f$$

and

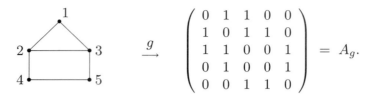

$$\xrightarrow{g} \begin{pmatrix} 0 & 1 & 1 & 0 & 0 \\ 1 & 0 & 1 & 1 & 0 \\ 1 & 1 & 0 & 0 & 1 \\ 0 & 1 & 0 & 0 & 1 \\ 0 & 0 & 1 & 1 & 0 \end{pmatrix} = A_g.$$

The second numbering yields a smaller deviation, or *bandwidth*, of the 1's in A_g from the main diagonal (namely, $b_g = 2$) than does the first numbering f ($b_f = 3$). To save computer memory, we shall prefer g. The bandwidth b_f of the graph G (numbered via f) is defined by

$$b_f = \max_{uv \in E} |f(u) - f(v)|.$$

This suggests an interesting question, the *bandwidth problem*: Determine for a graph G the **bandwidth** $b(G) = \min_f (b_f : f \text{ numbering})$.

We can easily determine the bandwidth for simple graphs. For a path P_n ($n \geq 2$) we have $b(P_n) = 1$ with the numbering

A circuit C_n has bandwidth $b(C_n) = 2$. Clearly, we have $b(C_n) \geq 2$, since the vertex with number 1 has a neighbor with number greater than or equal to 3. The following numberings are optimal:

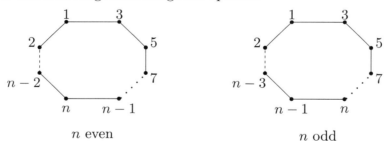

Clearly, we have $b(K_n) = n - 1$, and also for the graphs $K_{m,n}$, the bandwidth is not difficult to determine (see Exercise 6.28). For hypercubes Q_n, the determination of the bandwidth is not so easy, and indeed, the general bandwidth problem is a provably difficult, that is, an NP-complete, problem. We shall have more to say about this in Section 8.5.

6.3. Paths and Circuits

To learn more about a particular graph G, we should get to know something about its substructures. The following definition is straightforward: A graph

$H = (V', E')$ is called a **subgraph** of $G = (V, E)$ if $V' \subseteq V$ and $E' \subseteq E$. The subgraph H is called an **induced subgraph** if $E' = E \cap \binom{V'}{2}$, that is, if H contains all edges between the vertices in V' that are also present in G. In the following example, H_1, H_2 are subgraphs of G, but only H_2 is induced:

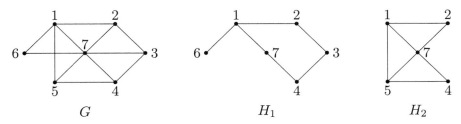

We obtain particularly important subgraphs by removing vertices or edges. Let $G = (V, E)$. Then $G \smallsetminus A$, $A \subseteq V$, denotes the graphs obtained by removing A and all edges incident with A, and $G \smallsetminus B$, $B \subseteq E$, denotes the graph $G = (V, E \smallsetminus B)$. The graphs $G \smallsetminus A$ and $A \subseteq V$ are thus always induced subgraphs. Let us remove $A = \{1, 3, 6\}$ in the example above, thereby obtaining

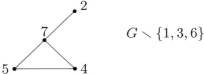

$$G \smallsetminus \{1, 3, 6\}$$

Our first investigations into graphs are motivated by their pictorial representation. Let us think again about the example of a system of roads. We are interested most of all in whether we can get from one nodal point to another, and if so, how many streets we must traverse in a *shortest* path. The following definitions arise from such considerations.

Let $G = (V, E)$. We say that $v \in V$ is *reachable* from $u \in V$ is there is a path P with start vertex u and end vertex v, for short, a u, v-path. Clearly, reachability is an equivalence relation on V. We call the induced graphs on the individual equivalence classes the **components** of G.

The following graph has four components:

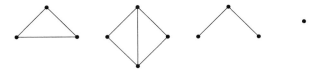

A graph G is said to be **connected** if it consists of a single component. We call an edge k of a graph $G = (V, E)$ a **cut edge** if the removal of k increases the number of components. The graph H_1 above contains a single cut edge, namely the edge $\{1,6\}$. The graph H_2 has no cut edges. An edge is a cut edge of G if and only if it is contained in no circuit of G (clear?).

Using paths, we can also introduce a notion of distance. The **distance** $d(u,v)$ between two vertices u and v is the length of the shortest path from u to v, with $d(u,u) = 0$, $u \in V$. If there is no such path, in which case u and v lie in different components, we set $d(u,v) = \infty$. We call $D(G) = [b] \max_{u,v \in V} d(u,v)$ the **diameter** of G.

In the following graph G we have $d(u,v) = 2$, $d(u,w) = 3$, $d(v,w) = 2$, and $D(G) = 3$:

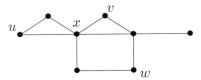

It should be clear that $d : V^2 \to \mathbb{N}_0$ satisfies the triangle inequality $d(u,v) \leq d(u,w) + d(w,v)$ and therefore represents a metric, as defined in function theory.

As an example, let us consider the hypercube Q_n. We see at once that $d(u,v)$ is equal to the number of coordinate places in which u and v differ, from which it follows that $D(Q_n) = n$.

Using circuits, we can derive a very useful characterization of bipartite graphs.

Theorem 6.4. *A graph G with $n \geq 2$ vertices is bipartite if and only if all circuits have the same length. In particular, G is bipartite if there are no circuits at all.*

Proof. We may assume that G is connected, since otherwise, we could simply consider the individual components separately. For the proof in one direction, suppose that G is bipartite with defining vertex sets S and T. Every circuit must run alternately between S and T, and therefore have even length. Conversely, let us suppose that all circuits have even length. We choose $u \in V$ arbitrarily and set $u \in S$. We create the sets S and T according to the following rules:

$$v \in \begin{cases} S & \text{if } d(u,v) \text{ is even,} \\ T & \text{if } d(u,v) \text{ is odd.} \end{cases}$$

It remains to show that vertices from S and from T are never neighbors. Let us suppose to the contrary that $v, w \in T$, $vw \in E$ (the case $v, w \in S$ is analogous). From $vw \in E$ it follows that $|d(u,v) - d(u,w)| \leq 1$, and therefore $d(u, v) = d(u, w)$, since both numbers are odd:

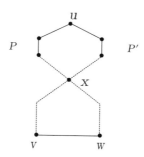

Let P be a u, v-path of length $d(u, v)$, P' a u, w-path of length $d(u, w)$, and x the last common vertex of P and P'. Then $d(x, v) = d(x, w)$, and we obtain the circuit $P(x, v), vw, P'(w, x)$ of odd length, which is a contradiction. $\square$

Example 6.5. The hypercubes Q_n are bipartite with defining vertex sets

$$S = \{u : \text{the number of 1's in } u \text{ is even}\}$$

and

$$T = \{v : \text{the number of 1's in } v \text{ is odd}\}.$$

The following figure shows the bipartition of Q_3, where the vertices of S are denoted by $\circ$, and those of T by $\bullet$:

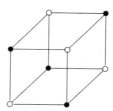

6.4. Directed Graphs

Up till now, we have considered graphs with edges uv. In many problems it is useful to consider the edges as directed. A system of roads including one-way streets is an obvious example.

A **directed** or **oriented graph** $\vec{G} = (V, E)$ consists of a vertex set V and a set $E \subseteq V^2$ of ordered pairs, which we call *directed* or *oriented* edges. If $k = (u, v)$, we call $u = k^-$ the **start vertex** and $v = k^+$ the **end vertex** of k. Every directed edge (u, v) appears at most once, and we require $u \neq v$. Otherwise, we speak of a **directed multigraph**.

A directed graph $\vec{G}$ is also called a **digraph** (for *directed graph*). To represent this visually, we use arrows to indicate directed edges $u \longrightarrow v$, with u the start vertex and v the end vertex.

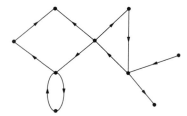

Note that in $\vec{G}$, both directed edges (u, v) and (v, u) can appear.

If we remove the orientation, we obtain the underlying undirected graph, where we draw parallel edges uv and vu twice. Most of our previous definitions can be generalized to directed graphs. For $\vec{G} = (V, E)$ we call $d^+(u) = |\{k \in E : k^+ = u\}|$ the **in-degree** and $d^-(u) = |\{k \in E : k^- = u\}|$ the **out-degree**. Thus $d^+(u)$ denotes the number of arrows that lead to u, and $d^-(u)$ the number that lead away from u. We clearly have $\sum_{u \in V} d^+(u) = \sum_{u \in V} d^-(u) = |E|$.

The incidence matrix $B = (b_{ij})$ of a directed graph $\vec{G} = (V, E)$ is the $n \times q$ matrix with

$$b_{ij} = \begin{cases} 1 & \text{if } u_i = k_j^+, \\ -1 & \text{if } u_i = k_j^-, \\ 0 & \text{if } u_i \notin k_j. \end{cases}$$

The row sums are $d^+(u) - d^-(u)$, and the columns sum to 0. For the $n \times n$ matrix BB^T, we therefore obtain

$$BB^T = \begin{pmatrix} d(u_1) & & 0 \\ & \ddots & \\ 0 & & d(u_n) \end{pmatrix} - A,$$

where A is the adjacency matrix of the underlying undirected (multi)graph.

Example 6.6.

$$B = \begin{pmatrix} -1 & 0 & 0 & 0 & 0 & 0 & 0 & 0 \\ 1 & 1 & -1 & -1 & 0 & 0 & 0 & 0 \\ 0 & 0 & 0 & 1 & 0 & 1 & 1 & 0 \\ 0 & 0 & 0 & 0 & 0 & 0 & -1 & -1 \\ 0 & 0 & 0 & 0 & -1 & -1 & 0 & 1 \\ 0 & -1 & 1 & 0 & 1 & 0 & 0 & 0 \end{pmatrix}$$

A **directed path** is a sequence of distinct vertices $u_1, u_2, \ldots, u_n$ with $u_i \longrightarrow u_{i+1}$ for all i, and the length of the path is again the number of directed edges. A **directed circuit** $u_1 \longrightarrow u_2 \longrightarrow \cdots \longrightarrow u_n \longrightarrow u_1$ is defined analogously.

A graph $\vec{G}$ that contains no directed circuits is called **acyclic**. Every undirected graph G can be made into an acyclic graph through a suitable orientation. We need only number the vertices somehow $u_1, u_2, \ldots, u_n$, and for $u_i u_j \in E$ always choose the orientation $u_i \longrightarrow u_j$ with $i < j$. A directed path $u_{i_1}, u_{i_2}, u_{i_3}, \ldots$ then satisfies $i_1 < i_2 < i_3 < \cdots$, and it can therefore never close into a circuit.

Acyclic directed graphs $\vec{G}$ play a fundamental role in transportation problems. We establish that $\vec{G}$ always contains vertices u with $d^+(u) = 0$; that is, all edges lead out from u. Such vertices are called *source vertices*. Analogously, there are *sink vertices* v with $d^-(v) = 0$; that is, all edges lead into v. Let P be a directed path in $\vec{G}$ of maximal length going from, say, u to v. If $(w, u) \in E$, then w would have to lie on the path (otherwise, P would not be maximal). But then we would have

$$\underbrace{u \to \cdots \to w}_{P} \to u,$$

a directed circuit, which does not exist. Therefore, u is a source, and v a sink. In a transportation system we generally want to transport as many sources (e.g., production sites) to sinks (processing sites). Later, we shall discuss some of the questions that arise in considering such systems.

A directed graph $\vec{G}$ is called **connected** if the underlying graph is connected. For applications, the following definition is useful: The directed graph $\vec{G}$ is said to be **strongly connected** if there exists a directed path from every vertex u to every other vertex v. In our example above, there is a directed path from 4 to 2, but not from 2 to 4. The graph is therefore not strongly connected. The sources are 1 and 4, while the vertex 3 is the only sink.

Having introduced all of these definitions, it is time for us to solve a concrete problem: the **maze problem**. We would like to enter a maze at a particular point, investigate all the paths in the maze, and find our way out again. Modeled as a graph, the maze presents the following situation: We are given a connected graph G. We begin at a vertex u_0, pass through each edge once in each direction, and return to u_0.

The following algorithm constructs a maze tour: Start at u_0 and traverse the edges according to the following rules:

(1) No edge may be traversed in the same direction more than once.

(2) When we reach $v \neq u_0$ the first time, we mark the edge (u, v) along which we reached v. On leaving v, we are allowed to traverse a marked edge (v, u) only after all other edges (v, x), $x \neq u$, have been traversed.

Example 6.7.

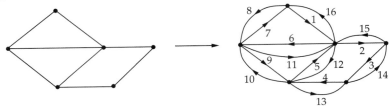

Let us prove the correctness of our algorithm. Let $u_0, u_1, \ldots, u_p$ be the tour through the edges W constructed by the algorithm. Clearly, we have $u_0 = u_p$ and $d_W^+(x) = d_W^-(x)$ for all vertices x in W, since wherever one arrives, one can also leave. We call v a *good* vertex if all edges incident to v appear in W in both directions. Clearly, u_0 is a good vertex, since otherwise, we could continue. Suppose that there exist bad vertices. Then let v be the first bad vertex on our tour W. We then have $d_W^+(v) = d_W^-(v) < d(v)$, and so from rule (2) the marked edge (u, v) in the direction (v, u) has not yet been used. But this means that $d_W^+(u) = d_W^-(u) < d(u)$, that is, the predecessor u is also bad, which contradicts the choice of v. Therefore, every vertex that we encounter on W is good. By the definition of "good," every neighbor of a good vertex is also good, that is, the good vertices form a connected graph, and since G is connected, every vertex must be good.

Exercises for Chapter 6

▷ **6.1** Let G be a graph with at least two vertices. Show that G always has two vertices of the same degree.

6.2 Determine the graphs with $n \geq 2$ vertices that have $n - 1$ different degrees.

6.3 Which of the three pictured graphs are isomorphic?

6.4 Determine the automorphism groups of the graphs P_n, C_n, and K_n.

▷ **6.5** Show that for every even integer $n \geq 4$ there is a 3-regular graph with n vertices.

6.6 Show that in a connected graph, every pair of paths of maximal length has a common vertex.

6.7 Suppose the graph G has the degree sequence $d_1 \leq \cdots \leq d_n$. Show that for the bandwidth, we have $b(G) \geq \max_j \max \left(d_j - \lfloor \frac{j-1}{2} \rfloor, \frac{d_j}{2}\right)$.

▷ **6.8** Show that a graph with n vertices and q edges has a least $n - q$ components.

6.9 Given $G = (V, E)$, for $A \subseteq V$ let $R(A)$ denote the set of vertices in $V \setminus A$ that are neighbors of at least one vertex of A. Show that for the bandwidth, we have $b(G) \geq \max_{1 \leq s \leq n} \min \left(|R(A)| : |A| = s\right)$.

▷ **6.10** A vertex set A in a graph G is called *independent* if no two vertices in A are joined by an edge. The integer $\alpha(G) = \max \left(|A| : A \text{ independent}\right)$ is called the *independence number* of G. Show that $\alpha(G) \geq \frac{|V|}{\Delta+1}$ with $\Delta = \max \left(d(u) : u \in V\right)$.

6.11 Calculate the independence number (see the previous exercise) $\alpha(G)$ for paths and circuits.

6.12 A *coloring* of a graph $G = (V, E)$ is an assignment $f : V \to C$ (the set of colors) such that $uv \in E$ implies $f(u) \neq f(v)$. The *chromatic number* $\chi(G)$ is the smallest number of colors necessary to color G. Determine $\chi(K_n)$, $\chi(K_{m,n})$, $\chi(Q_n)$, $\chi(\text{path})$, and $\chi(\text{circuit})$.

▷ **6.13** Prove that $\alpha(G)\chi(G) \geq n$, $n = |V|$, and give examples for each n for which equality holds.

6.14 Show that an edge k is a cut edge if and only if it is contained in no circuits. What graphs have only cut edges? Show further that G has no cut edges if all its degrees are even.

▷ **6.15** Suppose a graph $G = (V, E)$ with $|E| \geq 3$ and without isolated vertices has no induced subgraphs with exactly two edges. Show that $G = K_n$, $n \geq 3$.

6.16 The complete bipartite graph $K_{1,n-1}$ is called a *star*. Prove or disprove: (a) If G has diameter 2 on n vertices, then G contains a spanning star (that is, on n vertices). (b) If G contains a spanning star, then G has diameter 2.

6.17 Prove graph-theoretically: (a) $\binom{n}{2} = \binom{k}{2} + \binom{n-k}{2} + k(n - k)$ for $0 \leq k \leq n$. (b) Suppose $\sum_{i=1}^{t} n_i = n$. Then $\sum_{i=1}^{t} \binom{n_i}{2} \leq \binom{n}{2}$.

6.18 Let $G = (V, E)$ be a graph and $\bar{d} = \frac{1}{n} \sum_{u \in V} d(u)$ the average degree. Prove or disprove: (a) Deleting a vertex of maximal degree does not increase $\bar{d}$. (b) Deleting a vertex of minimal degree does not decrease $\bar{d}$.

6.19 Let the degree sequence of a graph be given in monotonic form by $d_1 \geq d_2 \geq \cdots \geq d_n$. Show that $\sum_{i=1}^{k} d_i \leq k(k - 1) + \sum_{j=k+1}^{n} \min(d_j, k)$ for all k. Hint: Consider the number of edges between the first k vertices and the remainder. Note: This condition characterizes the "graphical" sequences $d_1 \geq \cdots \geq d_n$.

▷ **6.20** Show that a triangle-free graph $G = (V, E)$ with n vertices has at most $\frac{n^2}{4}$ edges. Show further that the graph $K_{n/2,n/2}$ is the only triangle-free graph with n vertices and $\frac{n^2}{4}$ edges. Hint: Consider the smallest set $A \subseteq V$ of vertices that meet all edges.

6.21 Generalize the previous exercise: Let $G = (V, E)$ be a graph on n vertices that contains no K_{t+1}. Then $|E| \leq \frac{n^2}{2}(1 - \frac{1}{t})$. Hint: Induction on t.

▷ **6.22** Let S be a set of n points in the plane such that every pair of points are separated by a distance less than or equal to 1. Show that the maximal number of points separated by a distance greater than $\frac{1}{\sqrt{2}}$ is $\lfloor \frac{n^2}{3} \rfloor$. Hint: Construct a graph that contains no K_4.

6.23 The *complement* $\overline{G} = (V, \overline{E})$ of a graph $G = (V, E)$ has the same vertex set V with $uv \in \overline{E} \Leftrightarrow uv \notin E$. Let G be a k-regular graph. Show that the total number of triangles in G and $\overline{G}$ is exactly $\binom{n}{3} - \frac{n}{2}k(n - k - 1)$.

6.24 A graph G is called *self-complementary* if $G \cong \overline{G}$. Show that every self-complementary graph has $4m$ or $4m + 1$ vertices, and determine all such graphs with $n \leq 8$ vertices.

▷ **6.25** Let n, k be natural numbers with $2k \leq n$. The *Kneser graph* $K(n, k)$ has as vertices all k-subsets of an n-set, and two such k-sets A, B are joined by an edge if and only if $A \cap B = \emptyset$. Show that $\chi(K(n, k)) \leq n - 2k + 2$. What does $K(5, 2)$ look like?

▷ **6.26** Determine the automorphism group of the Petersen graph. Hint: Use the previous exercise.

6.27 Let G be a graph with n vertices and $n + 1$ edges. Show that G has a circuit of length $\leq \lfloor \frac{2n+2}{3} \rfloor$. Can equality hold?

▷ **6.28** Calculate the bandwidth of the graph $K_{m,n}$.

6.29 Derive from Exercise 6.9 a lower bound for the bandwidth of the hypercube Q_n.

6.30 Let G be a graph with n vertices and q edges. Show that G is connected if $q > \binom{n-1}{2}$. Is there a graph with $q = \binom{n-1}{2}$ edges that is not connected?

6.31 Generalize the previous exercise to strong connectivity in directed graphs.

▷ **6.32** Let G be a graph with maximal degree Δ. Show that $\chi(G) \leq \Delta + 1$.

6.33 For a graph H let $\delta(H)$ denote the minimal degree in H. Show that $\chi(G) \leq \max \delta(H) + 1$, where the maximum is taken over all induced subgraphs H of G.

▷ **6.34** Prove the inequalities $\chi(G) + \chi(\overline{G}) \leq n + 1$, $\chi(G)\chi(\overline{G}) \geq n$.

▷ **6.35** Prove that a graph G can be colored with k colors (that is, $\chi(G) \leq k$) if and only if the edges can be oriented in such a way that in every circuit C of G at least $\frac{|V(C)|}{k}$ edges are oriented in each of the two directions. Hint: Consider paths from a vertex u to a vertex v, and count 1 if the edge is oriented in the correct direction and $-(k - 1)$, if is oriented in the opposite direction.

6.36 Show that every self-complementary graph with at least two vertices has diameter 2 or 3.

▷ **6.37** The *girth* girth(G) of a graph G is the length of a shortest circuit in G (with girth$(G) = \infty$ if G has no circuits). A k-regular graph G with girth$(G) = g$ and smallest possible number of vertices $f(k, g)$ is called a (k, g)-graph, $k \geq 2$, $g \geq 3$.

Determine the (k, g)-graphs for: (a) $k = 2$, g arbitrary, (b) $g = 3$, k arbitrary, (c) $g = 4$, k arbitrary, (d) $k = 3$, $g = 5$, and show that for $k \geq 3$,

$$f(k, g) \geq \begin{cases} \frac{k(k-1)^r - 2}{k-2} & \text{if } g = 2r + 1, \\ \frac{(k-1)^r - 2}{k-2} & \text{if } g = 2r. \end{cases}$$

6.38 Suppose a graph $G = (V, E)$ on n vertices has independence number α. Prove that $|E| \geq \frac{1}{2}(\lceil \frac{n}{\alpha} \rceil - 1)(2n - \alpha \lceil \frac{n}{\alpha} \rceil)$. Hint: The extremal graphs consist of disjoint complete subgraphs.

6.39 Determine all graphs that contain no induced subgraphs with 3 or with 4 edges.

▷ **6.40** Let G be a connected graph on $n \geq 3$ vertices. A vertex u is called a *cut vertex* if $G \smallsetminus \{u\}$ (that is, G without u and the incident edges) is not connected. Show that G has at least two vertices that are not cut vertices, and if there are exactly two such vertices, then G is a path.

6.41 Let $G = (V, E)$ be a graph. For $k, k' \in E$, define $k \sim k'$ if either $k = k'$ or k, k' lie on some circuit. Show that $\sim$ is an equivalence relation. Show analogously that $k \approx k'$, if $k = k'$ or $G \smallsetminus \{k, k'\}$ has more components than G, is an equivalence relation. Is $\sim$ the same as $\approx$?

6.42 Let G be the following graph. The vertices are all $n!$ permutations $a_1 a_2 \ldots a_n$, where $a_1 \ldots a_n$, $b_1 \ldots b_n$ are neighbors if they differ by a single transposition $a_i \mapsto a_j$. Example: $134562 \sim 164532$. Show that G is bipartite.

▷ **6.43** A *tournament* T is a directed graph in which there is precisely one directed edge between each pair of vertices. Show that in a tournament there is always a vertex from which every other vertex can be reached by a directed path of length less than or equal to 2.

6.44 Show that a tournament (see the previous exercise) T is strongly connected if and only if T contains a spanning directed circuit (that is, a circuit of length $n = |V|$). Hint: Consider a longest directed circuit.

▷ **6.45** Let $\pi : v_1 v_2 \ldots v_n$ be an ordering of the vertices of a tournament. A *feedback edge* is a pair $v_i v_j$ with $i < j$ and $v_j \to v_i$. Let $j - i$ be the feedback length of this edge, and $f(\pi)$ the sum of all feedback lengths. Show that every ordering π with $f(\pi)$ minimal lists the vertices in nonincreasing out-degrees $d^-(v_1) \geq d^-(v_2) \geq \cdots \geq d^-(v_n)$. Hint: Consider what happens when in π two neighboring vertices are exchanged.

▷ **6.46** Show that the edges of a graph G can be oriented in such a way that the resulting directed graph $\vec{G}$ is strongly connected if and only if G is connected and has no cut edges.

6.47 Let G be a graph in which all vertices have even degree. Show that G can be oriented in such a way that in $\vec{G}$, one always has $d^+(u) = d^-(u)$.

Chapter 7

Trees

7.1. What Is a Tree?

The theory of trees originated in the study of hydrocarbons and isomers.[1]

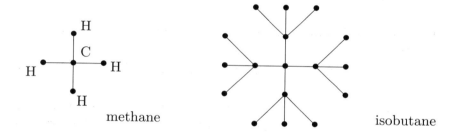

methane   isobutane

Toward the end of the nineteenth century, Cayley posed the question of determining the number of possible isomers of a given compound. This led to the theory of graph enumeration (see Chapter 4 and the suggestions for further reading at the end of Part 1). Trees are the fundamental building blocks for the construction of graphs. However, they are of interest not only in graph theory, for they also provide the optimal data structure for many discrete problems, in particular for searching and sorting problems, to which we shall return in Chapter 9.

Definition 7.1. A graph is called a **tree** if it is connected and contains no circuits. A graph all of whose components are trees is called a **forest**.

---

[1]Two chemical compounds with the identical composition but different arrangements of their atoms are called *isomers*.

The trees with five or fewer vertices are depicted below:

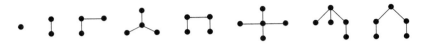

Let $G = (V, E)$ be a connected graph. A subgraph T that is a tree of order $n = |V|$ (i.e., contains all the vertices of G) is called a **spanning tree**. Clearly, every connected graph G contains at least one spanning tree. Either G is already a tree, or else G possesses a circuit C. If we remove an edge e from C, then $G_1 := (V, E \smallsetminus \{e\})$ is again connected. Either G_1 is a spanning tree or else G_1 possesses a circuit C_1. We remove an edge e_1 from C_1, and so on. After a finite number of steps we obtain a spanning tree.

Theorem 7.2. *The following are equivalent:*

a. $G = (V, E)$ *is a tree.*
b.   *Every pair of vertices in G are joined by exactly one path.*
c.   *G is connected, and $|E| = |V| - 1$.*

Proof. a $\Rightarrow$ b: If u and v were joined by two paths, these paths would yield a circuit.

b $\Rightarrow$ a: If C is a circuit, then two vertices of C are connected by more than one path.

a $\Rightarrow$ c: A tree contains one or more vertices of degree 1. Namely, let $P = u, u_1, u_2, \ldots, v$ be a longest path in G. Then all neighbors of u are in P, that is, $d(u) = 1$ (and also $d(v) = 1$), since G has no circuits. We now remove u and the incident edge uu_1, thereby obtaining a tree $G_1 = (V_1, E_1)$ on $n - 1$ vertices with $|V_1| - |E_1| = |V| - |E|$. After $n - 2$ steps we obtain a tree G_{n-2} on two vertices. That is, $G_{n-2} = K_2$, and we have $|V| - |E| = |V_{n-2}| - |E_{n-2}| = 1$.

c $\Rightarrow$ a: Let T be a spanning tree of G. From what we have just established,

$$1 = |V(G)| - |E(G)| \le |V(T)| - |E(T)| = 1 \,,$$

and so $E(G) = E(T)$, that is, $G = T$. $\square$

If a graph $G = (V, E)$ consists of t components, application of Theorem 7.2 to the individual components yields the result that every spanning forest possesses $|V| - t$ edges. There are some additional characterizations of trees. For example, G is a tree if and only if G is connected and every edge is a cut edge (proof?). Additionally, from Theorem 6.1 we immediately obtain the following corollary:

Corollary 7.3. *If T is a tree of order $n \geq 2$, and $(d_1, d_2, \ldots, d_n)$ is the degree sequence, then*

$$\sum_{i=1}^{n} d_i = 2n - 2.$$

How many spanning trees does a graph G possess? In general, this is a difficult problem. For complete graphs, however, we can easily provide an answer. Let K_n denote the complete graph on $\{1, 2, \ldots, n\}$. Let $t(n)$ denote the number of spanning trees. Let us examine the situation for small values of n:

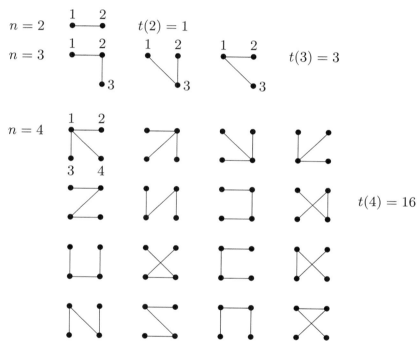

For $t(5)$ one may calculate $t(5) = 125$. Observe that the trees are not all distinct (in the sense of isomorphism). The following formula, suggested by the first few values, is one of the most astounding results in combinatorics.

Theorem 7.4. *The number of spanning trees in the complete graph K_n is given by $t(n) = n^{n-2}$.*

Proof. The expression n^{n-2} suggests that we use the rule of equality. Let $V = \{1, \ldots, n\}$ be the set of vertices. We begin by constructing a bijection from the set of all trees to the set of all sequences $(a_1, \ldots, a_{n-2})$ with $1 \leq a_i \leq n$, the number of which, as we know, is n^{n-2}. The mapping $T \rightarrow (a_1, a_2, \ldots, a_{n-2})$ is constructed as follows:

(1) From among all vertices of degree 1, find the one with minimal number v. The number of v's neighbor is a_1.
(2) Delete v and the incident edge. This yields a tree on $n-1$ vertices. Go to step (1) and execute the instructions $(n-2)$ times. This yields the sequence $a_1, a_2, \ldots, a_{n-2}$.

Example 7.5.

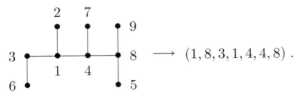

We must now show that conversely, for every sequence $(a_1, a_2, \ldots, a_{n-2})$ there exists precisely one tree T. What does the sequence tell us about the associated tree? Let d_i be the degree of the vertex i. Suppose the number i appears f_i times in the sequence. Since every time i is added to the sequence a neighboring vertex of i is deleted, we have $f_i \leq d_i - 1$ for all i. Note that $f_i \leq d_i - 1$, since i continues to reside in the part of the tree that remains. Therefore, its degree is at least 1. From Corollary 7.3 we conclude that

$$n - 2 = \sum_{i=1}^{n} f_i \leq \sum_{i=1}^{n}(d_i - 1) = 2n - 2 - n = n - 2 .$$

Thus $f_i = d_i - 1$ for all i. In particular, the numbers that never appear in the sequence are precisely those of the vertices of degree 1.

We thus obtain the inverse mapping:

(1) Find the minimal b_1 that does not appear in the sequence $(a_1, \ldots, a_{n-2})$; this yields the edge $b_1 a_1$.
(2) Find the minimal $b_2 \neq b_1$ that does not appear in the sequence $(a_2, \ldots, a_{n-2})$, and so on.

Observe that the last edge arises automatically from the above condition on the degree.

Example 7.6.

(a)   2 2 7 5 3 9 1 1
(b)   4 6 2 7 5 3 8 9

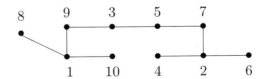

$\Box$

7.2. Breadth-First and Depth-First Search

How does one find a spanning tree, or more generally a spanning forest, in a graph G given by its adjacency matrix (or equivalently, by its neighborhood lists)? How can one tell whether G is connected?

Example 7.7. Let G be given by the following neighborhood lists:

a	b	c	d	e	f	g	h
b	a	b	a	b	g	c	a
d	c	d	b			f	g
h	d	g	c			h	
	e						

Is G connected? The following algorithm, **breadth-first search** (BFS), constructs a spanning tree (if one exists). The algorithm searches all the vertices neighboring a given vertex before proceeding, that is, the search goes first through the breadth of all neighboring vertices, whence the name of the algorithm.

Algorithm 7.8 (Breadth-first search).

(1) Select a start vertex and give it the number 1. Vertex 1 is now the *current vertex*.

(2) Suppose the current vertex is number i and that numbers $1, \ldots, r$ have been assigned to vertices. If $r = n$, stop: The spanning tree is complete. Otherwise, give the neighbors of i the numbers $r + 1, r + 2, \ldots$ and add the edges $i(r+1), i(r+2), \ldots$ to the spanning tree. If the number $i + 1$ has not been assigned, stop: The graph G is not connected (and so there is no spanning tree). Otherwise, make vertex $i + 1$ the current vertex and repeat step (2).

In our example, we obtain the following:

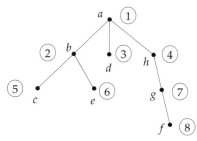

And indeed G is connected.

We would now like to prove the correctness of our algorithm. We will prove that if G is connected, then Algorithm 7.8 indeed creates a spanning tree and asserts that the graph is connected. A similar proof shows that if

G is not connected, the algorithm halts with the assertion that the graph is not connected.

Suppose, then, that G is connected. Since a numbered vertex is a neighbor of at most one vertex with a smaller number, the algorithm never creates a circuit, and since edges are always added to the graph that has been created thus far, the resulting graph T is a tree.

Suppose $v \notin V(T)$, with u the start vertex of the algorithm. Since G is connected, there is a path $u = v_0, v_1, \ldots, v$, and therefore an index i with $v_i \in V(T)$, $v_{i+1} \notin V(T)$. At some point in the algorithm, v_i was the current vertex. By step (2), all not-yet-numbered neighbors of v_i are added to T, and therefore v_{i+1} is after all in $V(T)$, a contradiction.

In a certain sense, the dual of breadth-first search is **depth-first search**, DFS. We move downward through the graph, edge by edge, until we can go no further. Then we take one step back and start again downward along another path. Here is the algorithm for depth-first search:

Algorithm 7.9 (Depth-first search).

(1) Choose a start vertex and assign it the number 1. This is now the current vertex, and vertex 1 is the *predecessor* vertex.

(2) Suppose the current vertex has number i, and the numbers $1, \ldots, r$ have been assigned. If $r = n$, stop: The graph is connected and the spanning tree is complete. Otherwise, choose a not-yet-numbered neighbor of i, give it the number $r + 1$, and add the edge $i(r + 1)$ to the graph under construction. The current vertex is now $r + 1$, and i is the predecessor vertex. If there is no unnumbered neighbor of i, go to the predecessor vertex of i if $i > 1$. This is now the current vertex. Repeat step (2). If $i = 1$ and there is no unnumbered neighbor, then G is not connected: stop.

In our example we obtain

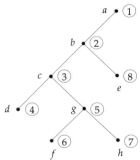

The proof of correctness of Algorithm 7.9 is similar to that of the previous algorithm.

We would like to know how long it takes our algorithms, say Algorithm 7.8, to run. Step (2) searches all unnumbered neighbors of the current vertex and therefore requires $O(\sum_{u \in V} d(u)) = O(|E|)$ steps. Since the algorithm also clearly requires $\Omega(|E|)$ operations, we have the running time $\Theta(|E|)$. The analysis of depth-first search is analogous.

7.3. Minimal Spanning Trees

Suppose we have created a communication network with a number of switches (these will be the vertices) and connections between the switches (the edges). To establish a connection between switches u and v costs $w(uv)$ units. We would like to arrange the switches in such a way that each switch can communicate with every other switch at minimal cost. A similar problem is the construction of a network of roads connecting a number of points with minimal cost of travel between points.

We model this problem with a **weighted graph**. Let there be given a weighted graph $G = (V, E)$ together with a weight function $w : E \longrightarrow \mathbb{R}$. We would like to find a spanning tree T with minimal weight $w(T) = \sum_{k \in E(T)} w(k)$.

Example 7.10.

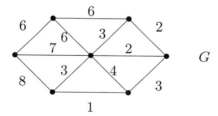

A naive approach is the following: Choose an edge of minimal weight. If one has already determined j edges, one chooses the next edge as that of minimal weight that does not lead to the creation of a circuit. After $n - 1$ steps a tree has been constructed.

For our graph, we obtain, for example,

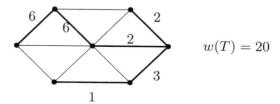

Is this tree optimal? In fact, our algorithm always produces an optimal tree, as we shall see shortly. Since our method always chooses the best edge available, we say that we have a **greedy algorithm**.

The structural elements of our algorithmic problem are the forests contained in G. Before we prove the optimality of the greedy algorithm, we would like to analyze the set-theoretic properties of forests and thereby develop the basic ideas of the algorithm.

Definition 7.11. Let S be a finite set and $\mathcal{U} \subseteq \mathcal{B}(S)$ a family of subsets of S. The pair $\mathcal{M} = (S, \mathcal{U})$ is called a **matroid** and $\mathcal{U}$ the family of **independent sets** of $\mathcal{M}$ if the following hold:

1. $\varnothing \in \mathcal{U}$,
2. $A \in \mathcal{U}$, $B \subseteq A \Rightarrow B \in \mathcal{U}$,
3. $A, B \in \mathcal{U}$, $|B| = |A| + 1 \Rightarrow \exists v \in B \setminus A$ with $A \cup \{v\} \in \mathcal{U}$.

A maximal independent set is called a **basis** of the matroid. From axiom 3 above, it follows that every basis of $\mathcal{M}$ has the same number of elements as any other basis (clear?). This number is called the **rank** $r(\mathcal{M})$ of the matroid. More precisely, axiom 3 ensures that every independent set can be extended to a basis by the inclusion of additional elements.

The term "matroid" suggests that we are dealing with a generalization of matrices. Consider n vectors $\boldsymbol{a}_1, \boldsymbol{a}_2, \ldots, \boldsymbol{a}_n$ in a vector space of dimension m over a field, for example, the real numbers $\mathbb{R}$. We may write the vectors $\boldsymbol{a}_j$ as columns $(a_{1j}, \ldots, a_{mj})^T$ and thereby obtain an $m \times n$ matrix. In this case, S is the set $\{\boldsymbol{a}_1, \ldots, \boldsymbol{a}_n\}$ and $A \subseteq S$ is independent if the vectors in A form a *linearly* independent set, where we declare that the empty set $\varnothing$ is linearly independent. Axiom 2 is clear, and axiom 3 is the Steinitz exchange theorem from linear algebra. The rank in this case is of course the dimension of the subspace spanned by the $\boldsymbol{a}_j$'s.

And now back to graphs. We consider all subgraphs $H = (V, A)$ on the complete set of vertices V and may therefore identify $H = (V, A)$ with the edge set $A \subseteq E$. The family of these subgraphs thus corresponds precisely to the family $\mathcal{B}(E)$ of all subsets of E. Now let $\mathcal{W} \subseteq \mathcal{B}(E)$ denote the family of edge sets of all *forests* of G.

Theorem 7.12. *If $G = (V, E)$ is a graph, then $\mathcal{M} = (E, \mathcal{W})$ is a matroid.*

Proof. Axioms 1 and 2 clearly are satisfied. Let $W = (V, A)$ and $W' = (V, B)$ be two forests with $|B| = |A| + 1$. Moreover, let $T_1, \ldots, T_m$ be the components of $W = (V, A)$ with vertex sets $V_1, \ldots, V_m$ and edge sets $A_1, \ldots, A_m$. From Theorem 7.2 we have $|A_i| = |V_i| - 1$, $i = 1, \ldots, m$, $V =$

$V_1 + \cdots + V_m$, $A = A_1 + \cdots + A_m$:

Since every forest on V_i has at most $|V_i| - 1$ edges, on account of $|B| > |A|$ there must be an edge $k \in B$ that connects two distinct sets V_s and V_t. But then $W'' = (V, A \cup \{k\})$ is a forest, and axiom 3 is satisfied. $\square$

We see that the bases of $\mathcal{M} = (E, \mathcal{W})$ are the spanning forests, and the rank of the matroid is $|V| - t$, where t is the number of components of G.

We now know that the forests form a matroid. But we would like to solve our minimal tree problem. The following theorem states that the greedy algorithm yields an optimal basis in *every* weighted matroid $\mathcal{M} = (S, \mathcal{U})$. Furthermore, this holds as well for forests, or in the case of connected graphs, for trees. The meaning of this theorem is now clear: Whenever we can prove a matroid structure in an optimization problem with weight function, the greedy algorithm works! And further, in Exercise 7.30 we shall see that in general, the greedy algorithm yields the optimum for every weight function *precisely* in the case of matroids.

Theorem 7.13. *Let $\mathcal{M} = (S, \mathcal{U})$ be a matroid with weight function $w : S \longrightarrow \mathbb{R}$. The following algorithm produces a basis of minimal weight:*

(1) *Let $A_0 = \varnothing \in \mathcal{U}$.*

(2) *If $A_i = \{a_1, \ldots, a_i\} \subseteq S$, then let $X_i = \{x \in S \setminus A_i : A_i \cup \{x\} \in \mathcal{U}\}$. If $X_i = \varnothing$, then A_i is the desired basis. Otherwise, choose some $a_{i+1} \in X_i$ of minimal weight and set $A_{i+1} = A_i \cup \{a_{i+1}\}$. Repeat step (2).*

Proof. Let $A = \{a_1, \ldots, a_r\}$ be the obtained set. That A is a basis follows at once from axiom 3. Because of the greedy construction, we see with the help of axiom 2 that $w(a_1) \le w(a_2) \le \cdots \le w(a_r)$ must hold. For $w(a_1) \le w(a_2)$ this is clear because of step (1). Let $2 \le i \le r - 1$. Since $\{a_1, \ldots, a_r\} \in \mathcal{U}$, it follows that $\{a_1, \ldots, a_{i-1}\} \in \mathcal{U}$ holds as well. Therefore, $a_i, a_{i+1} \in X_{i-1}$, and because of step (2), we have $w(a_i) \le w(a_{i+1})$. Suppose $B = \{b_1, \ldots, b_r\}$ were a basis with $w(B) < w(A)$, where we assume $w(b_1) \le \cdots \le w(b_r)$. There would then be a smallest index i with $w(b_i) < w(a_i)$, and on account of step (1), we would have $i \ge 2$. We consider the independent sets $A_{i-1} = \{a_1, \ldots, a_{i-1}\}$, $B_i = \{b_1, \ldots, b_i\}$. By axiom 3, there would then exist $b_j \in B_i \setminus A_{i-1}$ with $A_{i-1} \cup \{b_j\} \in \mathcal{U}$. Since now $w(b_j) \le w(b_i) < w(a_i)$, in the ith step, the greedy algorithm would have chosen b_j instead of a_i, which is a contradiction. $\square$

The specialization of the greedy algorithm to matroids on graphs was discovered by Kruskal and is therefore called Kruskal's algorithm for the

MST (minimal spanning tree) problem. How many computational steps does Kruskal's algorithm require? First we must order the edges k_i by weight: $w(k_1) \leq w(k_2) \leq \cdots \leq w(k_q)$, $q = |E|$. In other words, we must sort the q weights $w(k_i)$. We shall study how to do this effectively in Chapter 9, where we shall prove that $O(q \lg q)$ comparisons are necessary. Step (2) constructs the tree iteratively. After i steps we have a forest with $n - i$ components $V_1, V_2, \ldots, V_{n-i}$. Suppose k_h was the most recently added edge. On account of $w(k_1) \leq \cdots \leq w(k_h) \leq w(k_{h+1})$, we take the next edge $k_{h+1} = uv$ and test whether it is admissible, that is, whether a circuit is created by its addition. We have

$$k_{h+1} \text{ is admissible} \iff u, v \text{ are in different } V_j\text{'s}.$$

We determine in which sets V_u, V_v the vertices u, v are located. If $V_u \neq V_v$, we add k_{h+1} and merge $V_u \cup V_v \cup \{uv\}$ into one component. If $V_u = V_v$, we test the next edge. We must therefore execute at most n comparisons for each of u and v, and the total number of operations in step (2) is $O(nq) = O(q^2)$. Therefore, altogether our algorithm needs $O(q \lg q) + O(q^2) = O(q^2)$ operations, and the reader may determine that with a suitable data structure we need only $O(q \lg q)$ operations for step (2), and so altogether we require $O(|E| \lg |E|)$.

If we exchange "minimal" with "maximal" and $\leq$ with $\geq$, then the greedy algorithm gives a basis with maximal weight, or for graphs, a tree of maximal weight.

7.4. The Shortest Path in a Graph

Here is another optimization problem on weighted graphs. Suppose we have a street plan in front of us and find ourselves at location u. We would like to get from u to v in the shortest possible time. The roads k (the edges of the graph) have weight $w(k) \geq 0$, which gives the minimal time required to traverse the edge k (on foot, by car, tram, etc.).

Modeled as a graph, this means that we are given a connected graph $G = (V, E)$ and a weight function $w : E \to \mathbb{R}^+ = \{x \in \mathbb{R} : x \geq 0\}$. Let $u \in V$. For a path $P = P(u, v)$ from u to v, we denote by $\ell(P) = \sum_{k \in E(P)} w(k)$ the (weighted) *length* of P. We are seeking a *shortest* u, v-path for which $\ell(P)$ is minimal. The **distance** $d(u, v)$ is defined as the length of a shortest path. In the special case $w(k) = 1$ for all $k \in E$, it follows that $\ell(P)$ is precisely the previously defined length (the number of edges), and $d(u, v)$ the previously defined distance.

Let u be fixed. The following famous algorithm of Dijkstra constructs a spanning tree whose unique path from u to v is always the shortest, for *all* $v \in V$:

(1) Let $u_0 = u$, $V_0 = \{u_0\}$, $E_0 = \varnothing$, $\ell(u_0) = 0$.

(2) Suppose $V_i = \{u_0, u_1, \ldots, u_i\}$, $E_i = \{k_1, \ldots, k_i\}$. If $i = n - 1$, we are done. Otherwise, consider for all edges $k = vw$, $v \in V_i$, $w \in V \setminus V_i$, the expression $f(k) = \ell(v) + w(k)$ and choose $\overline{k}$ with $f(\overline{k}) = \min f(k)$. Let $\overline{k} = \overline{v}\,\overline{w}$. Then set $u_{i+1} = \overline{w}$, $k_{i+1} = \overline{k}$, $V_{i+1} = V_i \cup \{u_{i+1}\}$, $E_{i+1} = E_i \cup \{k_{i+1}\}$, $\ell(u_{i+1}) = f(\overline{k})$. Repeat step (2).

Example 7.14.

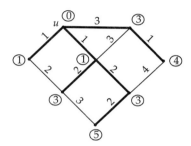

The circled numbers are the $\ell(v)$.

Theorem 7.15. *Let $G = (V, E)$ be a connected graph with weight function $w : E \longrightarrow \mathbb{R}^+$, $u \in V$. Dijkstra's algorithm gives a spanning tree T with the property that the unique path from u to v is always a minimal u, v-path in G with $d(u, v) = \ell(v)$ for all v.*

Proof. The algorithm certainly constructs a spanning tree. In the first step an edge of minimal weight $u = u_0$ to a neighbor is chosen; thus $k_1 = u_0 u_1$ is a minimal u_0, u_1-path with $\ell(u_1) = w(k_1) = d(u_0, u_1)$. Suppose the partial tree $T_i = (V_i, E_i)$ has the desired properties, and step (2) constructs $\overline{k} = \overline{v}\,\overline{w}$. We must show that $\ell(\overline{w}) = f(\overline{k}) = \ell(\overline{v}) + w(\overline{k})$ is equal to the (weighted) distance $d(u_0, \overline{w})$. For the $u_0, \overline{w}$-path P_0 just constructed, we have $\ell(P_0) = d(u_0, \overline{v}) + w(\overline{k}) = \ell(\overline{v}) + w(\overline{k}) = \ell(\overline{w})$. Let P be a shortest $u_0, \overline{w}$-path and v the last vertex of V_i in P, with $w \in V \setminus V_i$ as successor, $k = vw$. Then the partial paths $P(u_0, v)$, $P(w, \overline{w})$ are also shortest paths, and we obtain

$$
\begin{aligned}
d(u_0, \overline{w}) &= \ell(P(u_0, v)) + w(k) + \ell(P(w, \overline{w})) \\
&= (\ell(v) + w(k)) + \ell(P(w, \overline{w})) \\
&= f(k) + \ell(P(w, \overline{w})) \\
&\geq f(\overline{k}) = \ell(\overline{w}) = \ell(P_0) \,.
\end{aligned}
$$

Therefore, P_0 is a shortest path. $\qquad\square$

We see that our algorithm always constructs a shortest prolongation of the partial tree. We are therefore again dealing with a greedy algorithm.

Several variants of the shortest-path problem come at once to mind. Suppose we would like to determine a shortest path between only two given vertices u and v. We can use Dijkstra's algorithm with u as source, and this gives a shortest path from u to v. No algorithm is known that is asymptotically faster than Dijkstra's algorithm. Or perhaps we would like to find shortest paths for *all* pairs of vertices u, v. We can solve this problem by applying our algorithm to every vertex u as source, but there are generally faster algorithms available. See the cited literature at the end of this part, after Chapter 10.

Finally, we remark that it should be clear how the procedure may be modified for directed graphs. In this case, we seek shortest *directed* paths from u to all other vertices.

Exercises for Chapter 7

7.1 Prove the following characterization of trees: Let G be a graph on n vertices and q edges. Then G is a tree if and only if the following conditions are satisfied: (a) G has no circuits and $q = n - 1$. (b) G has no circuits and if any pair of nonneighboring vertices are joined by an edge, then the resulting graph has precisely one circuit. (c) G is connected ($G \neq K_n$ if $n \geq 3$), and if any two nonneighboring vertices are joined by an edge, the resulting graph has exactly one circuit.

▷ **7.2** Show that a connected graph with an even number of vertices always has a spanning subgraph in which all vertices have odd degree. Does this hold for disconnected graphs?

7.3 Let G be a connected graph. For $u \in V$ we set $r(u) = \max(d(u, v) : v \neq u)$. The parameter $r(G) = \min(r(u) : u \in V)$ is called the *radius* of G, and $Z(G) = \{u \in V : r(u) = r(G)\}$ is the *center* of G. Show that the center of a tree consists of either one vertex or two neighboring vertices.

7.4 Let $d_1 \geq \cdots \geq d_n > 0$ be a sequence of natural numbers. Show that $(d_1, \ldots, d_n)$ is the degree sequence of a tree if and only if $\sum_{i=1}^{n} d_i = 2n - 2$.

▷ **7.5** Determine among all trees with n vertices those for which the sum $\sum_{u \neq v \in V} d(u, v)$ is minimal and those for which it is maximal.

7.6 Carry out the correctness proof of depth-first search.

7.7 Generalize BFS or DFS by constructing an algorithm that determines the connected components of a graph.

7.8 Verify precisely that all bases of a matroid have the same size.

▷ **7.9** Suppose a connected graph G has a different weight on each edge. Show that G possesses a unique minimal spanning tree.

7.10 Consider the following graph G with cost function. Determine using Dijkstra's algorithm the shortest paths from u to all other vertices:

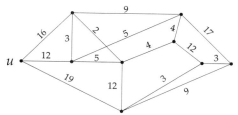

▷ **7.11** Show that a tree has at least as many vertices of degree 1 as the maximal degree. When does equality hold?

▷ **7.12** This exercise gives a recursive proof of the formula $t(n) = n^{n-2}$ for the number of spanning trees in K_n. Number the vertices $1, 2, \ldots, n$. Let $C(n, k)$ be the number of spanning trees in which the vertex n has degree k. Prove the recurrence $C(n, k) = \frac{k(n-1)}{n-1-k} C(n, k+1)$ and thereby conclude the desired formula.

▷ **7.13** Let B be the incidence matrix of a graph G. In every column we change arbitrarily one of the two 1's into -1 (this corresponds to an orientation of G) and call the new matrix C. Let $M = CC^T$. Prove that the number of spanning trees of G is given by $t(G) = \det M_{ii}$, where M_{ii} results from deleting the ith row and ith column from M (and this holds for every i). Hint: Let P be an $r \times s$ matrix and Q an $s \times r$ matrix, $r \leq s$. Then a theorem from linear algebra states that $\det(PQ)$ is equal to the sum of the products of determinants of the corresponding $r \times r$ submatrices.

7.14 Again verify, now using the previous exercise, $t(K_n) = n^{n-2}$.

7.15 Calculate $t(K_{m,n})$. Answer: $m^{n-1} n^{m-1}$.

7.16 Consider the complete graph K_n on $\{1, \ldots, n\}$ and let $d_1, \ldots, d_n$ be a sequence of natural numbers each greater than or equal to 1 with $\sum_{i=1}^{n} d_i = 2n - 2$. Show that the number of spanning trees in which the vertex i has degree d_i is equal to $\frac{(n-2)!}{(d_1-1)! \cdots (d_n-1)!}$, and derive from this an additional proof of the formula $t(K_n) = n^{n-2}$.

▷ **7.17** Let $G = K_n \smallsetminus \{k\}$ be the complete graph with an edge removed. Without using Exercise 7.13, calculate the number of spanning trees in G.

▷ **7.18** The lattice graph $G(2, n)$ consists of two paths of n vertices (let them be numbered 1 to n) whose vertices with the same number are joined by an edge. For example, $G(2, 2) \cong C_4$. Determine the number of spanning trees of $G(2, n)$ using generating functions.

7.19 Show that every automorphism of a tree leaves at least one vertex or one edge fixed. Hint: Exercise 7.3.

7.20 For n employees, each of whom drives his or her own car to work, there are n parking places available in the company parking lot. Each driver has a spot that he or she prefers, and in fact, driver i prefers $g(i)$, $1 \leq g(i) \leq n$. The drivers arrive at the parking lot one after the other: first 1, followed by 2, and so on. The ith driver parks in space $g(i)$ if it is free. Otherwise, he or she takes the next number $k > g(i)$ that is available if the space is empty. Otherwise, the driver goes home and never returns to work. Here is an example with $n = 4$:

$$\frac{\begin{array}{cccc} 1 & 2 & 3 & 4 \end{array}}{g \quad \begin{array}{cccc} 3 & 2 & 2 & 1 \end{array}}, \text{ then } 1 \to 3,\ 2 \to 2,\ 3 \to 4,\ 4 \to 1,$$

but

$$\frac{\begin{array}{cccc} 1 & 2 & 3 & 4 \end{array}}{g \quad \begin{array}{cccc} 2 & 3 & 3 & 2 \end{array}}, 1 \to 2,\ 2 \to 3,\ 3 \to 4,\ 4 \to \text{quits}$$

Let $p(n)$ be the number of functions g that allow each employee to obtain a parking space. Determine $p(n)$. Hint: $p(2) = 3$, $p(3) = 16$, $p(4) = 125$.

▷ **7.21** Let $G = (V, E)$ be a connected graph and let $w : E \to \mathbb{R}^+$ be a cost function. Show that the following algorithm constructs a minimal spanning tree: (1) Select an edge uv of minimal weight and set $S = \{u, v\}$, $T = V \smallsetminus S$. (2) If $T = \varnothing$, stop. Otherwise, choose from among the edges between S and T an edge $k = \overline{u}\,\overline{v}$ of minimal weight, $\overline{u} \in S$, $\overline{v} \in T$, and set $S \leftarrow S \cup \{\overline{v}\}$, $T \leftarrow T \smallsetminus \{\overline{v}\}$. Repeat step (2).

7.22 Estimate the running time of the algorithm in the previous exercise in terms of the number of vertices n.

7.23 Consider K_n on $\{1, \ldots, n\}$ with the cost function $w(ij) = i+j$. (a) Construct an MST tree T. (b) What is $w(T)$? (c) Is T uniquely determined?

▷ **7.24** Let $\mathcal{M} = (S, \mathcal{U})$ be a matroid, and $\mathcal{B}$ the family of bases. Show that $\mathcal{B}$ satisfies the following conditions: (a) $A \neq B \in \mathcal{B} \Rightarrow A \nsubseteq B, B \nsubseteq A$. (b) Let $A \neq B \in \mathcal{B}$. Then for each $x \in A$ there exists $y \in B$ with $(A \smallsetminus \{x\}) \cup \{y\} \in \mathcal{B}$. Show conversely that a family of sets $\mathcal{B}$ that satisfies conditions (a) and (b) is the family of bases of a matroid.

7.25 Let $\mathcal{M} = (S, \mathcal{U})$ be a matroid and $\mathcal{B}$ the family of bases. Show that the family $\mathcal{B}^* = \{S \smallsetminus B : B \in \mathcal{B}\}$ defines a matroid $\mathcal{M}^*$. The matroid $\mathcal{M}^*$ is called the *dual* of the matroid $\mathcal{M}$. What is the rank of $\mathcal{M}^*$?

7.26 Let $\mathcal{M} = (V, \mathcal{W})$ be the usual matroid induced by the graph $G = (V, E)$. Give a graph-theoretic description of the dual matroid $\mathcal{M}^*$. That is, what sets of edges are independent in $\mathcal{M}^*$?

▷ **7.27** As in the previous exercise, let $\mathcal{M} = (V, \mathcal{W})$ be given. A set of edges $A \subseteq V$ is said to be *minimally dependent* if $A \notin \mathcal{W}$ but $A' \in \mathcal{W}$ for every proper subset A' of A. Describe the minimally dependent sets in the graph and also the minimally dependent sets in $\mathcal{M}^*$. Hint: The minimally dependent sets in $\mathcal{M}^*$ are *minimal cut sets* A; that is, $G = (V, E \smallsetminus A)$ has one component more than G and is minimal with respect to this property.

7.28 Show that circuits and minimal cut sets always have an even number of edges in common.

7.29 Let B be the incidence matrix of $G = (V, E)$, interpreted as a matrix over the field $\{0, 1\}$. Show that $A \subseteq E$ is independent in the matroid $(E, \mathcal{W})$ if and only if the associated set of columns is linearly independent.

▷ **7.30** Let $(S, \mathcal{U})$ be a collection of sets satisfying axioms 1 and 2 of a matroid. Show that $(S, \mathcal{U})$ is a matroid (thus satisfies axiom 3 as well) if and only if the greedy algorithm yields the optimum for *every* weight function $w : S \to \mathbb{R}$.

7.31 Develop a Dijkstra algorithm for directed graphs.

7.32 Determine the shortest path from 1 to all i in the following directed graph, given by a weight matrix. A missing entry means that the corresponding edge is not present in the graph:

	1	2	3	4	5	6	7
1			4	10	3		
2			1	3	2	11	
3		9		8	3	2	1
4		4	5		8	6	3
5	1		1	2		3	1
6		1	1	3	2		
7	2	4	3			2	

▷ **7.33** Dijkstra's algorithm functions only for nonnegative edge weights $w(u, v) \in \mathbb{R}^+$. Suppose that there also exist negative weights on a graph. If directed circuits with negative total weight exist, then no shortest distance can exist (why?). Suppose, then, that we have a directed graph $\vec{G} = (V, E)$ with source $u \in V$, and a weight function $w : E \to \mathbb{R}$ without negative circuits, where we assume that all vertices $x \neq u$ from u can be reached by a directed path. Show that the following algorithm of Bellman–Ford creates a tree with shortest directed path from u: (1) Number $E = \{k_1, \ldots, k_q\}$; set $\ell(u) = 0$, $\ell(x) = \infty$ for $x \neq u$, $B = \emptyset$. (2) Run through the edges of E. Let $k_i = (x, y)$. If $\ell(x) + w(x, y) < \ell(y)$, set $\ell(y) \leftarrow \ell(x) + w(x, y)$ and $B \leftarrow (B \cup \{k_i\}) \smallsetminus \{k\}$, where k is the previous edge in B with end vertex y. Repeat $(|V| - 1)$ times. Then $\vec{G} = (V, B)$ is a "shortest" tree. Hint: Let $u, v_1, \ldots, v_k = v$ be a shortest path in $\vec{G}$. Show by induction that $\ell(v_i) = d(u, v_i)$ after the ith iteration.

7.34 Determine shortest paths from 1 in the following graph, given by its length matrix:

	1	2	3	4	5
1		6	5		
2			7	3	-2
3				-4	8
4		-1			
5	2			7	

▷ **7.35** Think about how the algorithm of Bellman–Ford should be extended to return the output "no solution" in the case of negative circuits.

7.36 A minimax or bottleneck tree is a spanning tree in which the maximum of the edge weights is as small as possible. Show that every MST tree is also a minimax tree.

Chapter 8

Matchings and Networks

8.1. Matchings in Bipartite Graphs

Recall the problem of associating jobs with candidates. We are given a set $S = \{P_1, \ldots, P_n\}$ of candidates and a set $T = \{J_1, \ldots, J_n\}$ of jobs. We set $P_i J_j \in E$ if P_i is suited for job J_j. We would like to find a mapping $P_i \longrightarrow J_{\varphi(i)}$ such that each person P_i finds a suitable job $J_{\varphi(i)}$. When is this possible? In general, we will have weights on the edges $P_i J_j$ (which we may interpret as suitability coefficients), and the mapping should be optimal (e.g., maximize the sum of suitabilities for the jobs assigned to the candidates).

Definition 8.1. For a bipartite graph $G = (S+T, E)$, a **matching** $M \subseteq E$ is a set of mutually nonincident pairs of edges.

We are interested in a maximally large matching M of $G = (S + T, E)$, and in particular, we would like to know the **matching number** $m(G)$ of G, which is the number of edges in a maximally large matching. A matching M is called a **maximum matching** if $|M| = m(G)$.

Example 8.2.

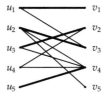

The heavy edges yield a matching with $m(G) = 4$. Why is $m(G) < 5$? Since u_3, u_4, u_5 together have as neighbors v_2, v_4, in any matching, one of the vertices u_3, u_4, u_5 must be left out.

We would like first to answer the question, when is $m(G) = |S|$? That is, when can *all* vertices of S be matched? For $A \subseteq S$ we set $N(A) = \{v \in T : uv \in E$ for some $u \in A\}$. Then $N(A)$ is the set of neighbors of A.

Theorem 8.3. *Let $G = (S + T, E)$ be a bipartite graph. Then $m(G) = |S|$ if and only if $|A| \leq |N(A)|$ for all $A \subseteq S$.*

Proof. The necessity is clear. If $|A| > |N(A)|$, then not all vertices in A can be matched simultaneously. Conversely, suppose that the condition of the theorem is satisfied. Given a matching $M \subseteq E$ with $|M| < |S|$, we will show that M is not a maximum matching. Let $u_0 \in S$ be a vertex that is not matched under M. Since $|N(\{u_0\})| \geq |\{u_0\}| = 1$, there exists a neighbor $v_1 \in T$. If v_1 is not matched in M, we add the edge $u_0 v_1$. Let us then assume $u_1 v_1 \in M$ for some $u_1 \neq u_0$. Since $|N\{u_0, u_1\}| \geq |\{u_0, u_1\}| = 2$, there is a vertex $v_2 \neq v_1$ that is a neighbor of u_0 or u_1. If v_2 is unmatched, then stop. Otherwise, there exists $u_2 v_2 \in M$ with $u_2 \notin \{u_0, u_1\}$. We continue in this fashion and finally reach an unmatched vertex v_r:

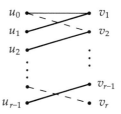

Each vertex v_j is the neighbor of at least one vertex u_i with $i < j$. If we proceed backward from v_r, then we obtain in reverse a path $P = v_r, u_a, v_a, u_b, v_b, \ldots, u_h, v_h, u_0$ with $r > a > b > \cdots > h > 0$. The edges $u_a v_a, u_b v_b, \ldots, u_h v_h$ are in M, while the edges $v_r u_a, v_a u_b, \ldots, v_h u_0$ are not in M. We now replace in P the M-edges by the edges that are not in M (of which there is one more), and we thereby obtain a matching M' with $|M'| = |M| + 1$. $\square$

Theorem 8.3 is also known as the "marriage theorem," due to the interpretation of S as a set of women, T a set of men, with $u_i v_j \in E$ if u_i, v_j are not averse to being married. The theorem gives the precise condition by which all the women can find acceptable marriage partners (a match) without having to commit bigamy.

In general, $G = (S + T, E)$ will not have a matching of order $|S|$. How large can $m(G)$ be? If $|A| - |N(A)| > 0$, then at least $|A| - |N(A)|$ vertices remain unmatched, that is, $m(G) \leq |S| - (|A| - |N(A)|)$. With

$\delta = \max_{A \subseteq S}(|A| - |N(A)|)$ we see that $m(G) \leq |S| - \delta$ must hold. Observe that $\delta \geq 0$, since for $A = \varnothing$, we have $|A| - |N(A)| = 0$. The following theorem shows that in $m(G) \leq |S| - \delta$ equality always holds.

Theorem 8.4. *Let $G = (S + T, E)$ be a bipartite graph. Then*

$$m(G) = |S| - \max_{A \subseteq S}(|A| - |N(A)|).$$

Proof. Let $\delta = \max_{A \subseteq S}(|A| - |N(A)|)$. We know already that $m(G) \leq |S| - \delta$. Let D be a new vertex set with $|D| = \delta$. We define the bipartite graph $G^* = (S + (T \cup D), E^*)$ by adding to G all edges between S and D. For $A \subseteq S$ we have $N^*(A) = N(A) \cup D$, where $N^*(A)$ are the neighbors of A in G^*. We thus have $|N^*(A)| = |N(A)| + \delta \geq |A|$, that is, G^* has a matching M^* of order $|S|$, from Theorem 8.3. If we now remove the edges of M^* that lead to D, we obtain a matching M in G with $|M| \geq |S| - \delta$, and thereby $m(G) = |S| - \delta$.   □

Theorem 8.4 can be interpreted in a different way. We call a vertex set $D \subseteq S + T$ a **vertex cover** of the bipartite graph $G = (S + T, E)$ if D meets every edge, or alternatively, if it *covers* every edge. Clearly, we have $|D| \geq |M|$ for every vertex cover D and every matching M, since D must cover every edge of M. In particular, we have $\min_D |D| \geq \max_M |M|$.

Theorem 8.5. *For a bipartite graph $G = (S + T, E)$, we have*

$$\max(|M| : M \text{ a matching}) = \min(|D| : D \text{ a vertex cover}).$$

Proof. From the previous theorem we know that $m(G) = \max|M| = |S| - |A_0| + |N(A_0)| = |S \setminus A_0| + |N(A_0)|$ for some $A_0 \subseteq S$. Since $(S \setminus A_0) + N(A_0)$ is clearly a vertex cover of G, we have the situation

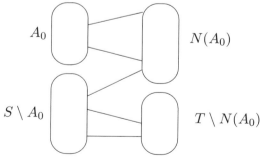

with $\min|D| \leq \max|M|$, and therefore $\min|D| = \max|M|$.   □

The result $\min|D| = \max|M|$ is one of the fundamental "equilibrium theorems" of graph theory, relating the two concepts of covering and matching. Later, we shall extend the theorem to networks and recognize it as a special case of the main theorem of linear optimization.

The marriage theorem can be interpreted in a number of ways combinatorially if we regard a given situation as a matching problem. The following example is a nice illustration:

Let $\mathcal{A} = \{A_1, \ldots, A_m\}$ be a family of sets on a ground set $T = \{t_1, \ldots, t_n\}$, where the sets A_i are not necessarily distinct. We construct a bipartite graph $G = (\mathcal{A} + T, E)$ by means of $A_i t_j \in E \Leftrightarrow t_j \in A_i$. As an example, we obtain for $A_1 = \{1, 2, 3\}, A_2 = \{1, 3\}, A_3 = \{2, 5, 6\}, A_4 = \{4, 6\}$ the following graph:

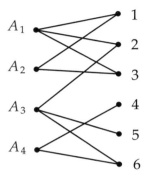

A matching M with $|M| = \ell$ corresponds to an *injective* assignment $\varphi \colon A_i \mapsto t_{\varphi i}$ $(i \in I \subseteq \{1, \ldots, m\}, |I| = \ell)$, with $t_{\varphi i} \in A_i$. We call $T_I = \{t_{\varphi i} : i \in I\}$ a **transversal** or **system of distinct representatives** of the subfamily $\mathcal{A}_I = \{A_i : i \in I\}$, and we call φ a **selection function** of $\mathcal{A}_I$. When does all of $\mathcal{A}$ possess a selection function? The answer is given by the marriage theorem, Theorem 8.3. For $I \subseteq \{1, \ldots, m\}$, by construction of the bipartite graph we have $N(\mathcal{A}_I) = \bigcup_{i \in I} A_i \subseteq T$. We thus obtain the following result:

Theorem 8.6. *Let $\mathcal{A} = \{A_1, \ldots, A_m\}$ be a family of sets. Then $\mathcal{A}$ has a transversal (or selection function) if and only if $|\bigcup_{i \in I} A_i| \geq |I|$ for all $I \subseteq \{1, \ldots, m\}$.*

A further surprising application of Theorem 8.5 yields an important theorem about matrices. Let $M = (m_{ij})$ be an $n \times n$ matrix with nonnegative real entries whose row and column sums are all equal to $r > 0$. We associate with M the bipartite graph $G = (Z + S, E)$, with rows $Z = \{z_1, \ldots, z_n\}$ and columns $S = \{s_1, \ldots, s_n\}$, by setting

$$z_i s_j \in E \iff m_{ij} > 0.$$

Thus a matching corresponds to a set of positive entries m_{ij} that appear in distinct rows and columns. We call such a set $\{m_{ij}\}$ a *diagonal* of the matrix M. A *vertex cover* D is a set of rows and columns that together cover all positive entries.

Example 8.7.

$$
\begin{array}{ccccc}
② & 0 & 1 & 3 & 0 \\
0 & 4 & 0 & 0 & ② \\
3 & ① & 2 & 0 & 0 \\
1 & 1 & ③ & 0 & 1 \\
0 & 0 & 0 & ③ & 3
\end{array}
$$

The circled elements form a diagonal.

Suppose that M has no diagonals of size n. Then by Theorem 8.5 there exist e rows and f columns with $e + f \leq n - 1$ that cover all nonzero entries. Thus $rn = \sum m_{ij} \leq r(e + f) \leq r(n - 1)$, which is a contradiction. Therefore there indeed exists a diagonal $m_{1j_1}, \ldots, m_{nj_n}$ with $\prod m_{ij_i} > 0$. Let $c_1 = \min(m_{1j_1}, \ldots, m_{nj_n}) > 0$. A **permutation matrix** is an $n \times n$ matrix that contains exactly one 1 in each column and each row, and zeros elsewhere. We consider the matrix $M_1 = M - c_1 P_1$, where P_1 is the permutation matrix that contains 1's at positions (i, j_i) and zeros elsewhere. Then M_1 again has nonnegative entries, equal row and column sums $r - c_1$, and *more* zeros than M. If we continue in this way, we finally obtain the zero matrix and thereby the representation $M = c_1 P_1 + \cdots + c_t P_t$ with $\sum_{i=1}^{t} c_i = r$. As a special case we obtain the famous theorem of Birkhoff and von Neumann. Let $M = (m_{ij})$ be a **doubly stochastic matrix**, that is, $m_{ij} \geq 0$, with all row and column sums equal to 1. Then $M = c_1 P_1 + \cdots + c_t P_t$, $\sum_{i=1}^{t} c_i = 1$, $c_i > 0$ $(i = 1, \ldots, t)$. In short, a doubly stochastic matrix is a convex sum of permutation matrices. We shall learn about the importance of this theorem in Section 15.6.

8.2. Construction of Optimal Matchings

Theorem 8.4 provides a formula for the matching number $m(G)$ in a bipartite graph. But how does one construct a maximum matching? For this, we use the idea of the proof of Theorem 8.3. The following considerations hold for arbitrary graphs $G = (V, E)$. A matching in $G = (V, E)$ is of course again a set of mutually nonincident edges, and $m(G)$ is the order of a maximum matching.

Let M be a matching in $G = (V, E)$. If the vertex u lies on an edge of M, we say that u is M-**saturated**. Otherwise, it is M-**unsaturated**. An M-**alternating path** P is a path in G that uses alternate edges of M and $E \smallsetminus M$, where the two end vertices of P are unsaturated. An M-alternating path thus contains exactly one more edge of $E \smallsetminus M$ than of M.

Example 8.8.

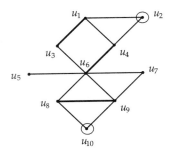

The M-edges are the heavy lines. The path $u_2, u_4, u_6, u_8, u_9, u_{10}$ is an M-alternating path.

Theorem 8.9. *A matching M in $G = (V, E)$ is a maximum matching if and only if there is no M-alternating path.*

Proof. Let M be a matching and $v_0, v_1, \ldots, v_{2k-1}, v_{2k}, v_{2k+1}$ an M-alternating path. Then $(M \smallsetminus \{v_1 v_2, v_3 v_4, \ldots, v_{2k-1} v_{2k}\}) \cup \{v_0 v_1, v_2 v_3, \ldots, v_{2k} v_{2k+1}\}$ is a matching M' with $|M'| = |M| + 1$. Conversely, let M and M' be matchings with $|M'| > |M|$. We set $N = (M \smallsetminus M') \cup (M' \smallsetminus M)$, the symmetric difference of M and M'. Let H be the subgraph spanned by the edge set N. Every vertex in H has degree 1 or 2, and thus every component of H is either an alternating circuit (alternating edges of M and M') or a path with alternating edges of M and M'. Since $|M'| > |M|$, there exists among these path components a path P that begins and ends with an M'-edge. Thus the path P is an M-alternating path, since the end vertices cannot be M-saturated (otherwise, they would be covered twice by M'). $\square$

Let us consider in our example the matching

$$M' = \{u_1 u_4, u_3 u_6, u_7 u_9, u_8 u_{10}\}.$$

We then obtain $N = (M \smallsetminus M') \cup (M' \smallsetminus M)$:

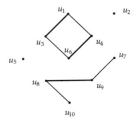

The vertices u_1, u_3, u_6, u_4 form an alternating circuit, and u_7, u_9, u_8, u_{10} is an M-alternating path. Exchanging the M- and M'-edges on this path leads to a new matching (see the left-hand figure below). Again there exists an M-alternating path u_5, u_6, u_4, u_2. Exchanging now leads to a maximal matching in the right-hand figure:

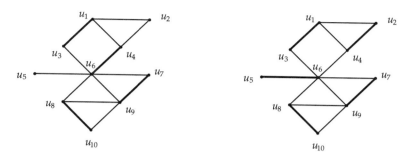

Theorem 8.9 yields for bipartite graphs the following method for construct-ing a maximum matching. Let $G = (S + T, E)$ be a graph. We ask the following question: Is $m(G) = |S|$? If yes, then the algorithm should return a maximum matching. If no, the algorithm should establish this fact.

Algorithm 8.10.

(1) Begin with a matching M (e.g., $M = \varnothing$).
(2) If every S-vertex is M-saturated, we are done. Otherwise, choose an M-unsaturated vertex $u \in S$ and begin the following construction of an "alternating" tree B. At the start, B consists of the vertex u.
   Suppose that B looks as follows during the course of the algorithm:

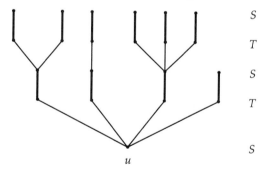

The unique path from each vertex to the "root" u alternates between M-edges and non-M-edges. The vertices of degree 1 are all in S. Let $S_B = S \cap V(B)$, $T_B = T \cap V(B)$.

(*)   Question: Is $N(S_B) \subseteq T_B$?

If yes, then $|N(S_B)| = |T_B| = |S_B| - 1$, and by Theorem 8.3, $m(G) < |S|$. Stop. If no, choose $y \in N(S_B) \smallsetminus T_B$. Let $xy \in E$, $x \in S_B$.
   Either y is M-saturated, $yz \in M$, in which case we grow the tree B by adding the edges $xy \notin M$, $yz \in M$. With this tree, go to (*).

Or y is M-unsaturated, in which case we have found an M-alternating path from u to y. We exchange edges as in the proof of Theorem 8.9. With the new matching M' we go to step (2) and allow the tree to grow.

The reader can adapt this algorithm without too much trouble to find in general a maximum matching and simultaneously a minimal vertex cover.

Let us return to our job-assignment problem, with S a set of persons, T a set of jobs with $u_i v_j \in E$ if person u_i is qualified for job v_j. We would now like to find not only an assignment $u_i \to v_{\varphi i}$ that is as large as possible, but among all such matchings, one with a maximal suitability coefficient.

Two variants are plausible: We are given a weight function $w : S \to \mathbb{R}^+$ (value of the persons) and we seek a matching M with $\sum_{u \in S \cap V(M)} w(u) = $ max. Or we can assume a weight function $w : E \to \mathbb{R}^+$ (suitability coefficient person/job) and are interested in a matching M with $\sum_{k \in M} w(k) = $ max.

We begin with the first variant. We shall call a subset $X \subseteq S$ that can be matched in $G = (S + T, E)$ a **transversal** (inclusive of $\varnothing$). We denote the set of transversals by $\mathcal{T} \subseteq \mathcal{B}(S)$. We are thus seeking a maximal transversal A with maximal weight $w(A)$. The following theorem shows that the transversals form a matroid, and therefore, by Theorem 7.13, the greedy algorithm leads to the desired goal.

Theorem 8.11. *Let $G = (S + T, E)$ be a bipartite graph, $\mathcal{T} \subseteq \mathcal{B}(S)$ the family of transversals. Then $(S, \mathcal{T})$ forms a matroid.*

Proof. Axioms 1 and 2 of a matroid are clearly satisfied. So let A and A' be transversals with $|A'| = |A| + 1$ and associated matchings M, M', that is, $A = S \cap V(M)$, $A' = S \cap V(M')$, $|M'| = |M| + 1$. As in the proof of Theorem 8.9, we see that there is an M-alternating path P whose edges lie alternately in M and M'. Since P has odd length, one of the two end vertices u of P must lie in S. Since $u \in A' \smallsetminus A$, the vertex u can be added to A. $\qquad\square$

We now consider the second variant. We are given a bipartite graph $G = (S + T, E)$ with a weight function $w : E \to \mathbb{R}^+$. We seek a matching M with maximal weight $w(M) = \sum_{k \in M} w(k)$. We first establish that we may assume that G contains all edges between S and T by providing any new edges with weight zero. Likewise, we may assume that $|S| = |T|$ by adding edges to S (or T) and again assigning weight zero to the additional edges. We may therefore assume that $G = (S + T, E)$ is a complete bipartite graph $K_{n,n}$. For a matching M of maximal weight, we may therefore assume $|M| = n$. Finally, we transform the given maximum problem into a minimum problem, which is easier to work with. Therefore, let $W = \max_{k \in E} w(k)$, and let $G' = (S + T, E)$ be the graph $K_{n,n}$ with edge weight $w'(k) = W - w(k) \geq 0$.

Then clearly, a maximum matching M of minimal weight in G' is a matching of maximal weight in G.

With $S = \{s_1, \ldots, s_n\}$, $T = \{t_1, \ldots, t_n\}$ we set $w_{ij} = w(s_i t_j)$, where we assume $w_{ij} \in \mathbb{N}_0$. The weights thus form an integer matrix (w_{ij}). We let x_{ij} denote the variable that assumes the value 1 if $s_i t_j \in M$ and 0 if $s_i t_j \notin M$. A matching M is thus determined by the choice of x_{ij}, so that we have

$$(8.1) \qquad \sum_{j=1}^{n} x_{ij} = 1 \ (\forall i), \qquad \sum_{i=1}^{n} x_{ij} = 1 \ (\forall j).$$

In other words, (x_{ij}) is a permutation matrix, and this implies precisely a matching in $K_{n,n}$.

Our problem thus has the following form: Given the matrix (w_{ij}), we seek an assignment (x_{ij}) satisfying (8.1) such that $\sum w_{ij} x_{ij}$ is minimal.

Suppose we subtract from the hth row of (w_{ij}) a fixed value p_h. This yields a matrix (w'_{ij}). Since on account of (8.1) we have

$$\sum_{i,j} w'_{ij} x_{ij} = \sum_{i,j} w_{ij} x_{ij} - p_h \sum_{j} x_{hj} = \sum_{i,j} w_{ij} x_{ij} - p_h,$$

we see that the optimal assignments for (w_{ij}) and (w'_{ij}) agree. The same holds of course for the columns, and we may carry out this subtraction multiple times without changing this optimal assignment. Likewise, we may add fixed numbers to a row or column without changing the optimal assignment. We would like to use this idea to obtain a new matrix $(\overline{w}_{ij})$ that in general contains more zeros than (w_{ij}). Let $p_h = \min_j w_{hj}$ for $h = 1, \ldots, n$. We subtract p_h from the hth row for all h and thereby obtain a new matrix (w'_{ij}). Let $q_\ell = \min_i w'_{i\ell}$, $\ell = 1, \ldots, n$. We then subtract q_ℓ from the ℓth column for all ℓ and obtain the *reduced* matrix $(\overline{w}_{ij})$, about which we know that it has the same optimal assignments (x_{ij}) as the original matrix (w_{ij}).

Let us take as an example the matrix

$$(w_{ij}) = \begin{pmatrix} 9 & 11 & 12 & 11 \\ 6 & 3 & 8 & 5 \\ 7 & 6 & 13 & 11 \\ 9 & 10 & 10 & 7 \end{pmatrix}.$$

We have $p_1 = 9$, $p_2 = 3$, $p_3 = 6$, $p_4 = 7$, and we obtain

$$(w'_{ij}) = \begin{pmatrix} 0 & 2 & 3 & 2 \\ 3 & 0 & 5 & 2 \\ 1 & 0 & 7 & 5 \\ 2 & 3 & 3 & 0 \end{pmatrix}.$$

We now have $q_1 = q_2 = q_4 = 0$, $q_3 = 3$, and we therefore obtain the reduced matrix

$$(\overline{w}_{ij}) = \begin{pmatrix} \textcircled{0} & 2 & 0 & 2 \\ 3 & 0 & 2 & 2 \\ 1 & \textcircled{0} & 4 & 5 \\ 2 & 3 & \textcircled{0} & 0 \end{pmatrix}.$$

A matching corresponds exactly to a diagonal in this matrix (every row and column contains an entry). If $(\overline{w}_{ij})$ contains a diagonal with all zeros, then we are done. This would be our optimal assignment $(\overline{x}_{ij})$, since we always have $\sum \overline{w}_{ij} x_{ij} \geq 0$. In our example the maximal length of a zero diagonal is equal to 3 (the circled entries). We know from the previous sections that the maximal length of a zero diagonal is equal to the minimum number of rows and columns that cover all zeros. (We simply exchange the roles of the zero and nonzero entries.) In our case these are, for example, rows 1 and 4 and column 2. We call such a set of rows and columns a **minimal covering** for short.

The last phase of the algorithm runs as follows: Either we have found a zero diagonal of length n, in which case we are done, or we change the matrix $(\overline{w}_{ij})$ to a new matrix $(\overline{\overline{w}}_{ij})$, without changing the optimal assignment, so that for the entire sum of weights, we have $\overline{\overline{W}} = \sum \overline{\overline{w}}_{ij} < \overline{W} = \sum \overline{w}_{ij}$. It is clear that the second possibility can occur only finitely often (on account of $w_{ij} \in \mathbb{N}_0$), so that the algorithm must eventually stop.

Suppose the second case occurs, and a minimal zero-covering contains e rows and f columns, $e + f < n$. Let $\overline{w} > 0$ be the minimum of the uncovered elements. We now subtract $\overline{w}$ from the $n - e$ rows that are not covered, and then add $\overline{w}$ to all f columns of the covering. The resulting matrix $(\overline{\overline{w}}_{ij})$ then has, according to our earlier observation, the same optimal assignment. In our example, $\overline{w} = 1 = w_{31}$:

$$(\overline{w}_{ij}) = \begin{array}{c|cccc} & 0 & 2 & 0 & 2 \\ & 3 & 0 & 2 & 2 \\ & 1 & 0 & 4 & 5 \\ & 2 & 3 & 0 & 0 \end{array} \longrightarrow \begin{array}{cccc} 0 & 2 & 0 & 2 \\ 2 & -1 & 1 & 1 \\ 0 & -1 & 3 & 4 \\ 2 & 3 & 0 & 0 \end{array} \longrightarrow \begin{array}{cccc} 0 & 3 & \textcircled{0} & 2 \\ 2 & \textcircled{0} & 1 & 1 \\ \textcircled{0} & 0 & 3 & 4 \\ 2 & 4 & 0 & \textcircled{0} \end{array} = (\overline{\overline{w}}_{ij}).$$

For the change $\overline{w}_{ij} \rightarrow \overline{\overline{w}}_{ij}$ we have

$$\overline{\overline{w}}_{ij} = \begin{cases} \overline{w}_{ij} - \overline{w} & \text{if } \overline{w}_{ij} \text{ is uncovered,} \\ \overline{w}_{ij} & \text{if } \overline{w}_{ij} \text{ is covered by a row or column,} \\ \overline{w}_{ij} + \overline{w} & \text{if } \overline{w}_{ij} \text{ is covered by a row and a column.} \end{cases}$$

In particular, we have $\overline{\overline{w}}_{ij} \geq 0$ for all i, j. The number of doubly covered entries $\overline{w}_{ij}$ is $e \cdot f$, and the number of uncovered entries is $n^2 - n(e + f) + ef$.

For the total weights $\overline{W}$ and $\overline{\overline{W}}$ we thus obtain

$$\overline{\overline{W}} - \overline{W} = (ef)\overline{w} - (n^2 - n(e + f) + ef)\overline{w}$$
$$= (n(e + f) - n^2)\overline{w} < 0$$

since $e + f < n$. This completes our analysis.

In our example, $(\overline{w}_{ij})$ now contains a zero-diagonal of length 4, and an optimal assignment is $x_{13} = x_{22} = x_{31} = x_{44} = 1$, and $x_{ij} = 0$ otherwise. The original job-assignment problem therefore has minimal weight $12 + 3 + 7 + 7 = 29$.

We assemble all the steps of the algorithm:

Algorithm 8.12.

Input: An $(n \times n)$ matrix (w_{ij}), $w_{ij} \in \mathbb{N}_0$.

Output: An optimal assignment (x_{ij}).

(1) Initialization.
(1.1) Subtract $p_i = \min_j w_{ij}$ from row i in (w_{ij}) for all $i = 1, \ldots, n$. Let the new matrix be (w'_{ij}).
(1.2) Subtract $q_j = \min_i w'_{ij}$ from column j in (w'_{ij}) for all $j = 1, \ldots, n$. Let the new matrix be $(\overline{w}_{ij})$.
(2) Find a minimal zero-covering in $(\overline{w}_{ij})$. If the covering contains fewer than n rows and columns, go to step (3). Otherwise, go to step (4).
(3) Modification of the reduced matrix: Let $\overline{w} > 0$ be the smallest uncovered entry. Transform $(\overline{w}_{ij})$ to $(\overline{\overline{w}}_{ij})$ by subtracting all uncovered entries $\overline{w}$ and adding $\overline{w}$ to all doubly covered entries. Go to step (2).
(4) Determine a zero diagonal of length n and set $x_{ij} = 1$ for the entries of the zero diagonal, and $x_{ij} = 0$ for the rest. Stop.

Two further observations: The running time of the algorithm is, as one can determine without too much difficulty, bounded by $O(n^3)$. A significant point in the analysis was that the lowering of the total weight $\overline{\overline{W}} < \overline{W}$ in (3) stops after finitely many steps because we are dealing with whole numbers. If the weights $w_{ij} \in \mathbb{Q}^+$ are merely rational, we can arrive at the previous case by multiplying by the least common multiple of the denominators. But if the weights w_{ij} are real, it is not at once clear that the algorithm stops (the amounts of decrease can get smaller and smaller). In this case we must bring continuity arguments into the picture; we will have more to say about that in the next section. Instead of bipartite graphs, we could also consider arbitrary graphs. A special case of particular importance is the determination of a maximum matching M in the complete graph K_{2m}. In this case as well there is an algorithm with running time $O(n^3)$.

8.3. Flows in Networks

The job-assignment problem is a special case of a general situation that we would now like to discuss.

Definition 8.13. A **network** over $\mathbb{N}_0$ from u to v consists of a directed graph $\vec{G} = (V, E)$, $u \neq v \in V$, together with a weight function $c : E \longrightarrow \mathbb{N}_0$. The vertex u is called the **source**, v the **sink** of the network, and c the **capacity**.

We may imagine $\vec{G} = (V, E)$ as a system of roads, with $c(k)$ as the capacity, namely, how much can be transported over road k. Our task is to determine how much we can transport from source u to sink v given the capacity restrictions c.

A function $f : E \longrightarrow \mathbb{N}_0$ is called a **flow** in the network $\vec{G} = (V, E)$. The *net flow* in the vertex $x \in V$ is $(\partial f)(x) = \sum_{k^+ = x} f(k) - \sum_{k^- = x} f(k)$; that is, $(\partial f)(x)$ measures the inflow minus the outflow in x.

Definition 8.14. Given a network $\vec{G} = (V, E)$ with source u, sink v, and capacity c, a function f is called an **admissible flow of u to v** if

a. $0 \leq f(k) \leq c(k)$ for all $k \in E$,
b. $(\partial f)(x) = 0$ for all $x \neq u, v$.

An admissible flow thus transports a certain quantity from the source to the sink, in such a way that the capacity of the roads is not exceeded (condition a) and nothing is left over at intermediate nodes (condition b).

The **value** $w(f)$ of an admissible flow is $w(f) = (\partial f)(v)$, that is, the net flow into the sink. Clearly, $\sum_{x \in V}(\partial f)(x) = 0$, since in the sum, every value $f(k)$ is counted once positively and once negatively. On account of condition b, it follows that $(\partial f)(u) + (\partial f)(v) = 0$, and therefore $w(f) = -(\partial f)(u) = \sum_{k^- = u} f(k) - \sum_{k^+ = u} f(k)$. Thus $w(f)$ is also equal to the net flow that flows out of the source. Our problem is to maximize the value $w(f)$ of an admissible flow. Let us consider the following network. At each edge k, the flow has been recorded on the left, the capacity on the right:

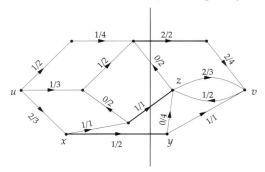

The value of the flow is 4. How large can $w(f)$ be? Theoretically, $2+3+3 = 8$ units can flow out of u along the three edges ux, and likewise, 8 units can flow into the sink v. We now think of the network as divided into two halves, to the right and to the left of the dotted line. It is clear that the value of a flow is limited by the capacity of the three "bridges" from the left to the right side (shown as heavy lines), in this case by $2 + 1 + 2 = 5$.

Definition 8.15. Given the network $\vec{G} = (V, E)$ with source u, sink v, and capacity c, a **cut** of the network is a partition $V = X + Y$ with $u \in X$, $v \in Y$. The **capacity** of the cut (X, Y) is $c(X, Y) = \sum c(k)$, where the summation is over all edges k with $k^- \in X$, $k^+ \in Y$.

Lemma 8.16. *Let $\vec{G} = (V, E)$ be a network with source u, sink v, and capacity c. Then for every admissible flow f and every cut (X, Y), we have*

$$w(f) \leq c(X, Y).$$

Proof. Set $S(A, B) = \{k \in E : k^- \in A,\ k^+ \in B\}$ for any partition $V = A + B$. Now let (X, Y) be a cut. Since $(\partial f)(y) = 0$ for $y \neq u, v$, we have

$$w(f) = (\partial f)(v) = \sum_{y \in Y} (\partial f)(y).$$

If both end vertices of an edge k are in Y, then in the right-hand sum, $f(k)$ is counted once positively, once negatively. We therefore obtain a nonzero value precisely for the edges k between X and Y, and from the condition on f, it follows that

$$w(f) = \sum_{k \in S(X,Y)} f(k) - \sum_{k \in S(Y,X)} f(k) \leq \sum_{k \in S(X,Y)} f(k) \leq \sum_{k \in S(X,Y)} c(k)$$

$$= c(X, Y),$$

which completes the proof of the lemma. $\qquad \square$

Our goal is to prove the fundamental equation

$$\max\ w(f) = \min\ c(X, Y),$$

the maximum flow, minimum cut theorem of Ford and Fulkerson. An admissible flow f_0 with $w(f_0) = \max w(f)$ is called a **maximum flow**, and analogously, a cut (X_0, Y_0) with $c(X_0, Y_0) = \min c(X, Y)$ is called a **minimum cut**. It is not clear a priori whether a maximum flow or minimum cut even exists. We shall prove this in our main result. How, then, do we find a maximum flow? The following method is a generalization of the alternating path in our solution of the matching problem in bipartite graphs. There, too, the result was a maximum–minimum theorem. We can increase the value of an admissible flow only by increasing the net outflow from the source without compromising the condition $(\partial f)(x) = 0$ for $x \neq u, v$.

An (undirected) path $P : u = x_0, x_1, x_2, \ldots, x_t = x$ is called an **increasing path** from u to x in the network $\vec{G}$ if

1. $f(k_i) < c(k_i)$ for every "forward edge" $k_i = (x_{i-1}, x_i)$.
2. $0 < f(k_j)$ for every "backward edge" $k_j = (x_j, x_{j-1})$.

If f is an admissible flow, we define the sets $X_f, Y_f \subseteq V$ by

$X_f = \{x : x = u$ or there exists an increasing path from u to $x\}$,

$Y_f = V \setminus X_f$.

Theorem 8.17. *Let $\vec{G} = (V, E)$ be a network over $\mathbb{N}_0$ with source u, sink v, and capacity c. The following conditions on an admissible flow f are equivalent:*

a. f is a maximum flow.
b. There is no increasing path from u to v.
c. (X_f, Y_f) is a cut.
For a maximum flow f, we have $w(f) = c(X_f, Y_f)$.

Proof. a $\Rightarrow$ b. Suppose P is an admissible path from u to v. We define an *elementary flow* f_P by

$$f_P(k) = \begin{cases} 1 & \text{if } k \in E(P), \ k \text{ a forward edge,} \\ -1 & \text{if } k \in E(P), \ k \text{ a backward edge,} \\ 0 & \text{if } k \notin E(P) . \end{cases}$$

We set $\alpha_1 = \min(c(k) - f(k) : k \in E(P)$ a forward edge$)$, $\alpha_2 = \min(f(k) : k \in E(P)$ a backward edge$)$. From the definition of an increasing path it follows that $\alpha = \min(\alpha_1, \alpha_2) > 0$. Clearly, $g = f + \alpha f_P$ is again an admissible flow, and for the value, we have $w(g) = w(f) + \alpha > w(f)$. Therefore, f was not a maximum flow.

b $\Rightarrow$ c. This follows from the definition of X_f, Y_f.

c $\Rightarrow$ a. We consider an edge $k \in S(X_f, Y_f)$ with start vertex $x = k^- \in X_f$ and end vertex $y = k^+ \in Y_f$. If we had $f(k) < c(k)$, we could prolong an increasing path from u to x by the edge k to y and thereby have an increasing path from u to y, in contradiction to $y \notin X_f$. We therefore conclude that

$$f(k) = c(k) \text{ for all } k \in S(X_f, Y_f),$$

and analogously,

$$f(k) = 0 \text{ for all } k \in S(Y_f, X_f).$$

It follows that

$$w(f) = \sum_{k \in S(X_f, Y_f)} f(k) - \sum_{k \in S(Y_f, X_f)} f(k) = \sum_{k \in S(X_f, Y_f)} c(k) = c(X_f, Y_f).$$

From Lemma 8.16 we conclude that f is a maximum flow with $w(f) = c(X_f, Y_f)$, whence everything has now been proved. $\qquad\Box$

Let us consider our starting example. The given flow f has value 4, and the cut has capacity 5. The path u, x, y, z, v is an increasing path with $\alpha = 1$, and so we can increase the flow by 1 along these edges and obtain thereby a maximum flow f_0 with value 5. The vertices of X_{f_0} are shown below as large dots:

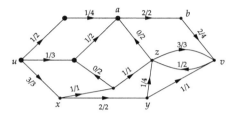

The "bridges" $k \in S(X_{f_0}, Y_{f_0})$ are ux, ab with $f_0(k) = c(k)$.

We showed in Theorem 8.17 that for a maximum flow f_0 we always have $w(f_0) = c(X_{f_0}, Y_{f_0})$, and so $\max w(f) = \min c(X, Y)$. But we do not yet know whether such a maximum flow even exists.

Corollary 8.18. *Let $\vec{G} = (V, E)$ be a network with integer capacities $c : E \to \mathbb{N}_0$. Then there exists an integer maximum flow $f : E \to \mathbb{N}_0$, and we have as well $\max w(f) = \min c(X, Y)$.*

Proof. Let us analyze our proof of Theorem 8.17. We begin with the integer flow $f = 0$. If f is not maximal in a step of the algorithm, then there exists an increasing path from u to v. We see now that the numbers $\alpha_1 = \min(c(k) - f(k))$, $\alpha_2 = \min(f(k))$ are integers, and therefore also $\alpha = \min(\alpha_1, \alpha_2)$. That is, at every step we always increase the value of the flow by an *integer*. Since the integer capacity of a cut is an upper bound, we obtain in finitely many steps an integer flow f_0 for which there cannot be another increasing path. By Theorem 8.17, it follows that f_0 is a maximum flow. $\qquad\Box$

Of course, there is also a maximum flow for rational capacities $c : E \to \mathbb{Q}^+$, as one can see by again multiplying by the greatest common multiple of the denominators. For the general case $c : E \to \mathbb{R}^+$, we use a continuity argument. A constructive proof is given by an algorithm of Edmonds–Karp. See the literature cited at the end of this part. Furthermore, in Chapter 15 we shall derive the complete theorem over $\mathbb{R}$ as a consequence of the main theorem of linear optimization.

The max-flow–min-cut theorem is the starting point for a sequence of interesting applications, a few of which we shall mention. Our first maximum–minimum result was Theorem 8.5, $\max(|M| : M$ a matching $) = \min(|D| : D$

a vertex cover) for bipartite graphs. We can derive this result easily from Corollary 8.18. We associate with $G = (S + T, E)$ a network $\vec{G}$ by orienting all edges of S toward T. Additionally, we define a source u^* and a sink v^*, where we add in all edges (u^*, x), $x \in S$, and (y, v^*), $y \in T$. Let the capacity be identically 1. Let the new vertex and edge sets be V^* and E^*:

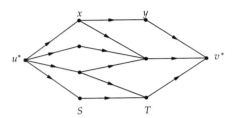

We know that there is a maximum flow f that takes on only the values 0 and 1. Let us consider an arbitrary admissible flow $f : E^* \to \{0, 1\}$. We see from the condition $(\partial f)(x) = 0$, $x \neq u^*, v^*$ that in the case $f(u^*, x) = 1$, precisely one edge (x, y) that leaves x in the direction T has the value 1. The same holds for $y \in T$. In other words, the edges xy, $x \in S$, $y \in T$, with $f(x, y) = 1$ constitute a *matching* in $G = (S + T, E)$, and the value of the flow f is $w(f) = |M|$. Conversely, a matching yields an admissible $0, 1$ flow if we assign the edges of M the flow value 1 as well as the connections to u^* and v^*. Thus maximum $0, 1$ flows are maximum matchings. In the same way, one sees that minimum cuts correspond to minimal vertex covers (check this!), and Theorem 8.5 indeed follows from Corollary 8.18. And that's not all. If we analyze our proof with the help of alternating paths, we see that we are dealing with the method of "increasing paths."

As a further example we will look at the **supply and demand problem**. We would like to transport some goods from a set S of production sites to a set T of buyer locations. Let us call S the sources, and T the sinks. With each source $x \in S$ we associate a *supply* $a(x)$ and with each sink $y \in T$ a *demand* $b(y)$. Our problem is to construct a flow that does not transport more out of $x \in S$ than the available supply, that is, at most $a(x)$, and brings at least $b(y)$ into $y \in T$.

Let $\vec{G} = (V, E)$ be the given directed graph with capacity $c : E \to \mathbb{R}^+$, $S, T \subseteq V$, $S \cap T = \varnothing$, with the functions $a : S \to \mathbb{R}^+$, $b : T \to \mathbb{R}^+$. We seek a flow $f : E \to \mathbb{R}^+$ that satisfies the following conditions:

a. $0 \leq f(k) \leq c(k)$.

b. $\sum_{k^- = x} f(k) - \sum_{k^+ = x} f(k) \leq a(x)$ for all $x \in S$.

c. $\sum_{k^+ = y} f(k) - \sum_{k^- = y} f(k) \geq b(y)$ for all $y \in T$.

d. $\sum_{k^+ = z} f(k) - \sum_{k^- = z} f(k) = 0$ for all $z \in V \smallsetminus (S \cup T)$.

Condition b measures the net outflow from the source x, while condition c measures the net inflow into the sink y. Otherwise, the usual conservation law $(\partial f)(z) = 0$ should hold.

Of course, such a flow does not necessarily have to exist. A simple example is

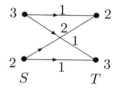

$$a(x) = 1 \qquad c=2 \qquad b(y) = 2$$

$$x \in S \qquad\qquad y \in T$$

It is clear that $\sum_{x \in S} a(x) \geq \sum_{y \in T} b(y)$ is a necessary condition, but it does not suffice, as the following example shows:

Here we have $\sum_{x \in S} a(x) = \sum_{y \in T} b(y) = 5$ to be sure, but only the quantity 2 can flow into the lower right-hand vertex z because of the limit on capacity, and therefore the demand $b(z) = 3$ cannot be met.

To solve our problem, we add a new source u^* and a new sink v^*, as well as all edges (u^*, x), $x \in S$, with $c(u^*, x) = a(x)$, and all edges (y, v^*), $y \in T$, with $c(y, v^*) = b(y)$. We denote the network arising in this way by $\vec{G}^* = (V^*, E^*)$.

We make the following claim: The original graph $\vec{G}$ possesses a flow that satisfies conditions a through d if and only if $\vec{G}^*$ has an admissible flow f^* in the previous sense that *saturates* all edges (y, v^*), that is, satisfies $f^*(y, v^*) = c(y, v^*) = b(y)$. The value of the flow f^* is therefore $w(f^*) = \sum_{y \in T} b(y)$.

If f^* if such an admissible flow for $\vec{G}^*$, then $f(u^*, x) \leq a(x)$ $(x \in S)$, and so the net outflow from x in $\vec{G}$ cannot amount to more than $a(x)$, that is, condition b is satisfied. Likewise, on account of $f^*(y, v^*) = b(y)$, the net inflow into $y \in T$ must be exactly $b(y)$ (condition c). The converse is just as easily proven. By our main theorem, Theorem 8.17, there exists a flow in $\vec{G}$ that satisfies conditions a through d if and only if $c(X^*, Y^*) \geq \sum_{y \in T} b(y)$ for every cut (X^*, Y^*) in $\vec{G}^*$. If we write $X^* = X \cup \{u^*\}$, $Y^* = Y \cup \{v^*\}$, where X or Y can be empty, we have the necessary and sufficient condition

$$c(X^*, Y^*) = c(X, Y) + \sum_{x \in S \cap Y} a(x) + \sum_{y \in T \cap X} b(y) \geq \sum_{y \in T} b(y),$$

and therefore

$$(8.2) \qquad c(X,Y) \geq \sum_{y \in T \cap Y} b(y) - \sum_{x \in S \cap Y} a(x) \quad \text{for all } V = X + Y.$$

One can imagine a host of further variants. For example, we could describe admissible flows f by two functions $c(k) \leq f(k) \leq d(k)$ or additionally introduce a *cost function* $\gamma(k)$ on the edges. We then seek a maximum flow with minimal cost.

8.4. Eulerian Graphs and the Traveling Salesman Problem

A familiar problem is to copy a given figure without lifting one's pencil from the page. For example, we can draw the left-hand figure in the following diagram (beginning with the topmost vertex) in one go by following the numbered edges in order, returning to the starting vertex. But the right-hand figure cannot be so drawn:

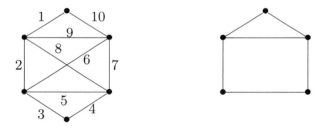

This type of problem goes back to Euler, who posed and solved the famous problem of the bridges of Königsberg (which can be found in every book on graph theory). In looking at this problem we shall make an exception and consider multigraphs $G = (V, E)$, that is, we allow loops and parallel edges.

Let $G = (V, E)$ be a multigraph with q edges. An edge sequence $k_1, \ldots, k_q$, with k_i incident to k_{i+1}, that contains each edge exactly once and returns to its starting point is called an **Euler circuit**. If we do not require that the start and end vertices are the same, we speak of an **open Euler circuit**.

Our second example possesses no (closed) Euler circuit, but it does contain an open Euler circuit. A graph is said to be **Eulerian** if it contains an Euler circuit.

We give ourselves two tasks:

1. Characterize all Eulerian multigraphs.
2. How does one find an Euler circuit?

Theorem 8.19. *A multigraph $G = (V, E)$ with $E \neq \varnothing$ is Eulerian if and only if G is connected and all vertices have even degree (where loops are counted twice).*

Proof. Clearly, the graph G must be connected, and since entering and leaving a vertex traverses two edges (or one loop), all the edges must have even degree. Conversely, suppose the conditions are satisfied. Since every degree is even, there must exist circuits (why?). Let C be a circuit with maximal number of edges. If $E = E(C)$, we are done. Otherwise, let $G' = (V, E \smallsetminus E(C))$. Since G is connected, G' and C must have a common vertex u with $d_{G'}(u) > 0$. In G' every vertex has even degree, and so there exists a circuit C' in G' that contains u, and C' can be added to a traversal of C when the vertex u is reached, in contradiction to the maximality of C. $\qquad\square$

How does one find an Euler circuit? One cannot simply start walking, as the example below shows. If we choose the edges in the order shown, we get stuck at vertex u. The reason is that removing edges numbered 1 through 6 from the graph breaks the remaining graph into two components:

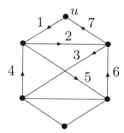

The following algorithm constructs an Euler circuit in every Eulerian multigraph:

Algorithm 8.20.

(1) Start at an arbitrary vertex u_0.
(2) Let $u_0, u_1, \ldots, u_i$ with edges $k_j = u_{j-1}u_j$ $(j = 1, \ldots, i)$ have already been constructed. Let $G_i = (V, E \smallsetminus \{k_1, \ldots, k_i\})$ be the remaining graph. If $E \smallsetminus \{k_1, \ldots, k_i\} = \varnothing$, stop. Otherwise, choose from among the edges in G_i incident to u_i an edge k_{i+1} that is not a cut edge (bridge) in G_i as long as this is possible. Repeat step (2).

Proof of the algorithm's correctness. Suppose the algorithm has constructed $u_0, u_1, \ldots, u_p$ with $W = \{k_1, \ldots, k_p\}$. Clearly, we have $u_p = u_0$. In G_p the vertex u_0 has degree zero, since otherwise, we could continue. Suppose $E(G_p) \neq \varnothing$. Then there exist vertices v with $d_{G_p}(v) > 0$. Let $S = \{v \in V : d_{G_p}(v) > 0\}$ and $T = \{v \in V : d_{G_p}(v) = 0\}$. Thus we have

$S \neq \varnothing$, $T \neq \varnothing$. Now let ℓ be the largest index such that $u_\ell \in S$, $u_{\ell+1} \in T$. There must exist such an ℓ, since otherwise, the edge tour W (which begins at $u_0 \in T$) would exhaust T without moving into S, in contradiction to the connectedness of G (see the figure below).

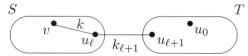

Since after $u_{\ell+1}$, the trip W does not leave the set T, and $d_{G_p}(x) = 0$ for all $x \in T$, it follows that $k_{\ell+1}$ is the only edge in G_ℓ between S and T and is therefore a cut edge in G_ℓ. But now $d_{G_p}(u_\ell) > 0$ is even, and so there must be an additional edge k in G_ℓ incident to u_ℓ. By rule (2), k is therefore also a cut edge in G_ℓ and thus also in G_p, since G_ℓ and G_p are identical on S. We now know that in G_p, every vertex has even degree, since in traversing a vertex in W, the degree is reduced by 2. If A and B are the two components of G_p that arise when k is removed (see the figure),

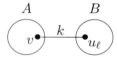

then u_ℓ would be the only vertex of odd degree in B, which by Corollary 6.2 is impossible. Thus k was not a cut edge after all, and we are done.   □

Remark 8.21. Since it can be determined in polynomial time whether an edge is a cut edge (clear?), our algorithm provides a polynomial-time procedure for constructing an Euler circuit.

As an application we shall solve the problem of the Chinese postman (the reason for the name is simply that the problem was first posed in this form by a Chinese mathematician). Given a connected multigraph $G = (V, E)$ and a cost function $w : E \to \mathbb{R}^+$, An Ya-Ren, the postwoman, begins at a vertex u_0 (the post office) and must travel each edge (street) at least once and return to u_0. Her task is to construct a tour W with minimal cost $\sum_{k \in W} w(k)$.

If G is Eulerian, then An Ya-Ren simply constructs an Euler circuit. If G is not Eulerian, she must traverse at least one street more than once, producing a multigraph $G^* = (V, E^*)$ that is Eulerian and for which $\sum_{k \in E^* \setminus E} w(k)$ is minimal.

Solution. Let U be the set of vertices of odd degree, $|U| = 2m$.

(1) For $u, v \in U$ determine the length $d(u, v)$ of a weighted shortest path (the shortest-path problem from Section 7.4).

(2) Consider the complete graphs K_{2m} on U with weights $w(uv) = d(u,v)$. Determine a weighted minimal matching $M = \{u_1v_1, \ldots, u_mv_m\}$ (the weighted matching problem from Section 8.2).

(3) Insert the optimal u_iv_i paths P_i. The multigraph $G^* = G \cup \bigcup_{i=1}^m P_i$ thus obtained is then a solution.

$\square$

Example 8.22. In the following graph G, we have $U = \{u, v\}$. A shortest weighted u, v path is u, x, y, z, v with $d(u, v) = 6$. We thus obtain the minimal multigraph G^* in the right-hand figure:

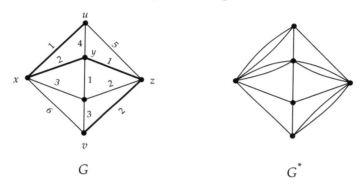

$$G \qquad\qquad G^*$$

Euler circuits can be defined on directed multigraphs $\vec{G} = (V, E)$, $W = \{k_1, \ldots, k_q\}$ with $k_i^+ = k_{i+1}^-$ and $k_q^+ = k_1^-$. The following theorem is the analogue of Theorem 8.19 and is proved similarly.

Theorem 8.23. *A directed multigraph $\vec{G} = (V, E)$ is Eulerian if and only if G is connected and $d^-(u) = d^+(u)$ for all $u \in V$.*

Up to now we have considered closed circuits containing each edge exactly once. We may consider as well closed sequences of edges that contain each vertex exactly once. In other words, we are looking for a circuit in G of length $n = |V|$. With this problem there is also associated a famous name, that of Hamilton, who posed the problem for the dodecahedral graph (see the figure). Circuits C_n are therefore called **Hamiltonian circuits**, and a graph G is said to be **Hamiltonian** if it contains such a circuit.

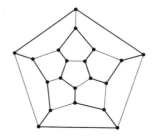

The heavy lines in the figure form a Hamiltonian circuit.

In contrast to the case of Eulerian graphs, the characterization of Hamiltonian graphs is a provably difficult (NP-complete) problem, about which we shall have more to say in the following section. We shall therefore have to be satisfied with necessary and sufficient conditions (see the exercises).

We turn now to one of the most famous algorithmic problems: the **traveling salesman problem** (TSP), which was mentioned earlier, in Section 5.3. A traveling salesman begins in his home city, visits each city on his list once, and returns to his starting point. What route should he take to minimize the total cost?

In graph-theoretic terms, we are given the complete graph K_n and a cost function $w : E \to \mathbb{R}^+$, or equivalently, a cost matrix (w_{ij}), $1 \le i, j \le n$, with

a. $w_{ij} \ge 0$,

b. $w_{ij} = w_{ji}$ (symmetry).

We seek a Hamiltonian circuit, that is, a cyclic permutation $\pi = (i_1 = 1, i_2, \ldots, i_n)$ such that

$$w(\pi) = \sum_{i=1}^{n} w(i, \pi(i))$$

is a minimum. Since there are $(n-1)!$ cyclic permutations, trying each one is impossible for large n. In the next section we shall see that the traveling salesman problem belongs to the class of NP-complete problems, for which no good (i.e., polynomial running time in n) algorithms are known and for which such algorithms likely do not exist. How is one to proceed? One seeks heuristic procedures in the hope of approaching the actual minimum as closely as possible.

Example 8.24 (Rhineland problem: M. Grötschel). The cost matrix is given by the following table of distances (in kilometers) between cities in the Rhineland region of Germany:

	A	B	D	F	C	W
Aachen	–	91	80	259	70	121
Bonn	91	–	77	175	27	84
Düsseldorf	80	77	–	232	47	29
Frankfurt	259	175	232	–	189	236
Cologne	70	27	47	189	–	55
Wuppertal	121	84	29	236	55	–

Our task is to find a shortest trip through these six cities. The procedure that probably first comes to mind is the "nearest-neighbor" (NN) algorithm:

Algorithm 8.25 (Nearest neighbor (NN)). Let $V = \{V_1, \ldots, V_n\}$. Choose an arbitrary start vertex V_{i_1}. If $V_{i_1}, \ldots, V_{i_m}$ has been constructed, search for a vertex $V_{j_0} \in V' = E \setminus \{V_{i_1}, \ldots, V_{i_m}\}$ with $w(V_{j_0}, V_{i_m}) = \min \ (w(V_j, V_{i_m}) : V_j \in V')$ and set $i_{m+1} = j_0$.

If we consider nearest neighbors at both ends of the constructed path, we obtain the **double nearest-neighbor** (DNN) algorithm:

Algorithm 8.26 (Double nearest neighbor (DNN)). As in NN, suppose the sequence $V_{i_1}, \ldots, V_{i_m}$ has already been chosen. In $V' = V \setminus \{V_{i_1}, \ldots, V_{i_m}\}$ let V_{j_0} satisfy $w(V_{j_0}, V_{i_1}) = \min \ (w(V_j, V_{i_1}) : V_j \in V')$ and V_{j_1} satisfy $w(V_{j_1}, V_{i_m}) = \min(w(V_j, V_{i_m}) : V_j \in V')$. If $w(V_{j_0}, V_{i_1}) \leq w(V_{j_1}, V_{i_m})$, add V_{j_0} to the tour before V_{i_1}. Otherwise, add V_{j_1} after V_{i_m}.

In our example, NN with start city Cologne and with start city Frankfurt yields the tours

$$\text{E} - \text{B} - \text{D} - \text{W} - \text{A} - \text{F} - \text{C} \qquad w = 702,$$
$$\text{F} - \text{B} - \text{C} - \text{D} - \text{W} - \text{A} - \text{F} \qquad w = 658.$$

Observe that the start city makes a significant difference in the length of the tour.

For DNN with start city Cologne, we begin our tour with

$$\text{C} - \text{B}.$$

Now, Düsseldorf is the closest remaining city to both Cologne (47) and Bonn (77). Since the distance to Cologne is less, we add Düsseldorf onto Cologne and obtain

$$\text{D} - \text{C} - \text{B}.$$

Continuing in this fashion, we end up with the circuit

$$\text{C} - \text{B} - \text{A} - \text{F} - \text{W} - \text{D} - \text{C} \qquad w = 689.$$

DNN will generally be better than NN, but in unlucky cases, both algorithms can be bad, with long stretches to travel at the end of the tour. To compensate for this disadvantage, let us attempt to enlarge *circuits* successively. We begin with an arbitrary circuit of length 3. Suppose that $C = (V_{i_1}, \ldots, V_{i_m})$ of length m has been constructed. If we wish to add $V_j \notin \{V_{i_1}, \ldots V_{i_m}\}$ to C, say between V_{i_k} and $V_{i_{k+1}}$, the net additional cost for the tour C is equal to $d(V_j, k) = w(V_{i_k}, V_j) + w(V_j, V_{i_{k+1}}) - w(V_{i_k}, V_{i_{k+1}})$. We now set $d(V_j) = \min_k d(V_j, k)$ and choose V_{j_0} with $d(V_{j_0}) = \min \ (d(V_j) : V_j \notin \{V_{i_1}, \ldots, V_{i_m}\})$. Now V_{j_0} is inserted at the proper place. This algorithm is called **cheapest insertion** (CI).

Up to now we have considered tours that are constructed step by step. We now would like to study two global methods.

Minimum spanning tree (MST) heuristic: The idea is simple: A tour is a Hamiltonian circuit. If we delete an edge, we obtain a spanning tree. MST is divided into three parts:

1. Construct a minimal spanning tree T (e.g., with the greedy algorithm from Section 7.3).
2. Double all edges in T. This gives an Eulerian multigraph T_D. Let $C = \{v_1, v_2, \ldots\}$ be an Euler circuit in T_D.
3. A Hamiltonian circuit is contained in C (by skipping over vertices that already have been traversed).

   In our example we have the following result:

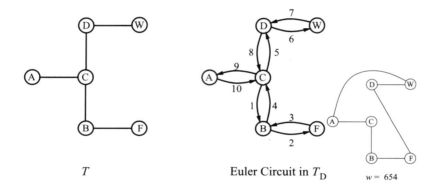

T Euler Circuit in T_D $w = 654$

Christofides heuristic:

(1) Construct a minimal spanning tree T.
(2) Let U be the set of vertices of odd degree in T, with $|U| = 2m$. Construct a maximum matching (with minimal cost) M on U, $M = \{u_1 v_1, \ldots, u_m v_m\}$, and add M to T. This yields an Eulerian multigraph T_D.
(3) A Hamiltonian circuit is contained in T_D.

   For the Rhineland problem this gives the following result:

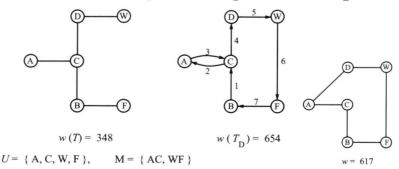

$w(T) = 348$ $w(T_D) = 654$

$U = \{A, C, W, F\}$, $M = \{AC, WF\}$ $w = 617$

One can show that this last tour is optimal for the Rhineland problem.

We mentioned earlier that the traveling salesman problem is provably hard. But is it possible to be able to assert that a particular (polynomial) algorithm at least constructs a tour that is not too far from the actual minimum? We say that an algorithm has a *performance guarantee* ε if for the tour C that it constructs,

$$w(C) \leq (1 + \varepsilon)w(C_{\text{opt}}).$$

For an arbitrary cost matrix (w_{ij}), this is also a difficult problem. However, for an important special case, namely, when the w_{ij}'s are interpreted as geometric lengths as in our example, we can say something more precise. In this case, the w_{ij}'s satisfy the triangle inequality $w_{ij} \leq w_{ih} + w_{hj}$ for all i, j, h, and we speak of a *metric* TSP. Both minimum spanning tree (MST) and the Christofides heuristic (CH) are fast algorithms, since the individual steps have polynomial running time. Let C_{MST} and C_{CH} be tours constructed by the MST heuristic and Christofides heuristic.

Theorem 8.27. *Let a metric TSP with cost matrix (w_{ij}) be given. Then the following hold:*

a. $w(C_{\text{MST}}) \leq 2\,w(C_{\text{opt}})$.
b. $w(C_{\text{CH}}) \leq \frac{3}{2}w(C_{\text{opt}})$.

Proof. a. Since C_{opt} minus an edge is a tree, we have $w(T) \leq w(C_{\text{opt}})$ for the tree T constructed in phase 1 of MST, and hence $w(T_{\text{D}}) \leq 2w(C_{\text{opt}})$. Let $C = \{v_1, v_2, \dots\}$ be the Euler circuit, and $C_{\text{MST}} = \{v_{i_1}, v_{i_2}, \dots\}$ the tour contained within it. From the triangle inequality we conclude that

$$w(v_{i_1}, v_{i_2}) \leq w(v_{i_1}, v_{i_1+1}) + \cdots + w(v_{i_2-1}, v_{i_2})$$

in the Euler circuit, and analogously for $w(v_{i_2}, v_{i_3})$, and so on. We thereby obtain $w(C_{\text{MST}}) \leq w(T_{\text{D}}) \leq 2\,w(C_{\text{opt}})$.

b. Again we have $w(T) \leq w(C_{\text{opt}})$. Let $\{v_1, v_2, \dots, v_n\}$ be a minimal tour, and $v_{i_1}, v_{i_2}, \dots, v_{i_{2m}}$ the vertices of U in this order. We consider the matchings

$$M_1 = \{v_{i_1}v_{i_2}, v_{i_3}v_{i_4}, \dots, v_{i_{2m-1}}v_{i_{2m}}\},$$
$$M_2 = \{v_{i_2}v_{i_3}, v_{i_4}v_{i_5}, \dots, v_{i_{2m}}v_{i_1}\}.$$

For the matching M determined in the Christofides heuristic, we have $w(M) \leq w(M_1)$, $w(M) \leq w(M_2)$, and from the triangle inequality, $w(M_1) + w(M_2) \leq w(C_{\text{opt}})$. We thereby obtain $w(M) \leq \frac{1}{2}(w(M_1) + w(M_2)) \leq \frac{1}{2}w(C_{\text{opt}})$, and so altogether,

$$w(C_{\text{CH}}) \leq w(T) + w(M) \leq \frac{3}{2}w(C_{\text{opt}}).$$

$\square$

Remark 8.28. The performance guarantee $\varepsilon = \frac{1}{2}$ determined by the Christofides heuristic is the best such guarantee known for the traveling salesman problem.

8.5. The Complexity Classes P and NP

We have become familiar with a variety of problems for which fast algorithms exist, for example the shortest-path problem and the characterization of Eulerian graphs together with the construction of an Euler circuit. We have also seen some "hard" problems, namely the characterization of Hamiltonian graphs and the traveling salesman problem. In this section we are going to make a few observations about the degree of difficulty of abstract problems.

We define an *abstract problem* as a family I of instances together with a family S of solutions. For example, in TSP, the instances are all the matrices (w_{ij}), and the solutions are the minimal tours. In the Hamiltonian problem HP, the instances are graphs, given, for example, by adjacency matrices, and the solutions are 1 and 0, the former for yes (the graph is Hamiltonian), and the latter for no (it is not Hamiltonian). We see that TSP is an **optimization problem**, and HP is a **decision problem**. The theory of complexity classes, which we shall sketch in the following pages, applies to decision problems. To apply it to optimization problems, we must transform these into decision problems. Typically, one does this in such a way that one specifies an additional bound M. For example, for TSP the associated decision problem would be, Is $w(C_{\text{opt}}) \leq M$? A solution of the optimization problem is then of course also a solution of the decision problem. The optimization problem is therefore at least as difficult as the associated decision problem.

How does one measure the degree of difficulty, known as the **complexity**, of a decision problem? It is done through the cost of obtaining the solution. This involves coding the input. Informally, this is an assignment of the input to words of 0's and 1's. For example, we can code the adjacency matrix of a graph with n^2 entries with the value 0 or 1. And we can code natural numbers using their binary representation. We say that an algorithm solves a problem in time $O(f(n))$ if given input of length n it produces a solution in $O(f(n))$ computational steps. That is a bit vague, but to be more precise would require the concept of formal languages and Turing machines. The book by Garey–Johnson provides an excellent overview of these matters. However, it should be clear at this point what is meant by a **running time** of $O(f(n))$.

A decision problem is said to be **polynomial** (more precisely, soluble in polynomial time) if there exists an algorithm with running time $O(n^s)$ for some constant s. The *complexity class* P contains all polynomial decision problems.

Let us consider the Hamilton problem HP. Suppose that our algorithm checks every possibility. That is, we write all $m!$ permutations of the m vertices and check for every permutation whether all the edges are in G. How large is the running time? If the graph is given by its adjacency matrix, the length of the input is $n = m^2$, or equivalently, $m = \sqrt{n}$. For the $m!$ permutations we therefore need $\Omega(m!) = \Omega(\sqrt{n}!) = \Omega(2^{\sqrt{n}})$ operations, and as we know from Chapter 5, this is never $O(n^s)$ for any s. We therefore say that the given algorithm is **exponential**. However, we may still ask whether some other algorithm for HP is polynomial, that is, whether indeed HP is in the class P. Likely it is not, for reasons that we shall now make clear.

The next fundamental complexity class is called NP, for "nondeterministic polynomial." As suggested by the name, this class was introduced in the context of nondeterministic problems. We will discuss this class using the equivalent notion of **verification**.

A decision problem is in the class NP if a *positive* solution (that is, a solution that purports to demonstrate that the answer is $1 = $ yes) can be verified in polynomial time. What exactly does this mean? Suppose someone asserts that he or she has, say, for the traveling salesman problem with input matrix (w_{ij}), constructed a tour C with $w(C) \leq M$. To say that the problem in this formulation is in NP means that we are able to determine in polynomial time whether C is first of all a Hamiltonian circuit, and if so, whether $w(C) \leq M$ is indeed the case.

And of course we can do this. We have merely to check that C is a circuit (one step per vertex) and then compare $w(C)$ with M. Thus the traveling salesman problem is indeed in NP. A quite different question is whether a *negative* solution (one with answer $0 = $ no) can be verified in polynomial time. This class is called co-NP.

It should be clear that $P \subseteq NP \cap$ co-NP. The two basic questions of complexity theory, which today belong among the most important open problems in all of mathematics, relate to these two classes: Is $P \neq NP$? Is $NP \neq$ co-NP? Most researchers in the field of complexity theory believe that the answer to both questions is yes. And with good reason. To see this, we shall look at a special class of problems in NP, the *NP-complete* problems. A decision problem Q is said to be **NP-complete** if it is in NP and if the solvability of Q in polynomial time would imply that *every* NP problem could be solved in polynomial time. In other words, for such a problem, $Q \in P \Rightarrow NP = P$. How does one prove that a problem Q is NP-complete? Suppose R is an arbitrary problem in NP. Then it would have to be possible to "transform" R into a special case of Q in *polynomial* time. The polynomial transformation of R into Q together with the polynomial solution of Q would then imply $R \in P$. Thus Q is "at least" as difficult as R. Thus the NP-complete problems are the "most difficult" problems in NP.

We shall now look at such a transformation. But first we must answer the question whether there even exist any NP-complete problems. The first example of such a decision problem was given by Cook in 1971, the *satisfiability problem* (SAT) for Boolean expressions. We shall look at this problem in Chapter 11. Another problem was proved to be NP-complete by Karp in 1972, namely our Hamilton problem HP.

As an example of what we mean by such a transformation as described above, we shall prove that TSP, the traveling salesman problem, is NP-complete. We know already that TSP is in NP. If we can transform HP into TSP in polynomial time, then TSP will have been shown to be NP-complete, since this holds already for HP, as proved by Karp. But such a transformation is simple: Given a graph $G = (V, E)$, $V = \{1, 2, \ldots, n\}$, we associate with G a *special* TSP in the following manner:

$$w_{ij} = \begin{cases} 0 & \text{if } ij \in E, \\ 1 & \text{if } ij \notin E . \end{cases}$$

Clearly, G possesses a Hamiltonian circuit if and only if the special version of TSP possesses a tour C with $w(C) \leq 0$ (and therefore equal to zero). Thus if we decide TSP in polynomial time, we can solve HP in polynomial time as well.

Most of the problems in NP that we have looked at thus far in this book either are in P or are NP-complete. One of the best-known problems whose complexity class is uncertain is the *graph isomorphism problem* GI. The input consists of a pair of graphs G and H, and the problem is to determine whether G and H are isomorphic. The problem GI is certainly in NP (see Exercise 8.46), and research to date suggests that GI is likely in P. Since the introduction of these concepts, hundreds of decision and optimization problems have been shown to be NP-complete. Yet despite much effort, not a single one of these problems has been shown to have a polynomial-time algorithm. And we know that if *one* such problem is polynomial, then they all are. This empirical fact speaks in favor of P $\neq$ NP. Stated informally, problems in P are easy, while those not in P, such as conjecturally the NP-complete problems, are difficult. Of course, one must attack NP-complete problems despite their likely intractability, that is, the probability that no fast algorithm exists. How one can approach such problems, for example using heuristics, was discussed in the previous section in reference to TSP.

Exercises for Chapter 8

8.1 Suppose the bipartite graph $G = (S + T, E)$ is k-regular, $k \geq 1$. Show that $|S| = |T|$ and G always contains a matching M with $|M| = |S| = |T|$.

8.2 A 1-*factor* in an arbitrary graph $G = (V, E)$ is a matching M that contains all vertices; hence $|M| = \frac{|V|}{2}$. The graph G is said to be 1-*factorizable* if E can be decomposed into disjoint 1-factors. Use the previous exercise to conclude that a k-regular bipartite graph, $k \geq 1$, is 1-factorizable.

▷ **8.3** Show that a bipartite graph $G = (S + T, E)$ with $|S| = |T| = n$ and $|E| > (m - 1)n$ contains a matching of size m. Is m the best possible?

8.4 Let $T = \{1, 2, \ldots, n\}$. How many distinct transversals does the family of sets $\mathcal{A} = \{\{1, 2\}, \{2, 3\}, \{3, 4\}, \ldots, \{n - 1, n\}, \{n, 1\}\}$ possess?

8.5 Show that a tree possesses at most one 1-factor.

8.6 Show that the Petersen graph is not 1-factorizable.

▷ **8.7** Let G be a graph on n vertices, n even, in which $d(u) + d(v) \geq n - 1$ for every pair of vertices u, v. Show that G possesses a 1-factor.

▷ **8.8** The $m \times n$ lattice $G(m, n)$ has as vertices the pairs (i, j), $1 \leq i \leq m$, $1 \leq j \leq n$, where (i, j), (k, ℓ) are joined by an edge if $i = k$, $|j - \ell| = 1$ or $j = \ell$, $|i - k| = 1$ (see Exercise 7.18 for $G(2, n)$). Show that $G(m, n)$ has a 1-factor if and only if mn is even. Calculate the number of 1-factors in $G(2, n)$.

8.9 Solve the optimal assignment problem for the following two matrices on $K_{4,4}$ and $K_{5,5}$, respectively:

$$
\begin{pmatrix} 8 & 3 & 2 & 4 \\ 10 & 9 & 3 & 6 \\ 2 & 1 & 1 & 5 \\ 3 & 8 & 2 & 1 \end{pmatrix}, \quad \begin{pmatrix} 8 & 7 & 5 & 11 & 4 \\ 9 & 7 & 6 & 11 & 3 \\ 12 & 9 & 4 & 8 & 2 \\ 1 & 2 & 3 & 5 & 6 \\ 11 & 4 & 2 & 8 & 2 \end{pmatrix}.
$$

8.10 Discuss in detail why Theorem 8.5 follows from Corollary 8.18.

8.11 Solve the Chinese postman problem for the following street system:

8.12 A pipeline transports oil from A to B. The oil can go by a northern route and by a southern route. Each route has an intermediate station with a pipeline from south to north. The first half of the northern route (up to the station) has a capacity of 300 barrels per hour, while the second half can carry 400 barrels per hour. For the southern route the analogous capacities are 500 and 300, while the capacity for the pipeline from south to north is 300 barrels. How many barrels per hour can be transported from A to B?

▷ **8.13** Prove the following statement: Let $G = (V, E)$ be Hamiltonian and $A \subseteq V$, $A \neq \emptyset$. Then $G \smallsetminus A$ has at most $|A|$ components.

8.14 Show that the Petersen graph is not Hamiltonian.

▷ **8.15** Show that all hypercubes Q_n, $n \geq 2$, are Hamiltonian.

8.16 Construct a non-Hamiltonian graph on 10 vertices for which $d(u) + d(v) \geq 9$ for each pair of nonneighboring vertices. (See Exercise 8.43 for the fact that this is best possible.)

▷ **8.17** A cube is formed out of $27 = 3 \times 3 \times 3$ cubes of cheese. A mouse wishes to travel from a corner cube through every other cube, ending at the middle cube, always traveling from one cube to an adjacent cube, in such a way that each cube is traversed exactly once. Is this possible? Discuss the analogous problem for $5 \times 5 \times 5$ and generally for $n \times n \times n$, n odd.

8.18 Test the method of cheapest insertion for the Rhineland problem in Section 8.4.

8.19 Another method of constructing a traveling salesman tour is "most distant insertion." Begin with an arbitrary circuit of length three. Suppose that $C = \{V_{i_1}, \ldots, V_{i_k}\}$ has already been constructed. For each V_{i_j} determine $V'_{i_j} \notin C$ with $w(V'_{i_j}, V_{i_j}) = \min(w(V_h, V_{i_j}) : V_h \notin C)$, and among the V'_{i_j} determine the "most distant" $V_{i_{j_0}}$ with $w(V'_{i_{j_0}}, V_{i_{j_0}}) \geq w(V'_{i_j}, V_{i_j})$ for all i_j. Now construct $V'_{i_{j_0}}$ at the most favorable location as in cheapest insertion. Test most distant insertion for the Rhineland problem.

▷ **8.20** The *vertex cover problem* has as input the graph G and $k \in \mathbb{N}$ with the question, does there exist a vertex cover D in G with $|D| \leq k$? The *clique problem* has as input a graph G and $k \in \mathbb{N}$ with the question, does there exist a complete subgraph $H \subseteq G$ (a clique) with $|V(H)| \geq k$? Show that both problems are in NP and are polynomially equivalent. Hint: Consider the complementary graph $\overline{G}$. Are both problems NP-complete?

8.21 Construct k-regular graphs $(k > 1)$ without 1-factors for every k.

▷ **8.22** Let $G = (S + T, E)$ be a bipartite graph. Prove the equivalence of the following statements: (a) G is connected and every edge is in a 1-factor. (b) $|S| = |T|$ and $|N(A)| > |A|$ for every $A \subseteq S$ with $\emptyset \neq A \neq S$. (c) $G \smallsetminus \{u, v\}$ has a 1-factor for every $u \in S, v \in T$.

8.23 In Exercise 6.10 we introduced the independence number $\alpha(G)$. Show that a graph G is bipartite if and only if $\alpha(H) \geq \frac{|V(H)|}{2}$ for every subgraph H of G if and only if $m(H) = d(H)$ for every subgraph H of G ($m(H)$ is the matching number, $d(H)$ is the vertex cover number, that is, the minimum size of a vertex cover).

8.24 Show that if G is a k-regular graph on n vertices, then $\alpha(G) \leq \frac{n}{2}$.

8.25 Show that the edge set of a bipartite graph G with maximal degree Δ is the union of Δ matchings.

▷ **8.26** Two friends are playing a game on a connected graph G. They alternately choose vertices $u_1, u_2, u_3, \ldots$ that have not yet been chosen with the condition $u_i u_{i+1} \in E$ $(i \geq 1)$. The first player who is unable to choose a vertex loses. Show that the first player has a winning strategy if and only if G contains no 1-factor.

8.27 Consider how the method of growing trees discussed in Section 8.2 could be modified to find a maximum matching.

▷ **8.28** Show that every k-regular bipartite graph $G = (S + T, E)$ contains at least $k!$ distinct 1-factors. Hint: Induction on $n = |S| = |T|$.

8.29 Consider the usual 8×8 chessboard. If two white (or two black) squares are removed, then the board can no longer be covered by 31 dominoes (clear?). Show that if one removes any white and any black square, then the board can indeed be covered by 31 dominoes.

8.30 Let M be a matching in a graph G, and suppose $u \in V$ is M-unsaturated. Show that if there is no M-alternating path that begins at u, then u is unsaturated in a maximum matching.

8.31 Let $G = (V, E)$ be a graph without isolated vertices, and let $m(G)$ be the matching number and $\beta(G)$ the smallest number of edges that meet all the vertices. Show that $m(G) + \beta(G) = |V|$.

▷ **8.32** Show that $\alpha(G) = \beta(G)$ is true for bipartite graphs G. Give an example of a nonbipartite graph for which the assertion is false.

8.33 Show that the running time of the optimal matching algorithm on a bipartite graph $G = (S + T, E)$ is $O(n^3)$ in the number of vertices n.

▷ **8.34** Generalized tic-tac-toe. A positional game is a pair $(S, \mathcal{F})$, $\mathcal{F} \subseteq \mathcal{B}(S)$. The sets in $\mathcal{F}$ are called *winning sets*. Two players alternately occupy positions, that is, elements of S. A player wins if he or she has completely occupied a winning set $A \in \mathcal{F}$. The usual 3×3 tic-tac-toe thus has $|S| = 9$, $|\mathcal{F}| = 8$. Suppose that $|A| \geq a$ for all $A \in \mathcal{F}$ and that every $s \in S$ is in at most b winning sets. Show that the second player can force a draw if $a \geq 2b$. Hint: Consider the bipartite graph on $S + (\mathcal{F} \cup \mathcal{F}')$ with s adjacent to A, A' if $s \in A$, where $\mathcal{F}'$ is a copy of $\mathcal{F}$. What is the result for $n \times n$ tic-tac-toe?

8.35 Determine a maximal flow from vertex 0 to vertex 11 in the following directed graph, where the weights represent capacities:

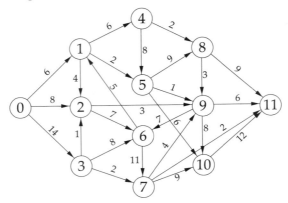

▷ **8.36** Let $(r_1, \ldots, r_m)$ and $(s_1, \ldots, s_n)$ be two sequences of nonnegative integers that satisfy the conditions $s_1 \geq \cdots \geq s_n$ and $\sum_{i=1}^m r_i = \sum_{j=1}^n s_j$. Prove the Gale–Ryser theorem: There exists an $m \times n$ matrix with $0, 1$ entries and row sums $r_1, \ldots, r_m$ and column sums $s_1, \ldots, s_n$ if and only if $\sum_{i=1}^m \min(r_i, k) \geq \sum_{j=1}^k s_j$ for all $k = 1, \ldots, n$. Hint: Transform the problem into a supply and demand problem on the bipartite graph $G = (Z + S, E)$, where Z is the number of rows and S the number of columns, and use the relevant theorem from Section 8.3.

▷ **8.37** Suppose the symmetric matrix (c_{ij}), $c_{ij} \geq 0$, satisfies the triangle inequality $c_{ik} \leq c_{ij} + c_{jk}$. Let c_{NN} be the cost of a "nearest neighbor" tour. Show that $c_{NN} \leq \frac{1}{2}(\lceil \lg n \rceil + 1)c_{opt}$. Hint: Let $\ell_1 \geq \cdots \geq \ell_n$ be the edge costs of the nearest neighbor tour. Show first that $c_{opt} \geq 2\sum_{i=k+1}^{2k} \ell_i$ for $1 \leq k \leq \lfloor \frac{n}{2} \rfloor$ and $c_{opt} \geq 2\sum_{i=\lceil n/2 \rceil+1}^n \ell_i$.

▷ **8.38** Let $\vec{G} = (V, E)$ be a directed graph, $u \neq v \in V$, $(u, v) \notin E$. A set $A \subseteq E$ is called a u, v-*separating edge set* if no directed path exists in $\vec{G} \smallsetminus A$ from u to v. A set $\mathcal{W}$ of directed u, v paths is called a u, v *path system* if every pair of paths in $\mathcal{W}$ are edge disjoint. Prove Menger's theorem: $\max |\mathcal{W}| = \min |A|$ over all path systems $\mathcal{W}$ and u, v-separating edge sets A. Hint: Set the capacity c equal to 1 and take u as source, v as sink.

8.39 Prove the vertex version of Menger's theorem: $u \neq v \in V$, $(u, v) \notin E$. The sets $A \subseteq V \smallsetminus \{u, v\}$ are separating vertex sets and the u, v paths in $\mathcal{W}$ are vertex disjoint (except for u, v). Hint: Transform the problem into the edge version.

8.40 Let $\vec{G} = (V, E)$ be a directed graph, $S, T \subseteq V$. An S, T-separating edge set A and an S, T vertex-path system $\mathcal{W}$ are defined in analogy to the definitions in the previous problem. Prove again $\max |\mathcal{W}| = \min |A|$. Hint: Adjoin additional vertices u^*, v^* to S and T.

▷ **8.41** Specialize the previous exercise to bipartite graphs $G = (S + T, E)$.

8.42 Estimate the running time of the algorithm from Section 8.4 for constructing an Euler circuit.

▷ **8.43** Let G be a graph with $n \geq 3$ vertices. Show that if $d(u) + d(v) \geq n$ for every pair of nonneighboring vertices, then G is Hamiltonian. Hint: Suppose the theorem is false for n. Then choose from among all counterexamples one with the maximal number of edges.

8.44 We consider in this problem the *asymmetric* traveling salesman problem ATSP, which permits $c_{ij} \neq c_{ji}$. Given a cost matrix (c_{ij}) on $\{1, \ldots, n\}$, we construct the following directed graph $\vec{G}$ on $\{1, \ldots, n, n+1\}$. The edges $(i, 1)$, $1 \leq i \leq n$, are replaced by $(i, n+1)$ with the same weight $c_{i,1}$, while all remaining edges remain unchanged. Show that ATSP is equivalent to determining a shortest directed path of length n from 1 to $n+1$ in $\vec{G}$.

8.45 Let (c_{ij}) be a symmetric cost matrix for TSP, $c_{ij} \geq 0$, that satisfies the triangle inequality $c_{ik} \leq c_{ij} + c_{jk}$. Begin with an arbitrary vertex v and write $C_1 = \{v\}$. Let the circuit $C_k = \{u_1, \ldots, u_k\}$ be already constructed. Determine $u \notin C_k$ with minimal distance to C_k and insert u before the corresponding vertex with shortest distance. Show that for the tour thus constructed, $c(T) \leq 2c(T_{\mathrm{opt}})$.

▷ **8.46** Show that the decision problem whether two graphs G and H are isomorphic is in NP by means of a polynomial description of the verification $G \cong H$.

8.47 Show that P $\subseteq$ co-NP.

▷ **8.48** Prove that NP $\neq$ co-NP implies P $\neq$ NP.

Chapter 9

Searching and Sorting

9.1. Search Problems and Decision Trees

A variant of the following game should be familiar to most readers. One player leaves the room. The remaining players agree on a particular word or object, and then the player who left the room has to guess what the others have chosen by asking questions that can be answered yes or no. If the item can be guessed in no more than twenty questions, the player wins.

This "twenty questions" game contains all the elements of a general search problem. One is given a search domain S and a set of tests by which the sought element x^* can be determined.

Let us consider a mathematical example: The questioner knows that the sought term is a number between 1 and 7. Therefore, the search domain is $S = \{1, 2, \ldots, 7\}$. The admissible tests are "$x^* < i$?" where i is some natural number. How many questions are required to determine x^*?

We can easily model an algorithm for posing questions using a **decision tree**:

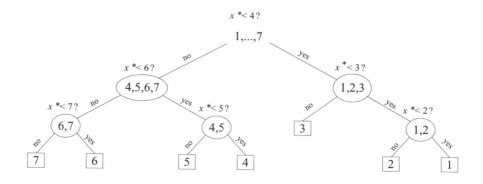

The round vertices contain the results that are still possible at that stage of the algorithm. The square vertices contain the unique result from following the given path through the tree. We say that the questioner requires three questions in the **worst case**, while the number required **on average** (assuming equiprobability) is $\frac{1}{7}(1 \cdot 2 + 6 \cdot 3) = \frac{20}{7}$.

Another well-known search problem deals with weighing coins. We have n coins, one of which is a counterfeit. All the genuine coins have the same weight, while the counterfeit has a different weight, which may be larger or smaller. The search domain is thus $\{1_L, 1_H, 2_L, 2_H, \ldots, n_L, n_H\}$, where i_L and i_H represent the ith coin as the counterfeit, being either lighter (L) or heavier (H). In a test, we take two sets A and B of coins, with $|A| = |B|$, place A in the left pan of a balance scale, B in the right pan, and observe the result. The test has three possible outcomes: $A < B$ (A is lighter than B), $A = B$ (they weigh the same), $A > B$ (A is heavier than B). The outcome $A < B$ gives us the information that the counterfeit coin is either in A (and is lighter) or in B (and is heavier). An analogous result holds for the case $A > B$. In the case $A = B$, we know that the counterfeit coin is in neither A nor B.

For $n = 12$ the following algorithm finds the counterfeit coin, where the right half of the decision tree is symmetric to the left and therefore omitted to save space:

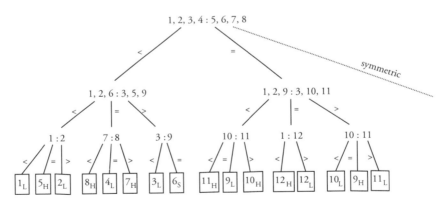

Every question algorithm can be thus represented as a decision tree, and the length of the algorithm corresponds precisely to the length of the path from the "root" of the tree to the relevant terminal vertex. We are ready now to introduce the ideas and terminology necessary for an understanding of general search problems.

A **rooted tree** (T, v) is a tree in the usual sense together with a distinguished vertex v, the **root** of T. We say also that T is rooted at v. We know that for every vertex x of T there is exactly one path from v to x.

We can picture these paths in the above examples as running from top to bottom; they are in fact *directed* paths. The following terminology is in line with this point of view. A vertex x is a **predecessor** of y if x is contained in the path $P(v, y)$ from v to y, and x is an *immediate predecessor* if $P(v, y) = v, \ldots, x, y$. Analogously, we have the terms **successor** and *immediate successor*. The vertices without successors (thus those of degree 1) are called **terminal vertices** or **leaves** of T. The remaining vertices are called **internal vertices**. It should be clear what is meant by a *subtree rooted at* x, namely, x is the root and the tree in question contains x and all of its successors.

The *length* $\ell(x)$ of a vertex x is the length of the unique path from the root v to x. Therefore, the root is the only vertex with length 0. Finally, the **length** $L(T)$ of the tree, $L(T) = \max_{x \in V} \ell(x)$, is the length of the longest path from the root to a leaf.

The following rooted tree has 22 vertices, of which 12 are leaves and 10 internal vertices. The length of the tree is 4. We shall consistently draw internal vertices round and leaves square:

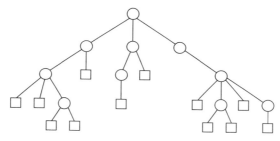

In a rooted tree T, every vertex other than the root has precisely one immediate predecessor. We say that T is an (n, q)-**tree** if T has n leaves and the maximal number of immediate successors of an inner vertex is q, where we always assume $q \geq 2$. A tree T is called a **complete** (n, q)-**tree** if every internal vertex has exactly q successors. It then follows that $q - 1 \mid n - 1$ (see Exercise 9.3). We denote by $\mathcal{T}(n, q)$ the class of (n, q)-trees.

Our first algorithm above returns a complete $(7, 2)$-tree, our weighing procedure an (incomplete) $(24, 3)$-tree, and our last example is a $(12, 4)$-tree.

We have thus represented each algorithm $\mathcal{A}$ for a given search problem as an (n, q)-tree T. The leaves of T correspond to the results, n is the size of the search domain, and q is the maximum number of outcomes of our test questions. The worst-case length $L(\mathcal{A})$ of the algorithm is precisely the length $L(T)$ of the associated decision tree T. For a given search problem, we are interested in the minimum worst-case length

$$L = \min L(\mathcal{A}),$$

over all possible algorithms $\mathcal{A}$, that is, over all possible decision trees. The following lower bound on L is called the *information-theoretic bound*.

Theorem 9.1. *Let* $T \in \mathcal{T}(n, q)$, $n \geq 1$, $q \geq 2$. *Then*

$$L(T) \geq \lceil \log_q n \rceil,$$

where $\log_q n$ *denotes the logarithm to the base* q

Proof. We shall show that for a tree $T \in \mathcal{T}(n, q)$ of length L, one always has $q^L \geq n$. For $L = 0$ this is clear. We now use induction on L. Suppose the immediate successors of the root v are $x_1, \ldots, x_t$, $t \leq q$. One of the subtrees T_i rooted at x_i must contain at least $\frac{n}{q}$ leaves. By induction, $q^{L(T_i)} \geq \frac{n}{q}$, and with $L \geq L(T_i) + 1$ we therefore have $q^L \geq q^{L(T_i)+1} \geq n$. Since $L(T)$ is an integer, it follows that $L(T) \geq \lceil \log_q n \rceil$. $\qquad\square$

The term "information-theoretic" bound comes from the interpretation of a search process as an increase of information. At the outset, we know nothing (information 0). With each question we increase our information about the actual result, until at the end, we know the whole story (complete information). The size $\lceil \log_q n \rceil$ thus gives a worst-case lower bound for the number of questions necessary for obtaining complete information.

Example 9.2. If every test is allowed, then we obtain all trees $T \in \mathcal{T}(n, q)$ as decision trees. For a complete (n, q)-tree T we clearly have $L(T) = \lceil \log_q n \rceil$, and thus the lower bound is satisfied in this case. This presents us immediately with the general problem: For what families $\mathcal{W}$ of (n, q)-trees is the lower bound achieved, that is, $\min(L(T) : T \in \mathcal{W}) = \lceil \log_q n \rceil$?

Let us return to our weighing problem. For 12 coins we have $n = 24$ possible results, and so we obtain $L \geq \lceil \log_3 24 \rceil = 3$, and thereby $L = 3$ from our previous algorithm. And what is the situation with 13 coins? Theoretically, $L = 3$ is again possible, since $\log_3 26 < 3$.

Suppose at the first weighing we place ℓ coins in each balance pan. If the left side is lighter, then any one of the ℓ coins on the left side could be lighter, or else any one of the coins on the right side could be heavier. Altogether, then, we obtain 2ℓ possible results. We obtain the same number 2ℓ if the right side is lighter. In the case of equality, we have $26 - 2\ell - 2\ell = 2m$ possible outcomes. From $26 = 2\ell + 2\ell + 2m$ we conclude that

$$\max(2\ell, 2m) \geq \frac{26}{3},$$

and hence $\max(2\ell, 2m) \geq 10$, since 2ℓ and $2m$ are both even. From the information-theoretic bound we see that $\max(2\ell, 2m)$ cannot be accomplished in two weighings, and so $L \geq 4$. That four weighings suffice for 13 coins is easy to see, and therefore $L = 4$ in this case. The reader can determine (or consult the exercises) that this is the typical exceptional case. In general, one has $L = \lceil \log_3(2n + 2) \rceil$ for the n-coin problem.

9.2. The Fundamental Theorem of Search Theory

Of greater interest than the determination of the length of a search process in the worst case is that of the average case. In most cases, the various possible outcomes do not occur with equal probability. If we know that an outcome x_i has high probability, we will construct a decision tree that assigns x_i a short length ℓ_i.

Our general problem then is as follows: Given an (n, q)-tree T with leaves $x_1, \ldots, x_n$ and a probability distribution $(p_1, \ldots, p_n)$, $p_i = p(x^* = x_i)$, let ℓ_i be the length of x_i. We are interested in the *average length*

$$\overline{L}(T) = \sum_{i=1}^{n} p_i \ell_i,$$

and in particular in $\overline{L} = \min \overline{L}(T)$, over all decision trees T. Thus $\overline{L}(T)$ is simply the expected value of the length of a randomly selected leaf.

First we must determine when a tree $T \in \mathcal{T}(n, q)$ with leaves of lengths $\ell_1, \ldots, \ell_n$ exists.

Theorem 9.3 (Kraft's inequality).

a. *Let $T \in \mathcal{T}(n, q)$ be given with leaves of length $\ell_1, \ldots, \ell_n$. Then $\sum_{i=1}^{n} q^{-\ell_i} \leq 1$, with equality if and only if T is complete.*
b. *Given $\ell_1, \ldots, \ell_n \in \mathbb{N}_0$ with $\sum_{i=1}^{n} q^{-\ell_i} \leq 1$, there exists a tree $T \in \mathcal{T}(n, q)$ with lengths $\ell_1, \ldots, \ell_n$.*

Proof. To prove assertion a, we first observe that any (n, q)-tree can be transformed into a complete (n', q)-tree T' with $n' \geq n$ by attaching leaves at "unsaturated" internal vertices. Since the sum $\sum q^{-\ell_i}$ thereby increases, it suffices to prove the equality for complete trees. We employ induction on n. For $n = 0$, the tree consists of the root only, and we have $q^0 = 1$. Now let $n > 0$. We replace a "branch" of leaves of length ℓ with a single leaf of length $\ell - 1$, as in the following diagram:

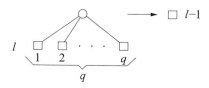

The new tree T' is a complete $(n-q+1, q)$-tree. By induction, we conclude that

$$\underbrace{\sum_{i=1}^{n} q^{-\ell_i}}_{T} = \sum_{i=q+1}^{n} q^{-\ell_i} + q \cdot q^{-\ell} = \underbrace{\sum_{i=q+1}^{n} q^{-\ell_i} + q^{-(\ell-1)}}_{T'} = 1 \, .$$

We now assume conversely that $\sum_{i=1}^{n} q^{-\ell_i} \leq 1$. Let $w_k = |\{i : \ell_i = k\}|$, $k = 0, 1, \ldots, L = L(T)$. That is, w_k is the number of leaves of length k in the tree T that we wish to construct. We may therefore write the inequality $\sum q^{-\ell_i} \leq 1$ as

$$\sum_{k=0}^{L} w_k q^{-k} \leq 1,$$

or equivalently, as

(9.1) $$w_0 q^L + w_1 q^{L-1} + \cdots + w_{L-1} q + w_L \leq q^L.$$

We construct the desired tree T inductively. If $w_0 = 1$, we have $L = 0$, and T consists of only the root. Suppose we have already determined T up to length k. Let $N_k = \{u \in E : \ell(u) = k\}$. We have already constructed $w_0, w_1, \ldots, w_k$ leaves of lengths $0, 1, \ldots, k$. Therefore, in N_k the vertices beneath these $w_0 + \cdots + w_k$ leaves are no longer available, and we conclude that in N_k there are still

(9.2) $$q^k - \sum_{i=0}^{k} w_i q^{k-i}$$

free internal vertices. From (9.1) we have

$$w_{k+1} q^{L-k-1} \leq q^L - \sum_{i=0}^{k} w_i q^{L-i} \, ,$$

and hence

$$w_{k+1} \leq q^{k+1} - \sum_{i=0}^{k} w_i q^{k+1-i} = q \left(q^k - \sum_{i=0}^{k} w_i q^{k-i} \right) .$$

From (9.2) it follows that we can place all w_{k+1} leaves of length $k+1$. $\square$

Example 9.4. Let $n = 6$, $q = 2$, $\ell_1 = 1$, $\ell_2 = 2$, $\ell_3 = 3$, $\ell_4 = \ell_5 = 5$, $\ell_6 = 6$. We have $w_0 = 0$, $w_1 = w_2 = w_3 = 1$, $w_4 = 0$, $w_5 = 2$, $w_6 = 1$, and $\sum_{k=0}^{6} w_k 2^{6-k} = 2^5 + 2^4 + 2^3 + 2 \cdot 2 + 1 = 61 \leq 2^6$. The construction is now as shown in the figure:

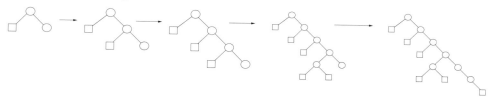

Before we arrive at our main result, we need a fact about logarithms. From calculus we know that for the natural logarithm, $\log x \leq x - 1$ for $x > 0$ with $\log x = x - 1$ only for $x = 1$.

Lemma 9.5. *Let $s_1, \ldots, s_n, y_1, \ldots, y_n$ be positive real numbers with $\sum_{i=1}^{n} s_i \leq \sum_{i=1}^{n} y_i$. Then $\sum_{i=1}^{n} y_i \log_q \frac{y_i}{s_i} \geq 0$ $(q > 1)$, with equality if and only if $s_i = y_i$ for all i.*

Proof. Since $\log_q x = \frac{\log x}{\log q}$, it suffices to consider only the natural logarithm. From $\log x \leq x - 1$ it follows that

$$\sum_{i=1}^{n} y_i \log \frac{s_i}{y_i} \leq \sum_{i=1}^{n} y_i \left(\frac{s_i}{y_i} - 1 \right) = \sum_{i=1}^{n} s_i - \sum_{i=1}^{n} y_i \leq 0,$$

and therefore $\sum_{i=1}^{n} y_i \log \frac{y_i}{s_i} \geq 0$. Equality can occur only if $\log \frac{s_i}{y_i} = \frac{s_i}{y_i} - 1$ for all i, that is, when $s_i = y_i$ for all i. $\qquad \square$

The following famous theorem of Shannon is known as the first fundamental theorem of information theory.

Theorem 9.6. *Let $n \geq 1$, $q \geq 2$, and let $p = (p_1, \ldots, p_n)$ be a probability distribution on the leaves of $T \in \mathcal{T}(n, q)$. Then*

$$-\sum_{i=1}^{n} p_i \log_q p_i \leq \overline{L} = \min \overline{L}(T) < \left(-\sum_{i=1}^{n} p_i \log_q p_i \right) + 1.$$

Proof. We first assume $p_i > 0$ for all i. To prove the left-hand inequality, we must show that $\overline{L}(T) = \sum_{i=1}^{n} p_i \ell_i \geq -\sum_{i=1}^{n} p_i \log_q p_i$ for *all* trees $T \in \mathcal{T}(n, q)$. From Kraft's inequality, Theorem 9.3, we have $\sum_{i=1}^{n} q^{-\ell_i} \leq 1 = \sum_{i=1}^{n} p_i$. If we set $s_i = q^{-\ell_i}$, $y_i = p_i$, in Lemma 9.5, we obtain $\sum_{i=1}^{n} p_i \log_q \frac{p_i}{q^{-\ell_i}} = \sum_{i=1}^{n} p_i \log_q (p_i q^{\ell_i}) \geq 0$, or

$$\overline{L}(T) = \sum_{i=1}^{n} p_i \ell_i \geq -\sum_{i=1}^{n} p_i \log_q p_i.$$

To prove the right-hand inequality, we define natural numbers ℓ_i by $-\log_q p_i \leq \ell_i < (-\log_q p_i) + 1$. Since $0 < p_i \leq 1$, the ℓ_i's are well defined. We therefore have $q^{-\ell_i} \leq p_i$ for all i and thus $\sum_{i=1}^{n} q^{-\ell_i} \leq \sum_{i=1}^{n} p_i = 1$.

From statement b of Kraft's inequality, there exists a tree $T \in \mathcal{T}(n,q)$ with leaf lengths $\ell_1, \ldots, \ell_n$, and we therefore obtain

$$\overline{L}(T) = \sum_{i=1}^{n} p_i \ell_i < \sum_{i=1}^{n} p_i(-\log_q p_i + 1) = \left(-\sum_{i=1}^{n} p_i \log_q p_i\right) + 1.$$

Finally, we may remove the assumption $p_i > 0$ by setting $0 \cdot \log_q 0 = 0$. The proof goes through in this case as well. □

The theorem determines the minimal length of (n,q)-trees with a distribution $(p_1, \ldots, p_n)$ up to an error less than 1. If $q = 2$, then $H(p_1, \ldots, p_n) = -\sum_{i=1}^{n} p_i \lg p_i$ is called the **entropy** of $(p_1, \ldots, p_n)$. In the sense of information theory, it is a measure of the average number of yes/no questions necessary to obtain complete information.

If all trees $T \in \mathcal{T}(n,q)$ are possible decision trees for the given search problem, then Theorem 9.6 gives upper and lower bounds for the average search length. In general, however, only certain trees come into play. For example, in the coin-weighing problem, since $|A| = |B|$, the left and right subtrees always have the same number of leaves. Therefore for general search problems we may use only the lower bound.

Let us return to the trees $T \in \mathcal{T}(n,q)$. We know that $\overline{L} = \min \overline{L}(T)$ is approximately $-\sum_{i=1}^{n} p_i \log_q p_i$. But what is the *exact* value $\overline{L}(p_1, \ldots, p_n)$? The following famous algorithm of Huffman determines this value. Furthermore, instead of a distribution $(p_1, \ldots, p_n)$ we may select weights $w_1, \ldots, w_n \in \mathbb{R}^+$. If we set $p_i = w_i / \sum_{i=1}^{n} w_i$ $(i = 1, \ldots, n)$, then the optimal trees are of course the same for both problems.

Therefore, let $(p_1, \ldots, p_n)$ be a distribution, where we assume $p_1 \geq p_2 \geq \cdots \geq p_n \geq 0$. Suppose the tree $T \in \mathcal{T}(n,q)$ is optimal for the distribution $(p_1, \ldots, p_n)$, that is, $\overline{L}(T) = \sum_{i=1}^{n} p_i \ell_i = \overline{L}(p_1, \ldots, p_n)$. We may assume that $q - 1$ is a divisor of $n - 1$ by including additional leaves with $p_j = 0$. This clearly has no effect on $\overline{L}(T)$. We shall now analyze T. Let $x_1, \ldots, x_n$ be the leaves of T with lengths $\ell_1, \ldots, \ell_n$ and probabilities $p_1 \geq \cdots \geq p_n$.

Step 1. We have $\ell_1 \leq \ell_2 \leq \cdots \leq \ell_n$. Suppose there exist indices i, j with $p_i > p_j$, $\ell_i > \ell_j$. If in T we exchange the places of x_i and x_j, we obtain the tree $T' \in \mathcal{T}(n,q)$ with

$$\overline{L}(T') = \sum_{k \neq i,j} p_k \ell_k + p_i \ell_j + p_j \ell_i = \overline{L}(T) - (p_i - p_j)(\ell_i - \ell_j) < \overline{L}(T),$$

contradicting the optimality of T.

Step 2. Let $L = L(T)$. From Step 1 it follows that $\ell_i = L$ implies $\ell_j = L$ for all j with $i \leq j \leq n$. From the condition $q - 1 \mid n - 1$ it follows moreover that T is complete. Indeed, if an internal vertex u with $\ell(u) \leq L - 2$ had fewer than q immediate successors, then we could attach a leaf of length L to u, and T would not be minimal. Let I be the set of internal vertices u with $\ell(u) \leq L - 2$, and let J be the set with $\ell(u) = L - 1$. Every vertex with the exception of the root has precisely one immediate predecessor. The vertices of I have q immediate successors, and suppose $v_j \in J$ has n_j successors (leaves of length L). By double counting, we obtain

$$|I|q + \sum_J n_j = |I| + |J| - 1 + n \,,$$

(-1 because of the root), and thereby

$$(n - 1) - |I|(q - 1) = \sum (n_j - 1) \,.$$

From $q - 1 \mid n - 1$ it follows that $q - 1 \mid \sum (n_j - 1)$. If $n_j = q$ for all j, then T is complete. Otherwise, we replace as many n_j as possible by q (this has no effect on $\overline{L}$). For the remaining n_j', since $q - 1 \mid \sum (n_j - 1)$, we have $n_j' = 1$. However, if an internal vertex u, $\ell(u) = L - 1$, has only one successor leaf x, then we can move up x to the place of u, contradicting the minimality of T.

Step 3. We may therefore assume that $x_{n-q+1}, \ldots, x_n$ with lengths $\ell_{n-q+1} = \cdots = \ell_n = L$ all have a common immediate predecessor v. We now substitute the branch

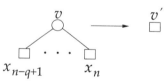

and obtain a complete tree $T' \in \mathcal{T}(n-q+1, q)$. If we assign v' the probability $p = p_{n-q+1} + \cdots + p_n$, then we obtain

$$\overline{L}(p_1, \ldots, p_{n-q}, p) \leq \overline{L}(T') = \overline{L}(T) - pL + p(L - 1) = \overline{L}(T) - p \,.$$

Step 4. Conversely, let U' be an optimal $(n - q + 1, q)$-tree for the distribution $(p_1, \ldots, p_{n-q}, p)$, and v' a leaf whose probability is p, $\ell(v') = \ell$. We make the substitution

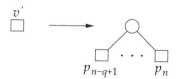

and obtain an (n, q)-tree U for the distribution $(p_1, \ldots, p_n)$, where we have

$$\overline{L}(p_1, \ldots, p_n) \leq \overline{L}(U') - p\ell + p(\ell + 1) = \overline{L}(U') + p.$$

Taking Steps 3 and 4 together, we draw the following conclusion:

Step 5. We have $\overline{L}(p_1, \ldots, p_n) = \overline{L}(p_1, \ldots, p_{n-q}, p) + p$, where $p = p_{n-q+1} + \cdots + p_n$. In other words, $T \in \mathcal{T}(n, q)$ is optimal for the distribution $(p_1, \ldots, p_n)$ if and only if $T' \in \mathcal{T}(n - q + 1, q)$ is optimal for the distribution $(p_1, \ldots, p_{n-q}, p)$, where $p = p_{n-q+1} + \cdots + p_n$ is the sum of the q smallest probabilities.

This is the basis for our algorithm. The substitution of a branch with the q smallest probabilities leads by Step 3 to an optimal $(n - q + 1, q)$-tree. We again substitute a branch, and so on, until we arrive at the trivial tree. Now, using Step 4, we develop the trivial tree step by step in the opposite direction until we finally obtain an optimal tree for the given distribution $(p_1, \ldots, p_n)$. We then eliminate any added leaves with $p_j = 0$. We note that the Huffman algorithm is once again an example of a greedy algorithm.

Example 9.7. Let $n = 8$, $q = 3$, $p_1 = p_2 = 22$, $p_3 = 17$, $p_4 = 16$, $p_5 = 15$, $p_6 = p_7 = 3$, $p_8 = 2$. To obtain $q - 1 \mid n - 1$, we add a leaf with $p_9 = 0$. Phase 3 of Huffman's algorithm looks like this:

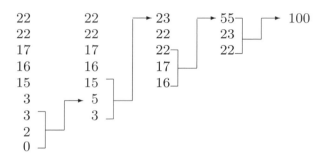

We now "develop" the trivial tree according to Step 4:

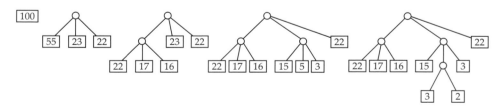

where we leave off the 0 leaf at the end. The average length of the optimal tree T is therefore $\overline{L} = 22 + 2(22 + 17 + 16 + 15 + 3) + 3(3 + 2) = 183$, or if we divide by $\sum_{i=1}^{8} p_i = 100$, we have $\overline{L}(T) = 1.83$. For the lower bound in Theorem 9.6 we obtain $-\sum \frac{p_i}{100} \log_3 \frac{p_i}{100} = 1.67$. Note that the length of T is equal to 3, and thus T is not optimal in the sense of the worst case, where we have $L(T) = \lceil \log_3 8 \rceil = 2$.

9.3. Sorting Lists

Many algorithms require as a first step that the items to be processed by the algorithm be sorted in some given order. An example is the greedy algorithm for matroids from Section 7.3. In this section we are going to analyze the problem of *sorting*.

Suppose, then, that we have a list $a_1, \ldots, a_n$ of n distinct integers, and our task is to rearrange the elements of the list so that the new list is ordered from the smallest to the largest element. Put another way, we must determine the unique permutation π for which $a_{\pi(1)} < a_{\pi(2)} < \cdots < a_{\pi(n)}$ holds. Instead of integers, we may consider any list of elements on which there exists a linear order, so that we have available the test of *pairwise comparison* $a_i : a_j$ with the possible answers $a_i < a_j$ and $a_j < a_i$.

Our search domain S therefore consists of all $n!$ possible permutations of the a_i, with each permutation yielding a single linear ordering and every test offering $q = 2$ possible answers. Let $S(n)$ denote the number of comparisons that an optimal algorithm will require (in the worst case). Then Theorem 9.1 yields

$$(9.3) \qquad\qquad S(n) \geq \lceil \lg n! \rceil.$$

How close can we get to the lower bound (9.3)? As an example, let us consider the case $n = 4$. We therefore have $S(n) \geq \lceil \lg 24 \rceil = 5$. Of course, we may sort $a_1, \ldots, a_4$ by carrying out *all* $\binom{4}{2} = 6$ comparisons $a_i : a_j$. The bound (9.3) tells us that we will need at least five comparisons. The following algorithm, represented as a decision tree, shows that indeed five comparisons suffice; that is, $S(4) = 5$:

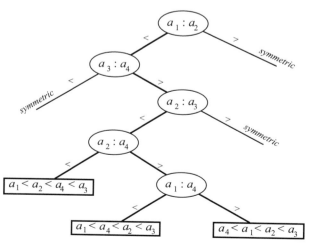

Let us trace through the execution of the algorithm. With the first comparison $a_1 : a_2$ we divided the $4! = 24$ possible linear arrangements into two halves, each of size 12, the first with $a_1 < a_2$, and the other with $a_2 < a_1$. We can therefore, without loss of generality, assume $a_1 < a_2$ (that is the significance of the word "symmetric" for the right-hand subtree). The same holds for $a_3 : a_4$, and we assume without loss of generality that $a_4 < a_3$. There are now six arrangements left that are compatible with $a_1 < a_2$ and $a_4 < a_3$. We now compare the two *maximal* elements a_2, a_3, and again without loss of generality assume $a_2 < a_3$. Our sorting to this point can be represented by a **Hasse diagram**:

Here we depict only the *immediate* order relations; the entire ordering constructed thus far is implied by transitivity; thus, for example, $a_1 < a_3$ is a consequence of $a_1 < a_2$ and $a_2 < a_3$. If we now compare $a_2 : a_4$, then in the case $a_2 < a_4$ we have determined the complete linear ordering $a_1 < a_2 < a_4 < a_3$. In the case $a_2 > a_4$ we obtain the Hasse diagram

and must then carry out a fifth comparison, $a_1 : a_4$.

For $n = 5$ we have $S(5) \geq \lceil \lg 120 \rceil = 7$. Since 120 is close to $2^7 = 128$, it is to be expected that a sorting algorithm with only 7 comparisons will require some delicacy. Nonetheless, it works, as the reader may convince him- or herself.

We are now going to consider three sorting algorithms, *insertion sort*, *merge sort*, and *quicksort*, and give a detailed analysis.

The **insertion sort** method is probably the idea that comes first to mind. We begin by comparing $a_1 : a_2$. Once we have achieved the correct sort order $a_{h_1} < \cdots < a_{h_i}$ for the first i elements, we then take the element a_{i+1} and proceed to insert it into the correct place in the sorted segment $a_{h_1} < \cdots < a_{h_i}$: We simplify the notation by setting $a_{h_j} = b_j$, and thus we have $b_1 < \cdots < b_i$. Now we compare a_{i+1} with the *median* element $b_{\frac{i+1}{2}}$ if i is odd, or with $b_{\frac{i}{2}}$ if i is even. In this way, we roughly halve the number of possible locations in which a_{i+1} can belong. We then compare with the median in this smaller interval and proceed until we know unambiguously where a_{i+1} belongs. How many comparisons will we need in the worst case for the insertion of a_{i+1} into its proper place? Since for a_{i+1} we have the $i + 1$ possibilities $a_{i+1} < b_1, b_1 < a_{i+1} < b_2, \ldots, b_i < a_{i+1}$, we will need at least, from Theorem 9.1, $\lceil \lg(i+1) \rceil$ comparisons. Induction now tells us at once that we can complete the insertion with this number of comparisons. For $i = 1$ this is clear. After the first comparison, there remain in the worst case $\lceil \frac{i+1}{2} \rceil$ possibilities. By induction, we can now insert a_{i+1} with $\lceil \lg \lceil \frac{i+1}{2} \rceil \rceil$ additional comparisons, and so we need altogether $1 + \lceil \lg \lceil \frac{i+1}{2} \rceil \rceil = \lceil \lg(i+1) \rceil$ comparisons.

We therefore obtain for $B(n)$, the total number of comparisons,

$$B(n) = \sum_{i=2}^{n} \lceil \lg i \rceil \,.$$

Let $n = 2^m + r$, $0 < r \le 2^m$. Since $\lceil \lg i \rceil = k$ for $2^{k-1} < i \le 2^k$, we conclude that

$$B(n) = \sum_{k=1}^{m} k2^{k-1} + r(m+1) \,.$$

In Section 2.1 we calculated $\sum_{k=1}^{m} k2^k = (m-1)2^{m+1} + 2$, and we therefore obtain

$$B(n) = (m-1)2^m + 1 + (n - 2^m)(m+1) = n(m+1) - 2^{m+1} + 1$$

and thus

$$(9.4) \qquad\qquad B(n) = n\lceil \lg n \rceil - 2^{\lceil \lg n \rceil} + 1 \,.$$

From Stirling's formula and (9.3) we conclude that $S(n) \ge \lg n! = \Omega(n \lg n)$, and from (9.4), we obtain $S(n) \le B(n) = O(n \lg n)$. Putting (9.4) and (9.5) together, we have solved our sorting problem asymptotically:

$$(9.5) \qquad\qquad S(n) = \Theta(n \lg n) \,.$$

The **merge sort** method (which we mentioned briefly in Section 5.3) is a recursive algorithm. First we divide the n elements as closely as possible into two even sublists, that is, into lists of size $\lfloor \frac{n}{2} \rfloor$ and $\lceil \frac{n}{2} \rceil$. We now sort the two sublists and merge the two sorted sublists together into a single list. Let $a_1 < \cdots < a_s$, $b_1 < \cdots < b_t$, with $s = \lfloor \frac{n}{2} \rfloor$, $t = \lceil \frac{n}{2} \rceil$, be the two sorted sublists. If we compare $a_1 : b_1$, we have determined the minimal element, which we put aside. We now compare the minimal elements of the new lists and thereby determine the second-smallest element. Continuing in this way, we obtain after $s + t - 1 = \lfloor \frac{n}{2} \rfloor + \lceil \frac{n}{2} \rceil - 1 = n - 1$ comparisons the complete sorted list. For the number $M(n)$ we thus obtain the recurrence

$$M(n) = M\left(\left\lfloor \frac{n}{2} \right\rfloor \right) + M\left(\left\lceil \frac{n}{2} \right\rceil \right) + (n - 1) \,.$$

One may easily convince oneself by induction that $M(n) = B(n)$ always holds, and so the merge-sort algorithm is also asymptotically optimal.

Let us look at the first few values of $S(n)$. As a lower bound we have $\lceil \lg n! \rceil$ from (9.3), and for an upper bound we have $B(n) = M(n)$ from (9.4):

n	2	3	4	5	6	7	8	9	10	11	12
$\lceil \lg n! \rceil$	1	2	5	7	10	13	16	19	22	26	29
$B(n)$	1	2	5	8	11	14	17	21	25	29	33

For $n \leq 11$ the lower bound is sharp, that is, $S(n) = \lceil \lg n! \rceil$, but for $n = 12$ a computer search yields the value $S(12) = 30$. Insertion sort, on the other hand, gives the sharp bound only for $n \leq 4$.

Let us analyze once again the merge-sort algorithm. It operates in three steps:

1. Decompose the list into two sublists (trivial).
2. Sort the sublists recursively.
3. Merge the sorted sublists together ($n - 1$ comparisons).

The following algorithm, quicksort, works in the opposite order:

Algorithm 9.8 (Quicksort).
1. Decompose the list into smaller and larger sublists R_1 and R_2, that is, $a_i < a_j$ for all $i \in R_1$, $j \in R_2$ ($n - 1$ comparisons).
2. Sort the sublists recursively.
3. Join the two sorted sublists together (trivial).

Let $a_1, \ldots, a_n$ be the given list. In step 1 we compare a_1 with all the remaining a_i and determine which a_i are less than or equal to a_1 ($i \in R_1$) and which a_j are greater than a_1 ($j \in R_2$). For this we require $n-1$ comparisons. If s is the correct location for a_1, we sort the smaller $s-1$ elements less than a_1 and the larger $n-s$ elements greater than a_1 recursively. Step 1 proceeds most easily with the help of two pointers i, j. At the start, $i = 1$, $j = n$. We now compare $a_i : a_j$. If $a_i > a_j$, the places of a_i, a_j are exchanged. Otherwise, j is decreased by 1 and we again compare $a_i : a_j$. After the first exchange, i is increased by 1 and again $a_i : a_j$ are compared. As soon as $a_i > a_j$, the locations of a_i and a_j are exchanged. We then decrease j by 1 and begin again the comparisons $a_i : a_j$. At the end of these $n - 1$ comparisons, a_1 is in the correct position, the elements to the left are less than a_1, and the elements to the right are greater than a_1.

The following example traces the execution of quicksort. The elements that are exchanged are underlined:

$$
\begin{array}{ccccccccc}
\underline{4} & 8 & 9 & 5 & 2 & 1 & 6 & 7 & \overset{\leftarrow}{\underline{3}} \\
3 & \overset{\rightarrow}{\underline{8}} & 9 & 5 & 2 & 1 & 6 & 7 & \underline{4} \\
3 & \underline{4} & 9 & 5 & 2 & \overset{\leftarrow}{\underline{1}} & 6 & 7 & 8 \\
3 & 1 & \overset{\rightarrow}{\underline{9}} & 5 & 2 & \underline{4} & 6 & 7 & 8 \\
3 & 1 & \underline{4} & 5 & \overset{\leftarrow}{\underline{2}} & 9 & 6 & 7 & 8 \\
3 & 1 & 2 & \overset{\rightarrow}{\underline{5}} & \underline{4} & 9 & 6 & 7 & 8 \\
3 & 1 & 2 & 4 & 5 & 9 & 6 & 7 & 8
\end{array}
$$

The number of comparisons is $1 + 1 + 3 + 1 + 1 + 1 = 8$. How good is quicksort? If the list $a_1, \ldots, a_n$ happens to be already in the correct order, $a_1 < \cdots < a_n$, then the first pass requires $n - 1$ comparisons with a_1 as the reference element, while the second pass requires $n - 2$ comparisons with a_2 as the reference element, and so on. Altogether, the algorithm executes all $(n - 1) + (n - 2) + \cdots + 2 + 1 = \binom{n}{2}$ comparisons, with the result that in the worst case, quicksort requires $O(n^2)$ comparisons, more than the optimal $\Theta(n \lg n)$. It is somewhat embarrassing that the worst case should occur precisely on the already presorted list. But now let us consider the average performance, where we assume that all $n!$ permutations are equally likely.

Let Q_n be the average number of comparisons. With probability $\frac{1}{n}$ we have that s is the correct location of a_1, $1 \leq s \leq n$. After the first pass, we have two sublists of lengths $s - 1$ and $n - s$. We thereby obtain the recurrence

$$
Q_n = n - 1 + \frac{1}{n} \sum_{s=1}^{n} (Q_{s-1} + Q_{n-s}), \quad Q_0 = 0,
$$

that is,

(9.6)
$$
Q_n = n - 1 + \frac{2}{n} \sum_{k=0}^{n-1} Q_k, \quad Q_0 = 0.
$$

With the method of summation factors from Section 2.1 we can solve the recurrence effortlessly. First we write

$$
nQ_n = n(n - 1) + 2 \sum_{k=0}^{n-1} Q_k \quad (n \geq 1),
$$

$$
(n - 1)Q_{n-1} = (n - 1)(n - 2) + 2 \sum_{k=0}^{n-2} Q_k \quad (n \geq 2).
$$

Subtraction yields

$$nQ_n - (n-1)Q_{n-1} = 2(n-1) + 2Q_{n-1} \quad (n \geq 2),$$

and thus

$$nQ_n = (n+1)Q_{n-1} + 2(n-1) \quad (n \geq 1),$$

since this also holds for $n = 1$ $(Q_1 = 0)$.

With the notation of (2.1) we have $a_n = n$, $b_n = n+1$, $c_n = 2(n-1)$. The summation factor s_n is therefore

$$s_n = \frac{(n-1)(n-2)\cdots 1}{(n+1)n\cdots 3} = \frac{2}{(n+1)n},$$

and we obtain

$$Q_n = \frac{n+1}{2} \sum_{k=1}^{n} \frac{4(k-1)}{k(k+1)} = 2(n+1) \sum_{k=0}^{n-1} \frac{k}{(k+1)(k+2)}.$$

How are we to determine the sum $\sum_{k=0}^{n-1} \frac{k}{(k+1)(k+2)}$? Clearly, we shall do so with our calculus of finite differences from Section 2.2. Partial summation yields

$$\sum_{k=0}^{n-1} \frac{k}{(k+1)(k+2)} = \sum_{0}^{n} x \cdot x^{\underline{-2}} = x \frac{x^{\underline{-1}}}{-1} \Big|_0^n + \sum_{0}^{n} (x+1)^{\underline{-1}}$$

$$= -\frac{n}{n+1} + H_{n+1} - 1 = H_n - \frac{2n}{n+1},$$

and we obtain

(9.7) $$Q_n = 2(n+1)H_n - 4n.$$

We now know that $H_n = \Theta(\log n)$, from which it follows that $Q_n = \Theta(n \lg n)$, and so quicksort is optimal on average.

In many problems we are interested not in the entire sorted list, but only a part. How many comparisons $W_1(n)$ do we need to determine, for example, only the largest of the elements $a_1, \ldots, a_n$? Since each of the n elements is a candidate, we have from Theorem 9.1 the lower bound $W_1(n) \geq \lceil \lg n \rceil$. In this case, the information-theoretic bound is of little help. Each of the elements a_i not equal to the maximum must yield the result $a_i < a_j$ in at least one comparsion $a_i : a_j$, since otherwise, a_i could not be excluded as a possible maximum. Therefore, we have $W_1(n) \geq n-1$. On the other hand, the sequence of comparisons $a_1 : a_2$, $\max(a_1, a_2) : a_3$, $\max(a_1, a_2, a_3) : a_4, \ldots$ certainly yields the maximum, and we obtain $W_1(n) = n-1$, in both the worst and average cases. The reader should think about how many comparisons one would need to determine both the maximum and the minimum. By what we have just proved, we need no more than $(n-1) + (n-2) = 2n - 3$ comparisons, but one can do better (see Exercise 9.30).

9.4. Binary Search Trees

In the previous sections we have studied decision trees as a tool for solving search problems. We are now going to look at **binary search trees**—rooted trees with at most two successors for every element—as *data structures* for ordered lists. We associate with each vertex v a *left* and *right* immediate successor v_L and v_R; if v_L or v_R does not exist, we leave the corresponding field empty. The *left subtree* of v is the subtree rooted at v_L, with the *right subtree* defined analogously.

Let T denote a binary tree with n vertices, and let $A = \{a_1, \ldots, a_n\}$ be a list of n distinct elements linearly ordered by $<$. We store A in the vertices V of T using the ordering $\kappa : V \to A$, and we say that T is a **binary search tree** for A if the following property is satisfied: Let $v \in V$. Then $\kappa(x) < \kappa(v)$ for all x in the left subtree of v and $\kappa(x) > \kappa(v)$ for all x in the right subtree of v.

The tree in the following diagram is a binary search tree for the set $\{2, 4, 5, 7, 8, 10, 11, 12, 13, 15, 16, 18, 20\}$:

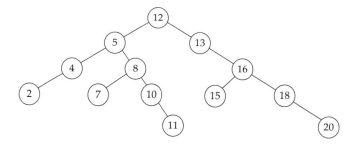

Why don't we simply store the elements in an ordered list? The advantage of a binary search tree as a data structure is that a number of elementary operations such as searching for an element, insertion, and deletion can be executed very quickly. We shall see that all these operations have a running time of $O(L)$ where L is the length of the tree.

Let the binary search tree T for A be given. For an element $a \in A$ we would like to determine the vertex x with $\kappa(x) = a$. It is clear what we have to do. We compare a with $\kappa(v)$, where v is the root. If $a = \kappa(v)$, we are done, and $x = v$. If $a < \kappa(v)$, we compare a with $\kappa(v_\mathrm{L})$, and if $a > \kappa(v)$, we compare a with $\kappa(v_\mathrm{R})$. In this manner we traverse a unique path from the root to the desired vertex x for which $\kappa(x) = a$, and the number of comparisons is $O(L)$.

We can just as easily determine the minimum or maximum. For the minimum, we always traverse the left branch, and for the maximum, the right branch. Again the number of operations is $O(L)$.

Suppose now that we would like to insert a new element $b \notin A$. That is, we must enlarge the tree T by a vertex z with $\kappa(z) = b$, in such a way that the new tree T' is a search tree for $A \cup \{b\}$. This is easily done. We compare b with $\kappa(v)$, where v is the root. If $b < \kappa(v)$, we move left, and otherwise right. If v_L is empty, we set $z = v_\mathrm{L}$, $\kappa(z) = b$; otherwise, we compare b with $\kappa(v_\mathrm{L})$. In the case $b > \kappa(v)$, we set $z = v_\mathrm{R}$ if v_R is empty, and otherwise, we compare b with $\kappa(v_\mathrm{R})$. In this way we reach a unique vertex y such that $b < \kappa(y)$ and y_L is empty ($z = y_\mathrm{L}$, $\kappa(z) = b$), or $b > \kappa(y)$ and y_R is empty ($z = y_\mathrm{R}$, $\kappa(z) = b$). In each case, the number of operations is again $O(L)$.

Suppose we wish to insert the number 9 in our example. We successively obtain $9 < 12$, $9 > 5$, $9 > 8$, $9 < 10$, and 10_L empty, and so we enlarge the tree thus:

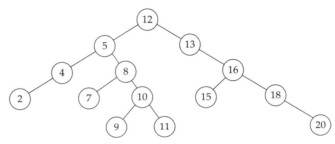

If T is a binary search tree, we can output the list stored in T in sorted form using the following recursive method, called **in-order**. The value $\kappa(x)$ of the root x of a subtree is output between the values of the left subtree (rooted at x_L) and those of the right subtree (rooted at x_R). This explains the name "in-order." The running time of this output procedure is clearly $\Theta(n)$.

How does one create a search tree for a list $\{a_1, \ldots, a_n\}$? A simple method is to insert the elements one at a time using the method just described. As an example, let us take $A = \{8, 2, 4, 9, 1, 7, 11, 5, 10, 12, 3, 6\}$. The tree is now built one element at a time:

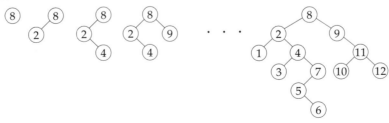

Each permutation $\pi = a_1 a_2 \ldots a_n$ determines by this method a unique binary search tree $T = T(\pi)$. We have seen that the most important parameter for the elementary operations that we have described is the length of the tree $L(T)$. Now we would like to determine two quantities. Let

$\ell_1(\pi), \ldots, \ell_n(\pi)$ be the lengths of the numbers $1, 2, \ldots, n$ in the tree $T(\pi)$. Then $\overline{L}(\pi) = \frac{1}{n} \sum_{i=1}^{n} \ell_i(\pi)$ is the *average* length, and $L(\pi) = \max_{1 \le i \le n} \ell_i(\pi)$ is the *maximum* length of the vertices i, that is, the length $L(T(\pi))$ of the tree. What are the expectations $E(\overline{L}(n))$ and $E(L(n))$ on the assumption that all $n!$ permutations are equally likely?

Let us take $n = 3$ as an example. The permutations π and their associated trees $T(\pi)$ look as follows:

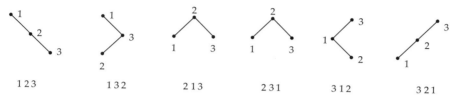

For $\overline{L}(\pi)$ and $L(\pi)$ we obtain the following associated values:

$$\overline{L}(\pi): \quad 1 \quad 1 \quad \tfrac{2}{3} \quad \tfrac{2}{3} \quad 1 \quad 1,$$

$$L(\pi): \quad 2 \quad 2 \quad 1 \quad 1 \quad 2 \quad 2.$$

Therefore we have $E(\overline{L}(3)) = \frac{1}{6}(1 + 1 + \frac{2}{3} + \frac{2}{3} + 1 + 1) = \frac{8}{9}$, $E(L(3)) = \frac{5}{3}$.

Let us turn our attention first to $E(\overline{L}(n))$. For a permutation π we have $\overline{L}(\pi) = \frac{1}{n} \sum_{i=1}^{n} \ell_i(\pi)$. If we define the random variable $X_n : \pi \to \sum_{i=1}^{n} \ell_i(\pi)$, then we have $E(\overline{L}(n)) = \frac{1}{n} E(X_n)$. We would now like to determine $f(n) = E(X_n)$. As starting values we have $f(0) = 0$, $f(1) = 0$, $f(2) = 1$, and from our example, $f(3) = \frac{8}{3}$. If we classify X_n according to the value i of the root, we obtain at once the recurrence

$$f(n) = \frac{1}{n} \sum_{i=1}^{n} [f(i-1) + (i-1) + f(n-i) + (n-i)],$$

since the left subtree has the expected length sum $f(i-1)$ plus 1 for each value less than i, and analogously for the right subtree. A rearrangement yields

$$(9.8) \qquad f(n) = \frac{2}{n} \sum_{i=0}^{n-1} f(i) + n - 1, \quad f(0) = 0,$$

and we have solved this recurrence in the previous section in relation to quicksort, with the result $f(n) = 2(n+1)H_n - 4n$. For the expected value $E(\overline{L}(n))$ we thus obtain

$$(9.9) \qquad E(\overline{L}(n)) = \frac{1}{n} f(n) = \left(2 + \frac{2}{n}\right) H_n - 4,$$

or

$$E(\overline{L}(n)) = \Theta(\log n).$$

If we check the result for $n = 3$, we again obtain

$$E(\overline{L}(3)) = \left(2 + \frac{2}{3}\right) H_3 - 4 = \frac{8}{3} \cdot \frac{11}{6} - 4 = \frac{8}{9}.$$

For the length L of a random tree, the analysis is somewhat more complicated. We define the random variable $Y_n : \pi \to \max \ell_i(\pi)$ over all *leaves i* and must calculate $E(Y_n)$. We are unable to derive a recurrence analogous to (9.8). Of course, we have $E(Y_n) \geq E(\overline{L}(n)) = \Theta(\log n)$. We shall therefore attempt to estimate $E(Y_n)$ from above by a more easily determined quantity.

We would like first to prove in general Jensen's inequality

$$(9.10) \qquad\qquad E(Y) \leq \lg E\left(2^Y\right)$$

for a real random variable Y, where we assume a uniform distribution on the probability space Ω. Let $\Omega = \{\omega_1, \ldots, \omega_m\}$, $Y(\omega_i) = y_i$. Then (9.10) means that

$$(9.11) \qquad\qquad \frac{1}{m} \sum_{i=1}^{m} y_i \leq \lg \frac{1}{m} \sum_{i=1}^{m} 2^{y_i}.$$

If we set $x_i = 2^{y_i}$, then (9.11) is equivalent to

$$\frac{1}{m} \sum_{i=1}^{m} \lg x_i \leq \lg \frac{1}{m} \sum_{i=1}^{m} x_i,$$

or

$$(9.12) \qquad\qquad \lg \left(\prod_{i=1}^{m} x_i\right)^{1/m} \leq \lg \frac{1}{m} \sum_{i=1}^{m} x_i.$$

Taking 2 to the power of each side of (9.12) results in the familiar arithmetic–geometric mean inequality. Thus (9.12) is correct, and therefore (9.10) as well. We now consider the random variable $Z_n : \pi \to \sum 2^{\ell_i(\pi)}$ over all leaves i. We have $E(2^{Y_n}) \leq E(Z_n)$ on account of $2^{\max \ell_i(\pi)} \leq \sum 2^{\ell_i(\pi)}$, and therefore from (9.10) we obtain

$$(9.13) \qquad\qquad E(Y_n) \leq \lg E(2^{Y_n}) \leq \lg E(Z_n).$$

We can now analyze the function $g(n) = E(Z_n)$ in detail. As initial values we have $g(0) = 0$, $g(1) = 1$, $g(2) = 2$. If we again classify according to the value i of the root, we obtain

$$g(n) = \frac{1}{n} \sum_{i=1}^{n} 2(g(i-1) + g(n-i)) \quad (n \geq 2),$$

since the expressions $\sum 2^{\ell_j}$ in the subtrees must be multiplied by 2 (the lengths are greater by 1). This yields

$$ng(n) = 4\sum_{i=0}^{n-1} g(i) \quad (n \geq 2),$$

$$(n-1)g(n-1) = 4\sum_{i=0}^{n-2} g(i) \quad (n \geq 3),$$

and by subtraction,

(9.14) $$ng(n) = (n+3)g(n-1) \quad (n \geq 3),$$

whence

$$g(n) = \frac{(n+3)(n+2)\cdots 6}{n(n-1)\cdots 3}g(2) = \frac{(n+3)(n+2)(n+1)}{30} \leq n^3 \quad (n \geq 2).$$

From this we obtain, with the help of (9.13),

$$E(L(n)) = E(Y_n) \leq \lg g(n) \leq \lg n^3 = 3\lg n.$$

Using $\lg n = \lg e \cdot \log n$, we finally obtain

(9.15) $$E(L(n)) \leq 3\lg e \cdot \log n \approx 4.34 \log n.$$

It is therefore not only the average length of a vertex, but the average *maximum* length of order of magnitude $\Theta(\log n)$.

To conclude, we ask how many binary search trees there are on n vertices. Let us call this number C_n. For $n = 1, 2, 3$ we have the following trees:

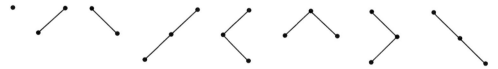

Therefore, $C_1 = 1, C_2 = 2, C_3 = 5$, and we set $C_0 = 1$. In considering the left and right subtrees of the root, we obtain at once the following recurrence:

$$C_n = C_0 C_{n-1} + C_1 C_{n-2} + \cdots + C_{n-1} C_0 \quad (n > 0),$$

or

(9.16) $$C_n = \sum_{k=0}^{n-1} C_k C_{n-1-k} + [n=0] \quad \text{for all } n.$$

This, of course, looks like a convolution, and so we employ the methods of Section 3.2. Let $C(z) = \sum_{n \geq 0} C_n z^n$ be the generating function. For $C(z)$, the convolution (9.16) is

$$C(z) = C(z) \cdot zC(z) + 1,$$

that is,

$$(C(z))^2 - \frac{C(z)}{z} + \frac{1}{z} = 0 \,.$$

Solving this quadratic equation yields

$$C(z) = \frac{1}{2z} \pm \frac{\sqrt{1-4z}}{2z} \,.$$

Since taking the plus sign would result in $C(0) = C_0 = \infty$, we must take the minus sign, and thus we have

$$C(z) = \frac{1 - \sqrt{1-4z}}{2z} \,.$$

We now have

$$\sqrt{1-4z} = \sum_{n \geq 0} \binom{1/2}{n} (-4z)^n = 1 + \sum_{n \geq 1} \frac{1}{2n} \binom{-1/2}{n-1} (-4z)^n \,.$$

We calculated the expression $\binom{-1/2}{n}$ in Exercise 1.40: $\binom{-1/2}{n} = (-\frac{1}{4})^n \binom{2n}{n}$. Therefore, we have

$$\sqrt{1-4z} = 1 + \sum_{n \geq 1} \frac{1}{2n} \binom{2n-2}{n-1} (-4)z^n = 1 - 2 \sum_{n \geq 1} \frac{1}{n} \binom{2n-2}{n-1} z^n$$

$$= 1 - 2 \sum_{n \geq 0} \frac{1}{n+1} \binom{2n}{n} z^{n+1} \,,$$

and hence

$$C_n = \frac{1}{n+1} \binom{2n}{n} \,.$$

The number C_n is called the nth **Catalan number**. It appears in counting problems almost as frequently as the binomial numbers and the Fibonacci numbers. A number of interesting examples are to be found in the exercises.

Exercises for Chapter 9

9.1 Solve the weighing problem on the assumption that one knows that the counterfeit coin is heavier than the true coins.

▷ **9.2** Consider the search problem with n coins, where one coin is lighter or heavier. Show that an optimal algorithm has length $L = \lceil \log_3(2n+2) \rceil$.

9.3 Show that for n, q ($q \geq 2$), there is exactly one complete (n, q)-tree if and only if $q - 1$ is a divisor of $n - 1$.

▷ **9.4** Let T be a complete binary tree with n leaves. Let $e(T)$ denote the sum of the lengths of the leaves, and $i(T)$ the sum of the lengths of the inner vertices. Show that $e(T) = i(T) + 2(n-1)$.

9.5 Let the set $S = \{1, \ldots, n\}$ be given, and let $x^* \in S$ be an unknown element. As tests one has available only $x^* < i$? for $i = 2, \ldots, n$ with yes/no answers. Show that $L = \lceil \lg n \rceil$ is the length of an optimal search algorithm.

9.6 Given the distribution $(30, 20, 15, 14, 11, 10)$ for the leaves $1, 2, \ldots, 6$, show that the following search trees are optimal. Only one of them is a Huffman tree. Which is it?

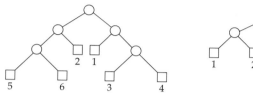 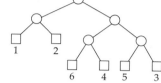

9.7 Prove that the binary insertion sort and the merge sort algorithms of Section 9.3 have the same running time $B(n) = M(n)$.

▷ **9.8** We are to determine the maximum value in a list of n elements, where in each round we may make $\lfloor \frac{n}{2} \rfloor$ disjoint comparisons in parallel (comparisons $a_i : a_j$ and $a_k : a_l$ are disjoint if $k, l \notin \{i, j\}$). Show that an optimal algorithm requires $\lceil \lg n \rceil$ rounds.

9.9 We would like to sort an (unordered) list of n elements in one round, where all comparisons are allowed (including nondisjoint ones). Show that we must execute all $\binom{n}{2}$ comparisons.

9.10 Determine optimal sorting algorithms for $n = 6, 7, 8$ elements. Hint: The search lengths are 10, 13, 16.

▷ **9.11** We are given the set $S = \{1, \ldots, n\}$ and a natural number $d \geq 0$. We are to determine an unknown number $x^* \in S$. The permitted tests are $x^* = i$? with the answers $x^* = i, |x^* - i| \leq d, |x^* - i| > d$. Let $L_d(n)$ be the optimal search length. (a) Prove that $L_0(n) = n - 1$, (b) Prove that $L_1(n) = \lceil \frac{n}{3} \rceil$ for $n \geq 3$ with $L_1(1) = 0$, $L_1(2) = 1$, (c) What is $L_2(n)$?

9.12 The next exercises deal with the Catalan numbers C_n from Section 9.4. Let $x_0, x_1, \ldots, x_n$ be variables that we wish to parenthesize in all possible ways with the condition that the order of x_0 to x_n remain intact. Example: $n = 1$, $(x_0 x_1)$; $n = 2$, $\big(x_0(x_1 x_2)\big)$, $\big((x_0 x_1) x_2\big)$; $n = 3$, $\big(x_0(x_1(x_2 x_3))\big)$, $\big((x_0 x_1)(x_2 x_3)\big)$, $\big(((x_0 x_1) x_2) x_3\big)$, $\big(x_0((x_1 x_2) x_3)\big)$, $\big((x_0(x_1 x_2)) x_3\big)$. Show that the number of possible parenthesizations is equal to C_n: (a) by setting up a recurrence; (b) by a bijection on the set of search trees.

▷ **9.13** There are $2n$ people standing in a ticket line somewhere in Europe. The cost of a ticket is 10 euros. Precisely n of the people have a € 10 note, while the other n each have only a € 20 note. At the start of this problem the cashier has no money in the till and can therefore make change only if at every transaction, the number of € 10 people who have bought tickets is at least as great as the number of € 20 people who have bought tickets. Show that the number of lines for which the cashier can always make change is the Catalan number C_n. Hint: Altogether, we have $\binom{2n}{n}$ lines. Now determine the number of lines for which making change fails.

9.14 We decompose a regular n-gon, $n \geq 3$, into triangles by drawing diagonals. Example: For $n = 4$,

Show that the number of possible triangulations of the n-gon is equal to C_{n-2}.

▷ **9.15** Suppose the numbers from 1 to 100 have been stored in a binary search tree. Which of the following sequences cannot be a search sequence for the element 37?

(a) $2, 7, 87, 83, 30, 31, 81, 37$;

(b) $75, 11, 67, 25, 55, 33, 34, 37$;

(c) $10, 80, 48, 60, 22, 70, 43, 30, 37$.

9.16 We can sort n numbers by first creating a binary search tree using the insertion algorithm of Section 9.4 and then outputting the numbers "in-order." What is the running time for the best and worst cases?

▷ **9.17** We are given a graph $G = (V, E)$, $|V| = n$, $|E| = m$. We are searching for an unknown edge k^*, and every test consists of a question, Is $u \in k^*$, $u \in V$, with possible answers yes and no. Let L be the minimum length of a search process. Show that (a) $|E| \leq \binom{L+1}{2} + 1$; (b) $|V| \leq \binom{L+2}{2} + 1$. (c) From this, derive lower bounds for L in terms of n and m. Hint: Induction on L.

9.18 Show that the bounds in the previous exercise can be satisfied with equality.

9.19 We are given a graph G. We are to determine whether the unknown edge k^* belongs to G or to the complementary graph $\overline{G}$ using the same test as in Exercise 9.17. Show that for the minimum search length L, one has $\min\left(n - \alpha(G), n - \omega(G)\right) \leq L \leq \min\left(n - \frac{\alpha(G)}{2}, n - \frac{\omega(G)}{2}\right)$, where $\alpha(G)$ is the independence number of G and $\omega(G)$ is the clique number (that is, $\omega(G) = \alpha(\overline{G})$). Hint: Consider a test sequence $v_1, \ldots, v_k$, where the answer is always no.

▷ **9.20** Suppose we have a set S with n elements containing an unknown element x^*. The admissible tests are $x^* \in A$ for $|A| \leq k$, where the number $k \leq n$ has previously been specified, with the possible answers yes and no. Let $L_{\leq k}(n)$ be the length of an optimal search algorithm. Show that (a) $L_{\leq k}(n) = \lceil \lg n \rceil$ for $k \geq \frac{n}{2}$; (b) $L_{\leq}(n) = t + \lceil \lg(n - tk) \rceil$, $t = \lceil \frac{n}{k} \rceil - 2$ for $k < \frac{n}{2}$. Hint: The function $L_{\leq k}(n)$ is monotonically increasing in n for fixed k. Consider the situation after the first test A_1, $|A_1| \leq k$.

9.21 For some reason, $m \cdot n$ people have been arranged in an $m \times n$ rectangle. We are to locate an unknown person x^* by questions of the form, "Is x^* in the ith row?" and "Is x^* in the jth column?" How many questions do we need?

9.22 We are given a set S of n people, each of whom is ill with the same probability $p > 0$. Our goal is to determine who are the sick persons $X^* \subseteq S$, where each test set $A \subseteq S$ determines the information $A \cap X^* \neq \varnothing$ or $A \cap X^* = \varnothing$. What is the probability distribution of the sets X^*? Determine the optimal search length L and give a lower bound for $\overline{L}$ with the help of Theorem 9.6. Hint: $\overline{L} \geq nH(p, 1 - p)$.

▷ **9.23** Let us take $n = 2$ in the previous exercise. An elementwise search yields $\underline{L} \le L = 2$. Show that $\overline{L} = 2$ holds precisely for $p \ge \frac{3-\sqrt{5}}{2}$.

9.24 Let the entropy of the distribution $(p_1, \ldots, p_n)$ be given by $H(p_1, \ldots, p_n) = -\sum_{i=1}^{n} p_i \lg p_i$. Show that

(a) $H(p_1, \ldots, p_n) \le H(\frac{1}{n}, \ldots, \frac{1}{n}) = \lg n$ for all $(p_1, \ldots, p_n)$, with equality only for the uniform distribution.

(b) $H(p_1, \ldots, p_n) = H(p_1, \ldots, p_k, s) + sH(p'_{k+1}, \ldots, p'_n)$ with $2 \le k \le n - 1$, $s = \sum_{i=k+1}^{n} p_i$, $p'_i = \frac{p_i}{s}$ $(i = k + 1, \ldots, n)$.

▷ **9.25** Let $h(n) = \min_T (T; \frac{1}{n}, \ldots, \frac{1}{n})$, that is, $h(n)$ is the optimal average length in the case of a uniform distribution (determined by the Huffman algorithm), $q = 2$. Show that $h(n)$ is an increasing function of n; more precisely, show that $h(n) + \frac{1}{n} \le h(n + 1)$. When does equality hold? Hint: Consider an optimal $(n + 1, 2)$-tree and remove a leaf.

9.26 Let $(p_1, \ldots, p_n)$ be a distribution, $q \ge 2$. Show that

$$\overline{L}(p_1, \ldots, p_n) \le \overline{L}\left(\frac{1}{n}, \ldots, \frac{1}{n}\right).$$

Hint: Set $p_1 \ge \cdots \ge p_n$ and show first that $\sum_{i=1}^{k} p_i \ge \frac{k}{n}$ for all k.

9.27 Let $\overline{L}(p_1, \ldots, p_n)$ be the optimum for the distribution $(p_1, \ldots, p_n)$, $q \ge 2$. Show that $\overline{L}(p_1, \ldots, p_n) = -\sum_{i=1}^{n} p_i \log_q p_i$ holds if and only if all p_i are of the form $p_i = q^{-\ell_i}$ for some $\ell_i \in \mathbb{N}_0$.

9.28 Consider the distribution $(p_1, \ldots, p_n)$ with $p_i = i/\binom{n+1}{2}$, $i = 1, \ldots, n$. Calculate $\overline{L}(p_1, \ldots, p_n)$ for $q = 2$.

▷ **9.29** We are given a list $y_1 < y_2 < \cdots < y_n$, and we wish to insert an unknown element x^*. That is, we must determine where x^* belongs. As tests we have $x^* = y_i$ with possible answers $<, =, >$. Determine the optimal search length. Hint: Represent a search algorithm as a binary tree (even though you are presented with three possible answers).

▷ **9.30** Show that an optimal algorithm for determining the maximum and minimum of a list of n elements requires $\lceil \frac{3n}{2} \rceil - 2$ comparisons. Hint: For the lower limit, construct an oppositional strategy that at each comparison reduces as little as possible the number of possible candidates for the maximum and minimum.

9.31 Determine the optimal average search length for finding the maximum of a list, assuming equal probabilities.

▷ **9.32** Let $M(n, n)$ be the minimal number of comparisons needed to merge an n-list $\{x_1 < \cdots < x_n\}$ with an n-list $\{y_1 < \cdots < y_n\}$. Determine $M(n, n)$.

9.33 Suppose we may make arbitrary comparisons in one round, but at most n of them. At the beginning, we have an unordered list. Show that after the first round, the number of possible candidates for the maximum is at least $\lceil \frac{n}{3} \rceil$, and that it is possible to limit this number to be less than or equal to $\lceil \frac{4}{11}n \rceil$. Consider the example $n = 16$ and verify that both bounds are sharp.

9.34 The following sorting algorithm is called "bubble sort." Let the input be $a_1, \ldots, a_n$. We compare $a_1 : a_2$, and if $a_1 < a_2$, the elements remain unchanged. Otherwise, we exchange the places of the two elements. Now $a_2 : a_3$ are compared, and so on up to $a_{n-1} : a_n$. In this way, the largest element moves to the end of the list. We now begin the next round, again with the first two elements, and so on, and then the next round, until finally the list is sorted. In the example we write the list from top to bottom, so that the larger elements "bubble up" to the top:

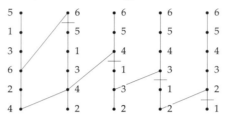

Execute bubble sort on the list $6, 10, 9, 2, 8, 4, 11, 3, 1, 7, 5$ and show that in general, all elements above the last element to be exchanged in a round are in the correct order (indicated with the dotted lines in the example).

$\triangleright$ **9.35** To analyze bubble sort, we investigate the quantities X, the number of exchanges of elements; R, the number of rounds and C, the number of comparisons. In our example in the previous exercise, we have $X = 8$, $C = 12$, $R = 5$. Let $a_1, \ldots, a_n$ be a permutation of $\{1, \ldots, n\}$, and $a'_1, \ldots, a'_n$ the permutation after one round. Let $b_1, \ldots, b_n$ and $b'_1, \ldots, b'_n$ denote the associated inversion tables (see Exercise 1.32). Show that $b'_1, \ldots, b'_n$ results from $b_1, \ldots, b_n$ by reducing each $b_j \neq 0$ by 1. Conclude that if $b_1, \ldots, b_n$ is the inversion table of the starting permutation, then $X = \sum_{i=1}^{n} b_i$, $R = 1 + \max(b_1, \ldots, b_n)$, $V = \sum_{i=1}^{R} c_i$ with $c_i = \max(b_j + j : b_j \geq i - 1) - i$.

$\triangleright$ **9.36** Suppose that all permutations of $\{1, \ldots, n\}$ are equally likely as input into bubble sort. Show that the probability that $R \leq k$ is equal to $\frac{1}{n!} k^{n-k} k!$ and conclude that one has the expectation $E(R) = n + 1 - \sum_{k=0}^{n} \frac{k^{n-k} k!}{n!}$. We have already calculated the expectation $E(X) = \frac{n(n-1)}{4}$ in Exercise 3.28. (Note: X and R are defined in the previous exercise.)

9.37 Develop a method to delete an element from a binary search tree.

9.38 Let $\ell_1(\pi), \ldots, \ell_n(\pi)$ be the lengths of elements $1, \ldots, n$ in the binary search tree induced by the permutation π. Show that $E(\ell_1(\pi)) = H_n - 1$ on the assumption that all permutations are equally likely.

$\triangleright$ **9.39** Under the same assumptions, calculate $E(\ell_k(\pi))$. Hint: Use a method similar to that of Section 9.4.

9.40 Describe the binary search trees in which the average length of the vertices is $\Theta(\lg n)$ but the maximal length is $\succ \lg n$.

$\triangleright$ **9.41** Show that Jensen's inequality (9.10) holds for arbitrary $c > 1$, that is, $E(Y) \leq \log_c E(c^Y)$. From this, derive a better upper bound for $E(L(n))$ (see (9.15)).

9.42 Show that every rooted binary tree with n leaves has a subtree with k leaves, $\frac{n}{3} \leq k \leq \frac{2n}{3}$.

9.43 Let T be a rooted tree. We travel at random from the root v_1 to a leaf $u_1, u_2, u_3, \ldots$, that is, from each internal vertex we go with equal probability to one of the x_i successors. Let S be the random variable $X = x_1 + x_1 x_2 + x_1 x_2 x_3 + \cdots$. Show that the expectation EX is equal to one less than the number of vertices of T.

Chapter 10

General Optimization Methods

In the previous chapters of Part 2 we have encountered a number of algorithms for solving important problems such as the job-assignment problem and the traveling salesman problem while investigating the underlying questions that arise in the design and analysis of algorithms: How do we describe an algorithm? What data structures should we use? How fast is the algorithm? Does there exist an efficient algorithm for the problem?

In this chapter we are going to look at some general methods for solving optimization problems and discuss which procedures are best adapted to which problems.

10.1. Backtracking

Suppose we enter a labyrinth at point A and must find our way to point E. We know neither the paths inside the labyrinth nor the location of the exit E. How should we proceed? Since we have no information, we must try all of the paths, which we shall do using the following set of rules:

(1) Walk, if possible, from the current position to a neighboring position that has not yet been visited.

(2) If no such position exists, go one position backward along the path that led to the current position. Go to step (1).

This **backtracking** is the essence of the method. It allows us to find our way out of every dead end, and furthermore, we will never traverse a false path twice, and therefore eventually, we will find our way to the exit E.

In the following example, the possible continuations of the path are shown as arrows:

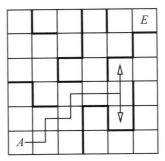

Backtracking is not a fast method for finding the optimal solution, but it organizes the exhaustive search in such a way that no loops arise in the search process. We extend a partial solution as long as we can, and when we can progress no further, we backtrack and try a new path. An example of this method that we have seen is depth-first search, from Section 7.2.

In general, we may model backtracking with a tree diagram. We are given sets $A_1, A_2, \ldots$, and we are seeking a sequence $(a_1, a_2, a_3, \ldots)$ with $a_i \in A_i$ satisfying certain conditions. We begin with the empty word $(\ )$ as root. Let S_1 be the set of candidates from A_1, from which we select the first element a_1. At this point, (a_1) is our partial solution. If we have already obtained $(a_1, \ldots, a_i)$, we select from the set of candidates $S_{i+1} \subseteq A_{i+1}$ (those that satisfy the conditions) the first element a_{i+1} and extend the partial solution to $(a_1, a_2, \ldots, a_i, a_{i+1})$. If S_{i+1} is empty, we backtrack one step and test the next candidate from S_i, and so on.

Consider the well-known queen-placement problem on an $n \times n$ chessboard. We seek to place as many queens as possible on the board in such a way that no queen can attack any of the others. Since there can be at most one queen in each row and in each column, we can place at most n queens. Can we always place the maximum n queens? For $n = 2, 3$ it is clear at once that only one and two queens can be placed.

In general, a solution $a_1, a_2, \ldots, a_n$ is a permutation of $\{1, \ldots, n\}$, where a_i specifies the position of the queen in row i. The diagonal condition says that we must have $|a_i - a_j| \neq |i - j|$ for $i \neq j$. Let us try $n = 4$. We see first by symmetry that we may assume $a_1 \leq 2$. Our backtracking tree therefore has the following form:

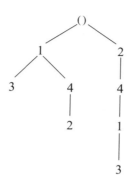

With the symmetry $a_1 = 2 \longleftrightarrow a_1 = 3$ we have found precisely two placements:

Even for the familiar 8×8 chessboard one can still solve the queen problem "by hand." Altogether, there are 92 possible placements, including the following:

For large n, backtracking will fail on account of the large number of branchings in the search tree. But there is always a solution for $n \geq 4$, and with a bit of intuition one can easily find one.

To make backtracking more efficient, we shall try to identify entire subtrees as superfluous. In our queen problem we have seen that we may use symmetry to assume $a_1 \leq \lceil \frac{n}{2} \rceil$: The subtrees with $a_1 > \lceil \frac{n}{2} \rceil$ do not have to be traversed.

A particularly interesting variant of this idea is the **branch and bound** procedure. Suppose we have a minimization problem to solve. With each vertex $(a_1, \ldots, a_k)$ of the backtracking tree we associate a lower bound $c(a_1, \ldots, a_k)$ for the desired minimum c_{opt}. We now *branch*, compare the new *bounds*, and proceed with the vertex that yields the smallest lower bound. If in this way we have obtained a leaf with the value c, we may then ignore all subtrees with $c(a_1, \ldots, a_\ell) > c$.

The traveling salesman problem (TSP) will help to clarify the idea of branch and bound. Consider the cost matrix

	1	2	3	4
1	∞	2	8	6
2	2	∞	6	4
3	4	5	∞	5
4	8	7	3	∞

The entry $c_{ij} \geq 0$ gives the costs of a trip from i to j. It is clear that we are dealing with an asymmetric TSP. In general, we have $c_{ij} \neq c_{ji}$, and we write ∞ along the main diagonal, since $i \to i$ is impossible. Since an edge must go out from each vertex, we may subtract the smallest value from each row and then from each column. This lowers the cost of the tour but does not alter the form of the optimal tour. The reader will recall that we used this idea in the weighted matching problem in Section 8.2.

The new matrices are now

	1	2	3	4
1	∞	0	6	4
2	0	∞	4	2
3	0	1	∞	1
4	5	4	0	∞

$\longrightarrow$

	1	2	3	4
1	∞	0	6	3
2	0	∞	4	1
3	0	1	∞	0
4	5	4	0	∞

all tours
bound ≥ 12

The sum of the reduced numbers is 12, and so 12 is a lower bound for the root, which contains all tours. We now search for a 0 entry, e.g., $(1, 2)$, and branch. The left subtree contains all tours with $1 \to 2$, the right subtree all tours with $1 \nrightarrow 2$. Since in the tours with $1 \to 2$ there are no edges coming from vertex 1 and no edge leads to vertex 2, we delete the first row and second column and set the $(2, 1)$ entry to ∞. In this way we obtain a new cost matrix in which we may again reduce the smallest entries. This yields

	1	3	4
2	∞	4	1
3	0	∞	0
4	5	0	∞

$\longrightarrow$

	1	3	4
2	∞	3	0
3	0	∞	0
4	5	0	∞

tours with $1 \to 2$
bound ≥ 13

In the tours with $1 \nrightarrow 2$ we set the $(1,2)$ entry to ∞ and obtain

	1	2	3	4
1	∞	∞	6	3
2	0	∞	4	1
3	0	1	∞	0
4	5	4	0	∞

$\longrightarrow$

	1	2	3	4
1	∞	∞	3	0
2	0	∞	4	1
3	0	0	∞	0
4	5	3	0	∞

tours with $1 \nrightarrow 2$
bound ≥ 16

We therefore continue with the subtree $1 \to 2$. We next try $(3,4)$, and thus for $3 \to 4$ we have the bound 21 and for $3 \nrightarrow 4$ the bound 13. The choice $(2,4)$ finally yields the following backtracking tree:

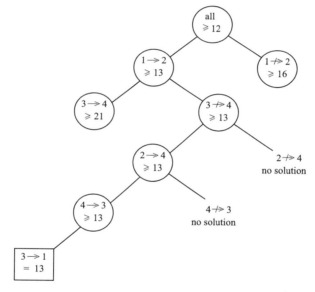

The tour $1 \longrightarrow 2 \longrightarrow 4 \longrightarrow 3 \longrightarrow 1$ with cost 13 is thus optimal.

10.2. Dynamic Programming

Dynamic programming, like the divide and conquer method, is used in optimization problems by reducing the given problem to a number of small subproblems and then computing the solution recursively. Before we discuss the general steps in dynamic programming, let us look at an example.

We are given a convex n-gon in the plane. That is, the closed boundary consists of n line segments, and the line joining any two edge points is entirely contained in the n-gon. In the figure, the left-hand 5-gon is convex, but the right-hand one is not:

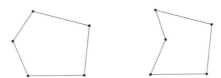

We are interested in triangulations of n-gons. That is, we decompose the n-gon into $n-2$ triangles by inserting $n-3$ diagonals. The following figure shows a triangulation of an 8-gon:

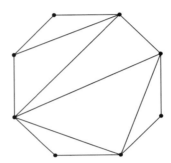

We are not interested in finding just any old triangulation, but an optimal one in relation to some weight function. Let $v_1, \ldots, v_n$ be the vertices. We define a weight function

$$w(\Delta(v_i v_j v_k)) = w_{ijk} \in \mathbb{R}$$

on the $\binom{n}{3}$ triangles. One natural candidate is $w = $ area, and another is $w = |v_i - v_j| + |v_i - v_k| + |v_j - v_k|$, the sum of the side lengths. We don't wish to specify the function w further. Any real-valued function is allowed.

If T is a triangulation, let $w(T) = \sum w(\Delta)$, extended over all $n-2$ triangles Δ of the triangulation, and our optimization task consists in determining a triangulation of minimal weight.

We might first attempt (for example with backtracking) to list all triangulations and then see which is the optimal one. But before we do that, we should determine (or at least estimate) the number of triangulations. Let $R_n (n \geq 3)$ be this number, where clearly the particular shape of the convex n-gon plays no role.

The starting value is $R_3 = 1$, and for $n = 4$ we obtain $R_4 = 2$. Consider now an $(n+2)$-gon. We number the vertices in clockwise order $v_1, v_2, \ldots, v_{n+2}$. The edge $v_1 v_{n+2}$ appears in precisely one triangle, say in $v_1 v_k v_{n+2}$, $2 \le k \le n+1$:

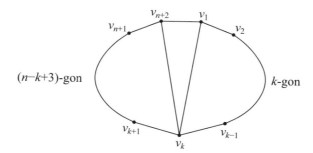

How many triangulations are there in which the triangle $v_1 v_k v_{n+2}$ appears? The convex k-gon $v_1 v_2 \ldots v_k$ can be triangulated in every possible way, and so can the convex $(n - k + 3)$-gon $v_k v_{k+1} \ldots v_{n+2}$. We thereby obtain the recurrence

$$R_{n+2} = \sum_{k=2}^{n+1} R_k R_{n-k+3} \qquad (n \ge 1) \, ,$$

where we set $R_2 = 1$ in order to include the case $v_{n+2} v_1 v_2$. An index transformation yields

$$R_{n+2} = \sum_{k=0}^{n-1} R_{k+2} R_{(n-1-k)+2} \qquad (n \ge 1) \, ,$$

and we recognize this recurrence from Section 9.4; it defines the Catalan numbers. Thus we have the result

$$R_{n+2} = \frac{1}{n+1} \binom{2n}{n} \qquad (n \ge 0) \, .$$

By induction one sees at once that $\frac{1}{n+1}\binom{2n}{n} \ge 2^{n-1}$, that is, $R_n \ge 2^{n-3}$. The number of triangulations therefore grows exponentially, and thus an exhaustive search is out of the question.

How, then, does **dynamic programming** function? Suppose we have a convex n-gon with vertices $v_1, v_2, \ldots, v_n$ and weight function w on the triangles. As a first step, we consider the *structure* of the optimal solution. Assume that in an optimal triangulation the triangle $v_1 v_k v_n$, $2 \le k \le n-1$, appears. It is then clear that the triangulations on the parts $v_1, \ldots, v_k$ and $v_k, v_{k+1}, \ldots, v_n$ must also be optimal. And that is the basis of our procedure: An optimal solution is composed of optimal subsolutions.

If $c(i,j)$, $i < j$, denotes the minimal weight of the triangulations of the $(j - i + 1)$-gon $v_i, v_{i+1}, \ldots, v_j$, then the above considerations lead us to

(10.1) $\qquad c(i,j) = \min_{i < k < j} (c(i,k) + c(k,j) + w_{ikj}) \qquad (j - i \ge 2)$

with the initial condition

$$c(i, i+1) = 0 \quad \text{for all } i.$$

From (10.1) we can calculate the desired optimum $c(1, n)$. Equation (10.1) is called, after its discoverer, **Bellman's optimality equation**.

It is now clear how a recursive procedure for determining $c(1, n)$ can be developed from (10.1). But it can be easily shown (see Exercise 10.5) that this method again has exponential running time. Was all our work in vain? No, for we proceed in precisely the opposite direction. Instead of decomposing $c(1, n)$ into smaller subproblems "top down," we build $c(1, n)$ from below "bottom up," beginning with

$$c(1, 2) = c(2, 3) = \cdots = c(n-1, n) = 0.$$

From this, using (10.1), we obtain

$$c(1, 3), \quad c(2, 4), \quad \ldots, \quad c(n-2, n),$$

then $c(i, j)$ with $j - i = 3$, and so on, until we arrive at $c(1, n)$. This bottom-up method is the second component of dynamic programming. We create the table of $c(i, j)$ recursively. Altogether, there are $\binom{n}{2}$ pairs $i < j$, and the calculational step (10.1) requires $O(n)$ operations, whence the total running time is $O(n^3)$.

Example 10.1. Let us consider the case $n = 6$ with weight function

ikj	w_{ikj}	ikj	w_{ikj}
123	3	234	6
124	1	235	1
125	6	236	3
126	1	245	7
134	2	246	1
135	4	256	4
136	4	345	4
145	5	346	4
146	1	356	2
156	2	456	5

We now construct two tables $c(i, j)$ and $p(i, j)$, where $k = p(i, j)$ is an index $i < k < j$ that yields the minimum in the calculation of $c(i, j)$ in (10.1). To clarify the bottom-up nature of the method, we rotate the triangle by 90 degrees:

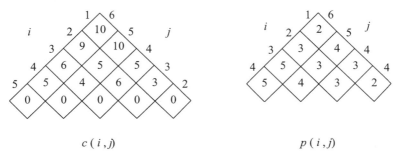

$$c(i,j) \qquad\qquad p(i,j)$$

The starting values are $c(i, i+1) = 0$, $1 \leq i \leq 5$. The values $c(i, i+2)$ are precisely $w_{i,i+1,i+2}$ and $p(i, i+2) = i+1$. We now proceed recursively. For example, we have

$$c(1,5) = \min \left\{ \begin{array}{l} c(2,5) + w_{125} = 5 + 6 = 11 \\ c(1,3) + c(3,5) + w_{135} = 3 + 4 + 4 = 11 \\ c(1,4) + w_{145} = 5 + 5 = 10 \end{array} \right\} = 10$$

and thus $p(1, 5) = 4$. The desired value is $c(1, 6) = 10$, and we can read off an optimal decomposition from the table $p(i, j)$. Since $p(1, 6) = 2$, we have the triangle 126. From $p(2, 6) = 3$ we obtain the triangle 236, and finally from $p(3, 6) = 5$, the triangles 356 and 345. Thus an optimal triangulation is

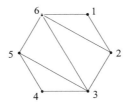

The sum is $w_{126} + w_{236} + w_{356} + w_{345} = 1 + 3 + 2 + 4 = 10$ as we have seen.

As an additional example we would like to attack a search problem. We are given a list $y_1 < y_2 < \cdots < y_n$ and a new element z, which is to be inserted into the list. We compare z with elements y_i and receive as answer $z < y_i$, $z = y_i$, or $z > y_i$. The final position of z is one from among $2n + 1$ possibilities: $z = y_i$ $(1 \leq i \leq n)$ or $z < y_1$ or $y_i < z < y_{i+1}$ $(1 \leq i \leq n)$, where we set $y_{n+1} = \infty$. Our search domain is therefore a set $X \cup Y$, $X = \{x_0, x_1, \ldots, x_n\}$, $Y = \{y_1, \ldots, y_n\}$, where the outcome x_j means $y_j < z < y_{j+1}$ (or $z < y_1$ for x_0) and y_i means $z = y_i$.

Since we have three possible answers to a comparison $z : y_i$, this is a ternary problem. However, we can model it as a binary search tree.

Take, for example, $n = 4$. We first compare $z : y_3$. If $z > y_3$, we make the test $z : y_4$, and in the case $z < y_3$ the comparison $z : y_1$, and if now

$z > y_1$, the comparison $z : y_2$. If a test gives the result $z = y_i$, then of course
we stop.

The decision tree therefore looks like this:

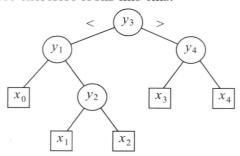

We see that the internal vertices correspond to the outcomes $z = y_i$ and the
leaves to the outcomes $x_j : y_j < z < y_{j+1}$. For the leaves, the length $\ell(x_j)$
is equal to the number of questions, while for the internal vertices we must
also include the question $z : y_i$, and so here we have $\ell(y_i) + 1$ as the number
of tests. The worst-case length of the algorithm is therefore

$$\max\left(\max_{0 \le j \le 4} \ell(x_j), \; \max_{1 \le i \le 4} \ell(y_i) + 1\right) = 3.$$

We pose now the following question, which generalizes our discussion from
Section 9.3. Suppose we are given a distribution $(p_0, \ldots, p_n, q_1, \ldots, q_n)$ on
$X \cup Y$, where $\sum_{j=0}^{n} p_j + \sum_{i=1}^{n} q_i = 1$. What is the minimal average length
of a sorting algorithm? That is, we are interested in the quantity

$$\overline{L}(p_0, \ldots, p_n, q_1, \ldots, q_n) = \min_T \left(\sum_{j=0}^{n} p_j \ell(x_j) + \sum_{i=1}^{n} q_i(\ell(y_i) + 1) \right),$$

taken over all trees with $n + 1$ leaves.

Suppose an optimal algorithm first tests $z : y_k$. Then the left subtree
must be optimal for the problem on the result set $S_{0,k-1} = \{x_0, y_1, x_1,$
$\ldots, y_{k-1}, x_{k-1}\}$, and the right subtree must be optimal on the set $S_{k,n} =$
$\{x_k, y_{k+1}, x_{k+1}, \ldots, x_n\}$. We thus have precisely the situation for dynamic
programming.

Let $c(i, j)$ be the optimal average length for the result set

$$S_{ij} = \{x_i, y_{i+1}, x_{i+1}, \ldots, y_j, x_j\}, \; i < j,$$

with $c(i, i) = 0$ for all i. Furthermore, we set $w_{ij} = \sum_{k=i}^{j} p_k + \sum_{k=i+1}^{j} q_k$.
Taking into account the comparison at the root (which increases the height
of the subtrees by 1), we obtain the Bellman optimality equation

(10.2) $c(i, j) = w_{ij} + \min_{i < k \le j} (c(i, k-1) + c(k, j)) \; (i < j),$

$c(i, i) = 0 \quad$ for all i.

With our bottom-up method we calculate the desired optimum $c(0, n)$ in time $O(n^3)$.

Example 10.2. Let $n = 4$ and suppose we have the following probabilities (multiplied by 100):

	0	1	2	3	4
p_k	12	6	16	16	10
q_k		5	15	8	12

We create the tables $c(i, j)$ and $p(i, j)$, where $k = p(i, j)$ indicates the root $z : y_k$ in the subproblem S_{ij}.

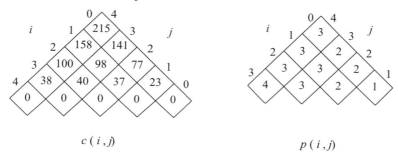

$$c(i, j) \qquad\qquad p(i, j)$$

An optimal tree with average length 2.15 is therefore given by the following:

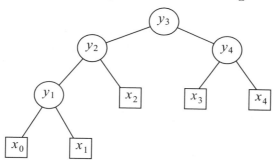

The traveling salesman problem can also be attacked with dynamic programming. As input we have the $n \times n$ matrix (w_{ij}). We fix 1 as the starting point of the tour. What do optimal tours T look like? Let k be the last vertex visited before returning to the starting point 1. If c^* is the length of an optimal tour T, then

$$c^* = c(k) + w_{k1},$$

where $c(k) = w_{1i_1} + w_{i_1 i_2} + \cdots + w_{i_{n-2} k}$ is equal to the cost of the path $1, i_1, \ldots, i_{n-2}, k$ in the tour T. And it is now clear that the path $1, i_1, \ldots, k$ must be optimal (that is, with minimal cost) among all paths from 1 to k.

From this we may at once apply the Bellman optimality equations. For $S \subseteq \{2, \ldots, n\}$ let $c(S, k)$, $k \in S$, denote the cost of an optimal path from 1 to k that beginning with 1 traverses precisely the vertices of S.

The equations are then

$$c(\{k\}, k) = w_{1k} \qquad (k = 2, \ldots, n),$$

$$c(S, k) = \min_{m \in S \smallsetminus \{k\}} (c(S \smallsetminus \{k\}, m) + w_{mk}) \quad (S \subseteq \{2, \ldots, n\}, |S| \geq 2),$$

$$c^* = \min_{k \in \{2, \ldots, n\}} (c(\{2, \ldots, n\}, k) + w_{k1}).$$

The operations we use are addition and comparison (for calculating the minimum). If $|S| = \ell$, $2 \leq \ell \leq n - 1$, then for determining $c(S, k)$ we require $\ell - 1$ additions, thus altogether $\ell(\ell - 1)$ for fixed S. Together with the $n - 1$ additions for calculating c^*, this gives

$$A(n) = \left(\sum_{\ell=2}^{n-1} \binom{n-1}{\ell} \ell(\ell - 1) \right) + (n - 1)$$

$$= \left(\sum_{\ell=2}^{n-1} (n-1)(n-2)\binom{n-3}{\ell-2} \right) + (n - 1)$$

$$= (n-1)(n-2)2^{n-3} + n - 1 = O(n^2 2^n)$$

for the total number of additions.

The total number of comparisons is of the same order of magnitude, and so the running time using dynamic programming is $O(n^2 2^n)$. Of course, this is exponential, but that was to be expected, since the traveling salesman problem is NP-complete. On the other hand, it is better than the algorithm using exhaustive search of all $(n - 1)!$ tours, since by Stirling's formula we have the asymptotic estimate $n! \sim \sqrt{2\pi n}(\frac{n}{e})^n$.

10.3. The Greedy Algorithm

We have encountered algorithms of greedy type in a number of situations, such as the by now almost classical problems of minimal spanning tree in Section 7.3 and minimal transversal in Section 8.2. In Section 7.3 we laid the theoretical foundations for when a **greedy algorithm** is appropriate, namely, when the sets that appear in the optimization procedure form a *matroid*.

Thus we are dealing here, as in the case of dynamic programming, with a situation in which the underlying structures are optimal. Let us consider an example, and then with an additional problem we will demonstrate how this method differs from that of dynamic programming.

Suppose we have n tasks to execute, each of which requires one time unit. In addition, we are given two sequences $b_1, \ldots, b_n$, $1 \leq b_i \leq n$, and $s_1, \ldots, s_n$, $s_i \geq 0$. The number b_i specifies that the ith task must end at the latest after b_i time units, unless we are willing to pay a penalty of s_i. Our problem consists in arranging the order of job execution in such a way that the total penalty for late completions is minimized.

Consider the following example:

i	b_i	s_i
1	3	2
2	2	3
3	6	1
4	1	2
5	2	4
6	2	2
7	4	5

We could proceed as follows. First, we arrange the tasks according to required completion time b_i:

i	4	2	5	6	1	7	3
b_i	1	2	2	2	3	4	6

Then we choose in a greedy fashion the greatest number of tasks that can be finished on time. Thus we choose first task 4, then 2; tasks 5 and 6 can't be accommodated; then tasks 1, 7, and 3. Thus the order will be $4, 2, 1, 7, 3, 5, 6$ with a total penalty of $s_5 + s_6 = 6$. Is this optimal?

We call an ordering of the n tasks a *plan* P and set $N = \{1, \ldots, n\}$. In P we let p_i denote the position of the ith task. A task is said to be *early* in P if it is completed on time, that is, if $p_i \leq b_i$. Otherwise, it is called *late*. Let $s(P)$ denote the penalty that must be paid in executing plan P. We first establish that we may assume that in P, all early tasks come before all late ones. Namely, if i is early and j is late with $p_i > p_j$, whence $b_i \geq p_i > p_j > b_j$, we may exchange the positions of i and j, and once again, i is early and j late. The penalty $s(P)$ remains the same. Moreover, we may arrange all the early tasks according to increasing end time. Namely, if i and j are early tasks with $p_i < p_j$, $b_i > b_j$, then we have $p_i < p_j \leq b_j < b_i$, and so we can exchange the positions of i and j, and as before, both tasks are early.

We may therefore assume that in any plan, first come the early tasks A and then the late tasks $S = N \smallsetminus A$. And now comes the crucial idea. The total penalty of P is $s(P) = \sum_{i \in S} s_i$. Therefore, $s(P)$ is minimal when $s(A) = \sum_{j \in A} s_j$ is *maximal*. We call a set $A \subseteq N$ *independent* (including $A = \varnothing$) if the tasks in A can be completed as early tasks in some plan. If we can show that $(N, \mathcal{A})$, with $\mathcal{A}$ the family of independent sets, forms a matroid, we may then apply our greedy algorithm of Theorem 7.13 to $(N, \mathcal{A})$ with the weight function s. We therefore see that we must order the tasks in increasing order of penalty and choose the next possible task. In hindsight it is clear: we must see to it that jobs with the greatest penalty are positioned first.

In our example we obtain

i	7	5	2	1	4	6	3
s_i	5	4	3	2	2	2	1
b_i	4	2	2	3	1	2	6

We take first task 7, then 5. Task 2 can be completed if we take the order $5, 2, 7$, and likewise 1 with the new ordering $5, 2, 1, 7$. Task 4 cannot be completed on time; nor can 6. But 3 can be. We thus obtain the optimal basis of the matroid $A = \{7, 5, 2, 1, 3\}$. Arranging A according to increasing b_i, we obtain the optimal plan P:

$$5, 2, 1, 7, 3, 4, 6$$

with $s(P) = s_4 + s_6 = 4$.

It remains to prove that $(N, \mathcal{A})$ is a matroid. The first two axioms are clear. Before we prove the exchange axiom, let us attempt to characterize the independent sets A. We assume that A is ordered by increasing end times b_i. For $k = 1, \ldots, n$ let $N_k(A)$ be the number of tasks in A with $b_i \leq k$, $N_0(A) = 0$. Clearly, we must have $N_k(A) \leq k$ for all k, and from this condition it follows clearly that A is independent. The condition $N_k(A) \leq k$ $(k = 0, \ldots, n)$ therefore characterizes independent sets.

To prove axiom 3 in Definition 7.11, let $A, B \in \mathcal{A}$ with $|B| = |A| + 1$. We have $N_0(A) = N_0(B) = 0$ and $N_n(A) = |A| < |B| = N_n(B)$. There thus exists a largest index t with $0 \leq t < n$ such that $N_t(B) \leq N_t(A)$. From $N_{t+1}(B) > N_{t+1}(A)$ it follows that there is a task $i \in B \smallsetminus A$ with $b_i = t + 1$. Let $A' = A \cup \{i\}$. For $k \leq t$ it follows that $N_k(A') = N_k(A) \leq k$, since A is independent, and for $j \geq t + 1$ we have $N_j(A') \leq N_j(B) \leq j$, since B is independent. From our observation above, we see that $A \cup \{i\}$ is independent, which completes the proof.

We note further that a greedy approach can work even if a matroid is not given (although one can often be found in retrospect). The Huffman algorithm from Section 9.2 is a greedy method—we always choose leaves of least probability. And so is Dijkstra's shortest-path algorithm from Section 7.4.

Let us now look at another famous problem: the knapsack problem. Ellie has found n objects of various weights and values. Each object i has value v_i and weight w_i, with $v_i, w_i \in \mathbb{N}$. Ellie would like to fill her knapsack with objects of the largest possible total value, but the total weight cannot exceed the number W, since that is the most that she can carry. Which objects should Ellie pack in her knapsack?

A greedy strategy would first arrange the objects according to their relative values $\frac{v_i}{w_i}$ in decreasing order and always pack the next possible object. Unfortunately, this approach fails already for $n = 3$. For example, let

i	1	2	3	
v_i	3	5	6	$W = 5$
w_i	1	2	3	
v_i/w_i	3	5/2	2	

The greedy method chooses 1 and then 2 for a total value of 8, but 2 and 3 are also possible, for a total value of 11.

Suppose that an optimal knapsack load L contains object m. Then $L \setminus \{m\}$ is an optimal load of the subproblem on $\{1, \ldots, n\} \setminus \{m\}$ with total weight $W - w_m$. The correct approach is therefore not the greedy algorithm, but dynamic programming (about which we shall have more to say in the exercises).

In the exercises we will also see that the greedy algorithm can veer arbitrarily far from the optimum. But let us consider the following variant, in which we may pack arbitrarily many exemplars of an object as long as we don't exceed the total weight W. We therefore seek a vector $(x_1, \ldots, x_n)$, $x_i \in \mathbb{N}_0$, with $\sum w_i x_i \leq W$ and a maximum for the sum $\sum v_i x_i$.

Again let us try the greedy algorithm with $\frac{v_1}{w_1} \geq \cdots \geq \frac{v_n}{w_n}$, where we may assume $w_1 \leq W$, since otherwise, object 1 plays no role. Using the greedy method, we pack x_1 exemplars of 1 with $x_1 = \left\lfloor \frac{W}{w_1} \right\rfloor$. Now supposing $(x_1, \ldots, x_n)$ to be an optimal solution, we have $v_j \leq \frac{v_1}{w_1} w_j$ for all j, and thus

$$v_{\text{opt}} = \sum_{j=1}^{n} v_j x_j \leq \frac{v_1}{w_1} \sum_{j=1}^{n} w_j x_j \leq v_1 \frac{W}{w_1} \leq 2 v_1 \left\lfloor \frac{W}{w_1} \right\rfloor \leq 2 v^*.$$

Thus the greedy value v^* achieves at least one-half of the optimum. In the exercises you will see that the bound $v^* \geq \frac{1}{2} v_{\text{opt}}$ can in general not be improved. For our final observation, we note that both versions of the knapsack problem are NP-complete.

Exercises for Chapter 10

▷ **10.1** Solve the king problem (analogous to the queen problem) on an $n \times n$ chessboard.

10.2 Using backtracking, determine all 3×3 magic squares (see Exercise 12.37). Hint: 1 cannot go in a corner.

▷ **10.3** Show that in the queen problem on the $n \times n$ chessboard we may assume $a_1 \neq 1$.

10.4 Discuss the usual 3×3 tic-tac-toe game using a game tree.

▷ **10.5** Show that determining $c(1, n)$ from the recurrence (10.1) has exponential running time.

10.6 Solve the following traveling salesman problem using branch and bound:

	1	2	3	4	5	6
1	∞	3	6	2	8	1
2	4	∞	3	4	4	5
3	3	2	∞	6	3	5
4	4	2	6	∞	4	4
5	3	3	2	6	∞	4
6	7	4	5	7	6	∞

10.7 Determine an optimal triangulation of the regular hexagon with weight function $w(v_i v_j v_k) = |v_i - v_j| + |v_j - v_k| + |v_k - v_i|$.

▷ **10.8** The *coin exchange problem* is as follows: You are given m coins with values $c_1, c_2, \ldots, c_m$, $c_i \in \mathbb{N}$. For $n \in \mathbb{N}$ determine (if possible) $k_1, \ldots, k_m$ with $\sum_{i=1}^{m} k_i c_i = n$ and $\sum_{i=1}^{m} k_i$ minimal. Show that a greedy algorithm works for the powers $c^0, c^1, \ldots, c^{m-1}$ ($c > 1$, $m \geq 1$).

10.9 Determine the number of comparisons in the dynamic program for the traveling salesman problem.

10.10 Solve, using backtracking, the following ferryboat problem: A ferryman must bring a wolf, a donkey, and a carrot across a river. He can take only one item at a time, and neither wolf and donkey nor donkey and carrot can be left unattended.

▷ **10.11** Construct an example in which the greedy algorithm (always take an allowable pair of greatest value) fails for the weighted job-assignment problem $c : K \to \mathbb{R}^+$ of Section 8.2.

10.12 Solve the queen-placement problem on the $n \times n$ chessboard.

10.13 You are given matrices $M_1, \ldots, M_n$ of type $r_i \times r_{i+1}$ ($i = 1, \ldots, n$). An evaluation of the product $M_1 \cdots M_n$ involves a sequence of matrix multiplications, in which the cost of multiplication of an $r \times s$ matrix A by an $s \times t$ matrix B is equal to rst. Determine using dynamic programming a cheapest sequence of multiplications. Hint: Look at the triangulation problem of Section 10.2.

10.14 Determine an optimal multiplication sequence for six matrices with dimensions 3×5, 5×11, 11×2, 2×9, 9×23, 23×6.

▷ **10.15** Three jealous men wish to transport themselves and their wives across a river. Their boat holds at most two persons. Determine a transportation strategy that brings all six parties across the river without a wife ever being left in the presence of a man not her husband—either on shore or in the boat—without her husband being present as well. Can it be done with four couples?

10.16 We fill an $n \times n \times n$ cube with n^3 unit cubes. The problem is to color each of the n^3 cubes white or black in such a way that the number of lines (horizontal, vertical, diagonal) of a single color is as small as possible.

▷ **10.17** The *stamp problem* is this: You have at your disposal stamps of n different denominations in $\mathbb{N}$, and at most m stamps can be affixed to any letter. Determine the largest value N for which all sums of stamp values from 1 to N can be realized. Example: For $n = 4$, $m = 5$ the four denominations $\{1, 5, 12, 28\}$ allow all values from 1 to 71 to be realized. Develop a backtracking program.

▷ **10.18** Let $\chi(G)$ be the chromatic number of the graph $G = (V, E)$; see Exercise 6.12. For $uv \notin E$ let $G^+ = G \cup \{uv\}$ and let G^- be the graph that results from identifying u and v as a new vertex $\overline{uv}$. Thus z is a neighbor of $\overline{uv}$ in G^- if and only if z was a neighbor of u or v in G. Show that $\chi(G) = \min(\chi(G^+), \chi(G^-))$. By iteration develop a branch and bound algorithm for determining $\chi(G)$. Hint: Use the largest complete subgraph as a bound.

10.19 Show that for $c_1 = 1$, $c_2 = 5$, $c_3 = 10$, $c_4 = 25$ the greedy algorithm works for the coin problem of Exercise 10.8. On the other hand, give an example $\{c_1, \ldots, c_m\}$ in which the greedy algorithm does not give the optimal solution (even though all n can be realized).

10.20 Show how dynamic programming can be applied to the knapsack problem with running time $O(Wn)$.

▷ **10.21** Show by means of an example that the greedy algorithm for the knapsack problem can stray arbitrarily far from the optimum. That is, for every $0 < \varepsilon < 1$ there exists a knapsack such that $v^* < (1 - \varepsilon)v_{\text{opt}}$, with v^* the greedy value. Hint: $n = 2$ will do.

▷ **10.22** In the knapsack problem let $\frac{v_1}{w_1} \geq \cdots \geq \frac{v_n}{w_n}$ be the relative values. Show that $v_{\text{opt}} \leq \sum_{j=1}^{k} v_j + \frac{v_{k+1}}{w_{k+1}}(W - \sum_{j=1}^{k} w_j)$ for all k, and conclude that $v^* \geq v_{\text{opt}} - \max(v_j : j = 1, \ldots, n)$, with v^* the greedy value.

10.23 A number of lectures $V_1, \ldots, V_n$ are to be held in different rooms, where lecture V_i requires a time of c_i. At a given time, only one lecture can take place in a given room. Develop a greedy algorithm that uses as few rooms as possible and such that all lectures are over at time $C \geq \max c_i$.

▷ **10.24** There are n jobs to be executed on a machine, where job J_i begins at time a_i and ends at time e_i. A set of jobs $J_{h_1}, \ldots, J_{h_m}$ is called *admissible* if $e_{h_i} \leq a_{h_{i+1}}$ for all i, that is, if they can be executed one after the other. Show that the following greedy algorithm returns a set of maximal size: Order the jobs by $e_1 \leq e_2 \leq \cdots \leq e_n$ and always choose the next possible job.

10.25 A variant of the knapsack problem allows partial objects to be packed. That is, we are seeking $(x_1, \ldots, x_n)$ with $0 \leq x_i \leq 1$ such that $\sum w_i x_i \leq W$ and $\sum v_i x_i$ is maximal. Show that the greedy algorithm returns an optimal solution. Hint: Show that there always exists an optimal solution with $x_1 = \text{greedy}(x_1)$.

10.26 Consider the variant of the knapsack problem given at the end of Section 10.3. Show by an example that the bound $v^* \geq \frac{1}{2}v_{\text{opt}}$ cannot be improved.

Bibliography for Part 2

There are quite a few good books on graph theory, such as those by Diestel and West. We have omitted entirely a classical subfield of graph theory: the embeddability of graphs in the plane. This problem goes back to the very beginnings of graph theory, in the form of the well-known four-color problem. The reader who wishes to learn more about the influence of the four-color problem on the development of graph theory may consult the book by Aigner. There is a large number of books on graph algorithms and combinatorial algorithms. The book by Corman, Leiserson, and Rivest is extensive and detailed; also recommended are the books by Even, Horowitz and Sahni, Papadimitriou and Steiglitz, and Jungnickel. The book by Lawler contains, among other things, an in-depth presentation of matroids and related algorithmic problems. For sorting algorithms, the classic text remains Knuth's book, and those interested in the relatively new area of search theory should consult the relevant book by Aigner. An excellent presentation of the complexity classes P and NP as well as complexity theory in general can be found in the book by Garey and Johnson.

M. Aigner: *Graph Theory: A Development from the 4-Color Problem.* BCS Associates.

M. Aigner: *Combinatorial Search.* Teubner-Wiley.

T. Corman, C. Leiserson, and R. Rivest: *Introduction to Algorithms.* MIT Press.

R. Diestel: *Graph Theory.* Springer-Verlag.

S. Even: *Algorithmic Combinatorics.* Macmillan.

M. GAREY AND D. JOHNSON: *Computers and Intractability: A Guide to the Theory of NP-Completeness.* Freeman.

E. HOROWITZ AND S. SAHNI: *Fundamentals of Computer Algorithms.* Computer Science Press.

D. JUNGNICKEL: *Graphs, Networks, and Algorithms.* Springer.

D. KNUTH: *The Art of Computer Programming III, Sorting and Searching.* Addison-Wesley.

E. LAWLER: *Combinatorial Optimization: Networks and Matroids.* Holt, Rinehart and Winston.

C. PAPADIMITRIOU AND K. STEIGLITZ: *Algorithms and Complexity.* Prentice-Hall.

D. WEST: *Introduction to Graph Theory.* Prentice-Hall.

Part 3

Algebraic Systems

In Part 1 we were concerned with counting sets, usually defined in terms of simple combinatorial properties, for example subsets and partitions. In Part 2 we studied graphs and solved a number of algorithmic problems by introducing graphs as a data structure. We repeatedly observed the utility of introducing algebraic concepts on sets that at the outset displayed no additional structure. An impressive example was that of generating functions $F(z) = \sum a_n z^n$. At first, the numbers a_n are simply the sizes of the given sets. But if we construe them as coefficients of the series $F(z)$, we can actually *calculate* with them: add them or take their convolution product.

A further example was the greedy algorithm in Section 7.2 for generating minimal spanning trees. It works because the forests form a matroid, and indeed, the proof of correctness became clearer and simpler when we lifted the discussion to the abstract realm of matroids. We represented graphs, and more generally, families of sets, as $0, 1$ matrices, which can be used for calculation: We can add matrices and multiply them. A matrix has a rank, and we may expect that the rank of an incidence matrix will tell us something about the underlying graph.

Stated informally, we have used algebraic methods in order better to understand discrete structures and to make problems easier to solve. In this third part we are going to take, in a certain sense, the opposite tack. We will investigate an important algebraic system, work out its structure, and then consider to what sorts of discrete problems it can be successfully applied.

For example, one may introduce a natural addition and multiplication to finite sets. We shall see that sets with such an addition or multiplication (or indeed both) exhibit a number of symmetries, and we shall use these symmetries to construct discrete configurations. A particularly important example is that of systems of inequalities, which we investigate in the last chapter. We arrive via an algebraic route at a fundamental result: the main theory of linear optimization, and we establish that this theorem encompasses all our "flow–cut" theorems, the more so in that the duality "maximum packing–minimum covering" is put in the proper context only when it is considered algebraically.

Chapter 11

Boolean Algebras

11.1. Definition and Properties

In this chapter we study the set $\mathcal{B}(n) = \{0, 1\}^n$. The elements of $\mathcal{B}(n)$ are sequences or vectors $\boldsymbol{x} = (x_1, \ldots, x_n)$ of 0's and 1's of length n, otherwise known as $0, 1$ words of length n. Initially, this is merely a set, but we may interpret $\mathcal{B}(n)$ in a number of ways and place various structures on it. And at that point things become interesting. We will always use $\boldsymbol{x}, \boldsymbol{y}, \boldsymbol{z}, \ldots$ to denote elements of $\mathcal{B}(n)$, while $x, y, z, \ldots$ will denote individual coordinates.

There is one interpretation that we know already. We consider the set $S = \{1, \ldots, n\}$ of indices and interpret

$$\boldsymbol{x} = (x_1, \ldots, x_n) \quad \text{as} \quad A_{\boldsymbol{x}} = \{i : x_i = 1\} \, .$$

Thus $\boldsymbol{x}$ is the **characteristic vector** of the set $A_{\boldsymbol{x}}$, and the mapping $\boldsymbol{x} \longleftrightarrow A_{\boldsymbol{x}}$ is bijective. From now on, $\boldsymbol{0} = (0, \ldots, 0)$ and $\boldsymbol{1} = (1, 1, \ldots, 1)$ will denote the zero word and the word of all ones. The associated sets are $A_{\boldsymbol{0}} = \varnothing$ and $A_{\boldsymbol{1}} = S$. Analogously, $\mathcal{B}(S)$ denotes the family of all subsets of S.

Another interpretation, one of fundamental importance for calculations with computers, results from regarding $\boldsymbol{x}$ as the **binary representation** of a natural number. In this case, we write $\boldsymbol{x} = (x_0, x_1, \ldots, x_{n-1})$ and obtain the relationship

$$\boldsymbol{x} = (x_0, x_1, \ldots, x_{n-1}) \longleftrightarrow z_{\boldsymbol{x}} = \sum_{i=0}^{n-1} x_i 2^i \, .$$

The zero word corresponds to $z_{\boldsymbol{0}} = 0$ and the word of all ones to $z_{\boldsymbol{1}} = 2^n - 1$.

Operations with sets and with numbers can also be carried out using $0, 1$ words, and that will be the guiding idea in what follows.

For sets we have three basic operations: union $\cup$, intersection $\cap$, and complement $^-$. We define the corresponding operations for $0, 1$ words coordinatewise as follows:

x	y	$x + y$
0	0	0
1	0	1
0	1	1
1	1	1

x	y	xy
0	0	0
1	0	0
0	1	0
1	1	1

x	$\overline{x}$
0	1
1	0

In general, we have

$$(x_1, \ldots, x_n) + (y_1, \ldots, y_n) = (x_1 + y_1, \ldots, x_n + y_n),$$
$$(x_1, \ldots, x_n)(y_1, \ldots, y_n) = (x_1 y_1, \ldots, x_n y_n),$$
$$\overline{(x_1, \ldots, x_n)} = (\overline{x}_1, \ldots, \overline{x}_n).$$

Clearly, $\boldsymbol{x} + \boldsymbol{y}$ corresponds to the *union* $A_{\boldsymbol{x}} \cup A_{\boldsymbol{y}}$, $\boldsymbol{xy}$ to the *intersection* $A_{\boldsymbol{x}} \cap A_{\boldsymbol{y}}$, and $\overline{\boldsymbol{x}}$ to the *complement* $\overline{A_{\boldsymbol{x}}}$. The rules for calculation are well known: The operations $+$ and $\cdot$ are commutative, associative, and distributive with zero element $\boldsymbol{0}$ and multiplicative identity $\boldsymbol{1}$, with well-defined complement $\boldsymbol{x} \mapsto \overline{\boldsymbol{x}}$, where $\overline{\overline{\boldsymbol{x}}} = \boldsymbol{x}$. Moreover, we have de Morgan's laws $\overline{\sum \boldsymbol{x}_i} = \prod \overline{\boldsymbol{x}}_i, \overline{\prod \boldsymbol{x}_i} = \sum \overline{\boldsymbol{x}}_i$, where $\sum$ and $\prod$ denote the sum and product just introduced.

The set $\mathcal{B}(n)$ equipped with the operations $+$, $\cdot$, and $^-$ (or equivalently, $\mathcal{B}(S)$ with $\cup$, $\cap$, $^-$) is called the **Boolean algebra** of order n. Note that neither $+$ nor $\cdot$ yields a group structure, since $\boldsymbol{1}$ has no additive inverse, and $(1, 0)$, for example, has no multiplicative inverse for $n = 2$.

We obtain an additional structure by defining the operations $\oplus$ and $\cdot$ as follows:

x	y	$x \oplus y$
0	0	0
1	0	1
0	1	1
1	1	0

x	y	xy
0	0	0
1	0	0
0	1	0
1	1	1

In this case, $\{0, 1\}$ with the operators $\oplus, \cdot$ is a *field*, which we also call the Galois field GF(2). (We shall discuss finite fields in detail in Section 12.2.) If we extend these operations coordinatewise, $\mathcal{B}(n)$ becomes a *vector space* of dimension n over GF(2). Interpreted as a set, $\boldsymbol{x} \oplus \boldsymbol{y}$ corresponds to the *symmetric difference* $(A_{\boldsymbol{x}} \setminus A_{\boldsymbol{y}}) \cup (A_{\boldsymbol{y}} \setminus A_{\boldsymbol{x}})$, which we therefore give the additional notation, for short, of $A_{\boldsymbol{x}} \oplus A_{\boldsymbol{y}}$.

Set inclusion $A \subseteq B$ provides the additional structure of an *order* $\boldsymbol{x} \leq \boldsymbol{y}$ on $\mathcal{B}(n)$: For $\boldsymbol{x} = (x_1, \ldots, x_n)$, $\boldsymbol{y} = (y_1, \ldots, y_n)$ we set

$$\boldsymbol{x} \leq \boldsymbol{y} \Longleftrightarrow x_i \leq y_i \quad \text{for all } i.$$

Clearly, we have $\boldsymbol{x} \le \boldsymbol{y}$ if and only if $A_{\boldsymbol{x}} \subseteq A_{\boldsymbol{y}}$. The order $\le$ has the additional property that every pair of elements $\boldsymbol{x}, \boldsymbol{y}$ has exactly one least upper bound and exactly one greatest lower bound, which we denote by $\boldsymbol{x} \lor \boldsymbol{y}$ and $\boldsymbol{x} \land \boldsymbol{y}$. This should all be clear. From the definitions, we obtain

$$\boldsymbol{z} = \boldsymbol{x} \lor \boldsymbol{y} \quad \text{with} \quad z_i = \max(x_i, y_i) \quad \text{for all } i,$$
$$\boldsymbol{w} = \boldsymbol{x} \land \boldsymbol{y} \quad \text{with} \quad w_i = \min(x_i, y_i) \quad \text{for all } i.$$

Since for $n = 1$ the defining tables for $+$ and $\cdot$ yield

$$x \lor y = x + y \quad \text{and} \quad x \land y = xy,$$

we have in general,

$$\boldsymbol{x} \lor \boldsymbol{y} = \boldsymbol{x} + \boldsymbol{y}, \quad \boldsymbol{x} \land \boldsymbol{y} = \boldsymbol{xy},$$

that is, $\boldsymbol{x} \lor \boldsymbol{y}$, $\boldsymbol{x} \land \boldsymbol{y}$ again correspond to the union and intersection of sets. This is of course the reason for the similarity in notation $\lor, \cup$ and $\land, \cap$. An order that possesses a *supremum* $\boldsymbol{x} \lor \boldsymbol{y}$ and an *infimum* $\boldsymbol{x} \land \boldsymbol{y}$ for all $\boldsymbol{x}, \boldsymbol{y}$ is called a *lattice*. Thus $\mathcal{B}(n)$ with the order $\le$ (or $\mathcal{B}(S)$ with $\subseteq$) is a lattice, called the **Boolean lattice**. The reason that we sometimes use $+$ and $\cdot$, and at other times $\lor$ and $\land$, will depend on our point of view. In the first instance, the algebraic aspect is in the foreground, while in the second, it is the order. We will use one or the other form depending on how a problem is formulated.

We will now define a notion of distance, the so-called *Hamming distance*:

$$\Delta(\boldsymbol{x}, \boldsymbol{y}) = |\{i : x_i \ne y_i\}|.$$

One sees immediately that $\Delta : \mathcal{B}(n) \to \mathbb{R}^+$ is a metric in the sense of calculus; that is, $\Delta(\boldsymbol{x}, \boldsymbol{y}) \ge 0$ with $\Delta(\boldsymbol{x}, \boldsymbol{y}) = 0$ if and only if $\boldsymbol{x} = \boldsymbol{y}$, $\Delta(\boldsymbol{x}, \boldsymbol{y}) = \Delta(\boldsymbol{y}, \boldsymbol{x})$, and the *triangle inequality* holds: $\Delta(\boldsymbol{x}, \boldsymbol{z}) \le \Delta(\boldsymbol{x}, \boldsymbol{y}) + \Delta(\boldsymbol{y}, \boldsymbol{z})$.

The reader may have noticed the connection with the hypercube graphs Q_n from Section 6.1. The vertices are the words $\boldsymbol{x} \in \mathcal{B}(n)$, and two vertices are neighbors if their Hamming distance is equal to 1. Furthermore, $\Delta(\boldsymbol{x}, \boldsymbol{y})$ corresponds precisely to the distance in Q_n in the graph-theoretic sense.

11.2. Propositional Logic and Boolean Functions

One of the historical starting points for the study of Boolean algebras was the connection to propositional logic. Consider the case $n = 2$. A mapping $f : \mathcal{B}(2) \to \{0, 1\}$ is called a (two-valued) **Boolean function**. Since $|\mathcal{B}(2)| = 4$, there exist $2^4 = 16$ such functions. The variables x and y will be interpreted as **propositions** with 1 meaning **true** and 0 **false**.

Let us write down once again the tables for our three basic operations
$\vee, \wedge, \oplus$:

x	y	$x \vee y$	$x \wedge y$	$x \oplus y$
0	0	0	0	0
1	0	1	0	1
0	1	1	0	1
1	1	1	1	0

The operation $x \vee y$ thus corresponds to the proposition $x \vee y$ is true whenever
x or y or both are true. We call $x \vee y$ in the sense of propositional logic
disjunction. The operator $x \wedge y$ corresponds to **conjunction**: $x \wedge y$ is
true whenever both x and y are true. In this way we can interpret all 16
functions in terms of propositional logic. For example, $x \oplus y$ corresponds to
"exclusive or" (either x is true and y is false or y is true and x is false). A
further example is

x	y	$x \rightarrow y$
0	0	1
1	0	0
0	1	1
1	1	1

This corresponds to **implication** $x \rightarrow y$. The proposition $x \rightarrow y$ is always
true except when x is true and y is false. The complement $x \rightarrow \overline{x}$ is naturally
interpreted as **negation** $\neg x$.

Let us look at an example for $n = 3$. What is the truth table for
the function $f(x, y, z) = (x \rightarrow y) \wedge [(y \wedge \neg z) \rightarrow (x \vee z)]$? We insert all
eight possible triples (x, y, z) in f and compute the right-hand side. For
$x = y = z = 0$, for example, we obtain

$$f(0, 0, 0) = (0 \rightarrow 0) \wedge [(0 \wedge 1) \rightarrow (0 \vee 0)] = 1 \wedge [0 \rightarrow 0] = 1 \wedge 1 = 1 \,,$$

and for $x = 1$, $y = 0$, $z = 1$, we have

$$f(1, 0, 1) = (1 \rightarrow 0) \wedge [(0 \wedge 0) \rightarrow (1 \vee 1)] = 0 \,,$$

since $1 \rightarrow 0 = 0$.

Two problems immediately present themselves: (1) Can we bring a given
Boolean function $f : \mathcal{B}(n) \rightarrow \{0, 1\}$ into a simple form that allows for
rapid evaluation? (2) Does there exist for f some $(x_1, \ldots, x_n)$ such that
$f(x_1, \ldots, x_n) = 1$? What we are asking is, can f be *satisfied*? That is, can
f ever be true?

Both questions are of the utmost significance for propositional logic.
One would like to develop a method by which compound propositions can
be recognized as true or false, or at least whether in principle they are
satisfiable.

To deal with the first question, one constructs *normal forms*. The best-known normal forms are **disjunctive normal form** (DNF), and by complementation, **conjunctive normal form** (CNF). As building blocks we employ $+$, $\cdot$, and $^-$. For every input $c = (c_1, \ldots, c_n)$ with $f(c) = 1$ we construct the **minterm function**

$$m_c(x_1, \ldots, x_n) = x_1^{c_1} x_2^{c_2} \cdots x_n^{c_n} \, ,$$

where $x^1 = x$ and $x^0 = \overline{x}$. By definition of the product, we therefore have $m_c(x) = 1$ if and only if $x = c$. (Clear?) The function f can therefore be written in the form

(DNF) $$f(x_1, \ldots, x_n) = \sum_{c : f(c)=1} x_1^{c_1} \cdots x_n^{c_n}.$$

The function f that is identically 0 is given by the empty sum.

For example, the function $f(x_1, x_2, x_3) = x_1 \oplus x_2 \oplus x_3$ has DNF

$$f(x_1, x_2, x_3) = x_1 \overline{x}_2 \overline{x}_3 + \overline{x}_1 x_2 \overline{x}_3 + \overline{x}_1 \overline{x}_2 x_3 + x_1 x_2 x_3 \, .$$

The DNF shows that *every* Boolean function can be expressed in terms of the operations $+$, $\cdot$, $^-$. We therefore say that the set $\{+, \cdot, ^-\}$ is a **basis** for all Boolean functions. Here we always stipulate that the constants (zero-valued functions) 0 and 1 are given. The significance of this will be seen later on.

By complementation we obtain the conjunctive normal form (see Exercise 11.4)

(CNF) $$f(x_1, \ldots, x_n) = \prod_{c : f(c)=0} (x_1^{\overline{c}_1} + \cdots + x_n^{\overline{c}_n}).$$

For our example this yields $x_1 \oplus x_2 \oplus x_3 = (x_1 + \overline{x}_2 + \overline{x}_3)(\overline{x}_1 + x_2 + \overline{x}_3)(\overline{x}_1 + \overline{x}_2 + x_3)(x_1 + x_2 + x_3)$.

The two operations $+$, $\cdot$ do not by themselves form a basis. For example, $x_1 \oplus x_2$ cannot be represented in terms of these two operators. Why? If a function f is given in the form

$$f(x_1, \ldots, x_n) = \sum_{(i_1, \ldots, i_\ell)} x_{i_1} \cdots x_{i_\ell},$$

we see at once that $f(x) \le f(y)$ holds for $x \le y$, since this follows from the tables for $+$ and $\cdot$. Thus only *monotone* functions can be represented in this way (and for such functions, $\{+, \cdot\}$ is in fact a basis; see the exercises). Our function $x_1 \oplus x_2$ is not monotone because of $(1,0) \le (1,1)$ and $1 \oplus 0 = 1$, $1 \oplus 1 = 0$, and therefore not representable in terms of $+, \cdot$. On the other hand, since $xy = \overline{\overline{x} + \overline{y}}$ and $x + y = \overline{\overline{x}\,\overline{y}}$, it follows that both $\{+, ^-\}$ and $\{\cdot, ^-\}$ are bases. Likewise, $\{\oplus, \cdot\}$ is a basis, as we see from $x + y = (x \oplus y) \oplus (xy)$ and $\overline{x} = x \oplus 1$. For example, we have $x_1 x_2 \overline{x}_3 = x_1 x_2 (1 \oplus x_3) = x_1 x_2 \oplus x_1 x_2 x_3$. For the basis $\{\oplus, \cdot\}$ the constants are necessary, since otherwise, the complement cannot be formed.

Before we progress any further, it is time to look at a typical example from everyday life. In the business section of a newspaper appears the following learned commentary:

> Either the euro will be devalued, or if exports do not decrease, there will have to be a price freeze. If the euro is not devalued, then exports will decrease and prices will not be frozen. However, if prices are frozen, then exports will not decrease and the euro will not be devalued.

What does it all mean? We let x stand for "the euro will be devalued," y for "exports will decrease," and z for "prices will be frozen." Then the above proposition may be written as

$$A_1 := x \vee (\neg y \rightarrow z), \ A_2 := \neg x \rightarrow (y \wedge \neg z), \ A_3 := z \rightarrow (\neg y \wedge \neg x).$$

This yields (check it!)

$$A_1 = x \vee y \vee z, \qquad A_2 = x \vee (y \wedge \neg z), \qquad A_3 = \neg z \vee (\neg y \wedge \neg x)$$

and therefore

$$\begin{aligned}
A_1 \wedge A_2 \wedge A_3 &= (x \vee y \vee z) \wedge (x \vee (y \wedge \neg z)) \wedge (\neg z \vee (\neg y \wedge \neg x)) \\
&= (x \vee (y \wedge \neg z)) \wedge (\neg z \vee (\neg y \wedge \neg x)) = (x \wedge \neg z) \vee (y \wedge \neg z) \\
&= \neg z \wedge (x \vee y) = \neg z \wedge (\neg x \rightarrow y).
\end{aligned}$$

The statement is thus equivalent to, "prices will not be frozen, and if the euro is not devalued, exports will decline," which seems much simpler.

Let us return to the normal forms. We have seen that disjunctive normal form expresses every Boolean function as a sum of products. Normally, however, the DNF is not the simplest expression of this form. Consider, for example, $f(x, y, z) = \overline{x}yz + \overline{x}y\overline{z} + x\overline{y}z$. Each of the three minterms contains all three variables. Since $z + \overline{z} = 1$ and $w1 = w$, we can also write

$$f(x, y, z) = \overline{x}y(z + \overline{z}) + x\overline{y}z = \overline{x}y + x\overline{y}z$$

as a sum of products with only two summands. We now state a general problem: For a Boolean function, say given in DNF, find a representation as a sum of products with the least possible number of summands. We thus have an additional representation: the SOPE (sums of products expansion) representation of Boolean functions. A well-known algorithm for constructing a SOPE representation is due to Quine and McCluskey. For small n one can realize the transformation from DNF to SOPE using a **Karnaugh diagram**.

Again let us consider a function $f(x, y, z)$ given in DNF. We know that the DNF contains eight possible minterms, and if (as in our example) two minterms differ in a single variable, we may simplify by adding these minterms and deleting the variable. This is the idea behind the Karnaugh diagram. We represent the eight possible minterms in a rectangular array in such a way that neighboring fields differ by a single variable:

	y	y	$\overline{y}$	$\overline{y}$
x				
$\overline{x}$				
	z	$\overline{z}$	$\overline{z}$	z

The upper left-hand field corresponds, for example, to the product xyz, and the lower right-hand field to $\overline{x}\,\overline{y}z$. In going from one field to a neighboring field, two variables remain the same, while the third is transformed into its complement. Here one must imagine that the fields are ordered cyclically, that is, if we move from the upper right-hand field $x\overline{y}z$ to the right, we land in the field xyz at the upper left, with an analogous situation for the columns.

We now write 1 in a field if the corresponding minterm appears in the DNF. If two 1's now appear in neighboring fields, we draw a rectangle around these 1's and eliminate the corresponding variable. In our example, this yields

	y	y	$\overline{y}$	$\overline{y}$
x				1
$\overline{x}$	1	1		
	z	$\overline{z}$	$\overline{z}$	z

$\overline{x}yz + \overline{x}y\overline{z} + x\overline{y}z$

$\longrightarrow$

	y	y	$\overline{y}$	$\overline{y}$
x				1
$\overline{x}$	1	1		
	z	$\overline{z}$	$\overline{z}$	z

$\overline{x}y + x\overline{y}z$

A SOPE representation thus corresponds to a minimal covering of all 1's with rectangles. Note that a square containing four 1's corresponds to the elimination of two variables. For example, let $f(x, y, z) = xy\overline{z} + xyz + x\overline{y}z + \overline{x}yz + \overline{x}\,\overline{y}z$. The Karnaugh diagram is

	y	y	$\overline{y}$	$\overline{y}$
x	1	1		1
$\overline{x}$	1			1
	z	$\overline{z}$	$\overline{z}$	z

and we obtain $f(x, y, z) = xy\overline{z} + z$, since at the right and left borders there is a compound square covering all four 1's.

We come now to the second question: When is a formula $f(x_1, \ldots, x_n)$, given for example, in disjunctive normal form, satisfiable? Here we have a decision problem in the sense of Section 8.5. This problem, SAT, was the first example for which NP completeness was proved, by Cook in 1971. The input length is the number of binary operations into which we decompose f, for example using DNF. Of course, SAT $\in$ NP, since for the x_i we have only to insert the values. To show that SAT is in fact NP complete is beyond the scope of this book. One may consult the book of Garey and Johnson. If we could solve SAT in polynomial time, we could do so for all problems in NP. The Boolean satisfiability problem thus stands at the center of the theory of algorithms.

11.3. Logical Nets

In the last section we saw that Boolean functions can be represented in terms of a few basis operations. We would like now to study some important functions more closely, and in particular, to derive the most efficient possible realizations.

Our model looks as follows: We are given a finite set $\Omega = \{g_i : \mathcal{B}(n_i) \to \{0,1\}\}$ of Boolean functions, which we call a *basis*. By a basis, we mean as before that all Boolean functions can be expressed in terms of Ω, say through $\Omega_0 = \{+, \cdot, ^-\}$. The input is $X = \{x_1, \ldots, x_n, 0, 1\}$, where $0, 1$ are the constants $0, 1$. A **calculation** or **realization** of $f \in \mathcal{B}(n)$ now follows step by step from the inputs and then from evaluations of functions g_i from our basis with arguments that we have already calculated. And finally, of course, we should end up with f. Formally, then, a calculation of f is a sequence $\alpha_1, \alpha_2, \ldots, \alpha_m$ such that for all α_j either $\alpha_j \in X$ or $\alpha_j = g_i(\alpha_{j_1}, \ldots, \alpha_{j_{n_i}})$ for some g_i and $j_1, \ldots, j_{n_i} < j$. Finally, $\alpha_m = f$.

We may represent a calculation $\alpha_1, \ldots, \alpha_m$ in a more suggestive way using a directed graph $\vec{G}$. The vertices are the α_j's, and we draw an arrow $\alpha_j \to \alpha_k$ if α_j goes into the calculation of α_k. For $\alpha_j \to \alpha_k$, we therefore always have $j < k$. This directed graph is called a **logical net** for the realization of f. The directed graph $\vec{G}$ has as sources the set X and as sink, $\alpha_m = f$.

Example 11.1. Let $\Omega_0 = \{+, \cdot, ^-\}$. A logical net for the calculation of $f(x_1, x_2, x_3) = x_1\overline{x}_2 + x_2\overline{x}_3 + \overline{x}_1 x_3$ is given in the following figure, where for the sake of clarity we have omitted the arrows:

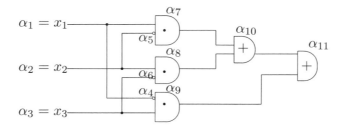

Formally, we set $\alpha_1 = x_1$, $\alpha_2 = x_2$, $\alpha_3 = x_3$, $\alpha_4 = \overline{\alpha}_1$, $\alpha_5 = \overline{\alpha}_2$, $\alpha_6 = \overline{\alpha}_3$, $\alpha_7 = \alpha_1\alpha_5$, $\alpha_8 = \alpha_2\alpha_6$, $\alpha_9 = \alpha_3\alpha_4$, $\alpha_{10} = \alpha_7 + \alpha_8$, $\alpha_{11} = \alpha_9 + \alpha_{10} = f$. We have three input steps, three negations, three multiplications, and two additions.

As indicated in the figure, we draw a small circle ∘ for negation, and for the binary operations we draw **gates**: AND gates ⊃ and OR gates ⊕. This yields a problem of both theoretical and practical interest: Given a basis Ω and $f \in \mathcal{B}(n)$, how many calculational steps do we need for the realization of f? By calculational steps we understand only compositions; the inputs are not counted. In our example, there are eight steps. Can we do better than eight? We will let $C_\Omega(f)$ denote the minimal number of calculational steps, and $C_\Omega^*(f)$ the number of *multivalued* operations, that is, negations are not counted. For our example we thereby obtain $C_{\Omega_0}(f) \leq 8$, $C_{\Omega_0}^*(f) \leq 5$.

This example suggests an additional measure of complexity. The three multiplications can be executed in *parallel*, and so they may be counted as a single step. Then come the two additions. We say that the *depth* of the net is 3 (without negations). The definition of depth is suggested by the directed graph $\vec{G}$. The *sources* are the inputs (vertices with in-degree zero), the *sinks* are the results (out-degree zero). The depth is then precisely the length of a longest directed path in $\vec{G}$, in our case 4 if we count negation, and 3 if we consider only the gates. We thus have two additional measures $D_{\Omega_0}(f) \leq 4$ and $D_{\Omega_0}^*(f) \leq 3$ for minimal realizations.

Let us again take the basis $\Omega_0 = \{+, \cdot,^- \}$. We would like to derive a lower bound for $C_{\Omega_0}^*(f)$, $f \in \mathcal{B}(n)$. To this end, we consider the graph $\vec{G}$ of the net. We say that f depends *essentially* on x_i if there exists $c_j \in \{0,1\}$, $j \neq i$, with $f(c_1, \ldots, c_{i-1}, 0, c_{i+1}, \ldots, c_n) \neq f(c_1, \ldots, c_{i-1}, 1, c_{i+1}, \cdots, c_n)$. It is clear that it is precisely the essential variables that must appear as sources in every realization of f. We may ignore the inessential variables. We therefore assume, without loss of generality, that f depends essentially on all x_i. Suppose an optimal net has ℓ negations (in-degree equals out-degree equals 1) and $C_{\Omega_0}^*(f)$ operations $+$ and $\cdot$. We now count the edges of $\vec{G}$ in two ways. There are $\ell + 2C_{\Omega_0}^*(f)$ edges that lead into vertices, and at least $n + \ell + C_{\Omega_0}^*(f) - 1$ that lead out of vertices (no edge leads from a sink, but inputs and gates may be used more than once), and thus we obtain $\ell + 2C_{\Omega_0}^*(f) \geq n + \ell + C_{\Omega_0}^*(f) - 1$, or

$$(11.1) \qquad\qquad C_{\Omega_0}^*(f) \geq n - 1 .$$

Example 11.2. Let us consider the *matching function* f_M for even n:

$$f_M(x_1, \ldots, x_{n/2}, y_1, \ldots, y_{n/2}) = \begin{cases} 1 & \text{if } x_i = y_i \text{ for all } i, \\ 0 & \text{otherwise.} \end{cases}$$

Here f_M depends on all variables. Namely, if we set all variables $\neq x_i$ equal to 1, then we have $f_M = 0$ for $x_i = 0$ and $f_M = 1$ for $x_i = 1$. By (11.1) we therefore have $C_{\Omega_0}^*(f_M) \geq n - 1$. In order to realize f_M, we first apply $g(x, y) = (x + \overline{y})(\overline{x} + y)$. The function g satisfies $g(x, y) = 1$ if and only if $x = y$. The matching function is accordingly realized by

$$f_M = g(x_1, y_1)g(x_2, y_2) \cdots g(x_{n/2}, y_{n/2}).$$

Every function $g(x_i, y_i)$ requires three binary operations from our basis, and so $\frac{3n}{2}$ altogether, and the successive multiplications of the $g(x_i, y_i)$ an additional $\frac{n}{2} - 1$. We thereby obtain

$$n - 1 \leq C_{\Omega_0}^*(f_M) \leq 2n - 1,$$

and it can be shown that $C_{\Omega_0}^*(f_M) = 2n - 1$ is the actual complexity. If instead of $\{+, \cdot, ^- \}$ we use the basis $\Omega_1 = \{\oplus, \cdot, ^- \}$, then we can realize $g(x, y) = \overline{x \oplus y}$ with one binary operation, and it follows that $C_{\Omega_1}^*(f_M) \leq \frac{n}{2} + \frac{n}{2} - 1 = n - 1$, that is, $C_{\Omega_1}^*(f_M) = n - 1$ based on the lower bound (11.1), which of course is valid as well for Ω_1.

A particularly important function arises from the interpretation of $0, 1$ words as numbers in their binary representation. Suppose we want to add $14 + 11$. In binary representation this is $14 = 0 \cdot 2^0 + 1 \cdot 2^1 + 1 \cdot 2^2 + 1 \cdot 2^3$, $11 = 1 \cdot 2^0 + 1 \cdot 2^1 + 0 \cdot 2^2 + 1 \cdot 2^3$:

$$
\begin{array}{llllll}
14: & 0 & 1 & 1 & 1, \\
11: & 1 & 1 & 0 & 1.
\end{array}
$$

We add in the binary system with sum $\boldsymbol{s}$ and carry $\boldsymbol{u}$:

$$
\begin{array}{l|cccc}
\boldsymbol{x} & 0 & 1 & 1 & 1 \\
\boldsymbol{y} & 1 & 1 & 0 & 1 \\
\hline
\boldsymbol{u} & & 0 & 1 & 1 & 1 \\
\boldsymbol{s} & 1 & 0 & 0 & 1 & 1
\end{array}
$$

The sum is $1 \cdot 2^0 + 0 \cdot 2^1 + 0 \cdot 2^2 + 1 \cdot 2^3 + 1 \cdot 2^4 = 25$.

We see that s_i and u_{i+1} are respectively functions of x_i, y_i, and u_i with the following table:

x_i	y_i	u_i	s_i	u_{i+1}
0	0	0	0	0
1	0	0	1	0
0	1	0	1	0
0	0	1	1	0
1	1	0	0	1
1	0	1	0	1
0	1	1	0	1
1	1	1	1	1

The individual building blocks of the logical net are called *full adders* (FAs). The full addition of $\boldsymbol{x} = (x_0, \ldots, x_{n-1})$, $\boldsymbol{y} = (y_0, \ldots, y_{n-1})$ is therefore represented as follows:

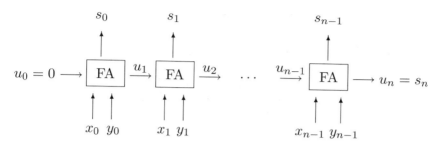

The reader may be easily convinced that the full adder is realized by the following net, where for the sake of brevity we also use $\oplus$ gates.

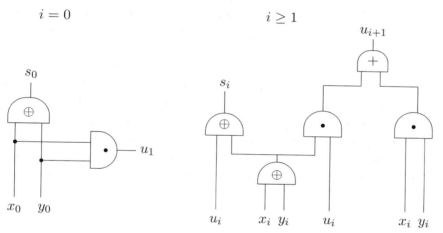

The total number of operations in the basis $\{+, \cdot, \oplus\}$ is therefore equal to $2 + 5(n - 1) = 5n - 3$. Since in $\Omega_0 = \{+, \cdot, ^-\}$ every $\oplus$ gate can be replaced by three additions and multiplications, we obtain $C^*_{\Omega_0}(f_A) \leq 4 + 9(n - 1) = 9n - 5$ for the addition function $f_A : \mathcal{B}(2n) \to \mathcal{B}(n + 1)$, $f_A(x_0, \ldots, x_{n-1}, y_0, \ldots, y_{n-1}) = (s_0, \ldots, s_{n-1}, s_n)$.

11.4. Boolean Lattices, Orders, and Hypergraphs

We now consider a Boolean algebra from the point of view of orders, and interpret $\mathcal{B}(n)$ as a **set lattice** $\mathcal{B}(S)$ on an n-set S with inclusion $A \subseteq B$ as the order relation. The lattice $\mathcal{B}(4)$ looks as follows, where we have omitted the usual braces in set notation:

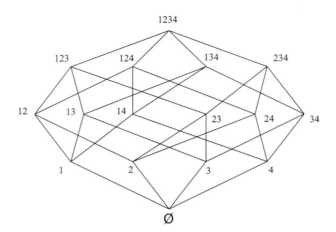

In the figure, the inclusions are indicated by edges going from the bottom to the top. We have not taken into account all inclusions, just the relations $A \subseteq B$ with $|B| = |A| + 1$; the remaining order relations are obtained by transitivity. For example, from $\{1\} \subseteq \{1,3\}, \{1,3\} \subseteq \{1,3,4\}$, it follows that $\{1\} \subseteq \{1,3,4\}$. This pictorial representation is called the **Hasse diagram** of $\mathcal{B}(4)$. (See Section 9.3, where we used Hasse diagrams in sorting problems.) We may interpret the Hasse diagram as a directed graph with edges directed upward, with $\varnothing$ as the unique source and $\{1,2,3,4\}$ as the unique sink.

Of course, a Hasse diagram can be drawn for arbitrary orders $(P, \leq)$. We connect x and y if $x < y$ and there is no z such that $x < z < y$. The following Hasse diagram represents the order on $P = \{a, b, \ldots, j\}$ with the relations $a < b, a < c, a < d, a < e, a < f, a < g, a < i, a < j, b < e, b < g, c < e, c < g, d < f, d < i, d < j, e < g, f < i, f < j$:

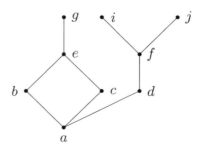

From the order $\leq$ on P certain concepts immediately suggest themselves. A **chain** in $(P, \leq)$ of *length* k is a sequence of $k + 1$ distinct elements $a_0 < a_1 < \cdots < a_k$. The length of a longest chain with terminal element a is called the *height* $h(a)$ of $a \in P$, and $h(P) = \max_{a \in P} h(a)$ is the height of P. In our example, we see that $h(a) = 0$, $h(b) = h(c) = h(d) = 1$, $h(e) = h(f) = 2$, $h(g) = h(i) = h(j) = 3$. An element a is said to be *minimal* if there is no b such that $b < a$, and analogously, it is *maximal* if there is no c for which $a < c$.

In a chain, every pair of distinct elements x, y are **comparable**, that is, either $x < y$ or $y < x$. Conversely, we say that x, y are **incomparable** if neither $x < y$ nor $y < x$ is the case. In our example, the pair b and c and the pair e and j are incomparable. An **antichain** is a subset $A \subseteq P$ of mutually incomparable elements. In other words, A contains no chains of length ≥ 1. In our example, $\{e, f\}$ and $\{b, c, i, j\}$ are antichains.

Let us return to the subject of Boolean lattices $\mathcal{B}(S)$, $|S| = n$. A chain $\mathcal{C}$ is a family of sets $A_0 \subset A_1 \subset A_2 \subset \cdots \subset A_k$, and an antichain is a family $\mathcal{A} \subseteq \mathcal{B}(S)$ with $A \not\subseteq B$ and $B \not\subseteq A$ for all $A, B \in \mathcal{A}$. What is the height of $A \subseteq S$? Clearly, it is $|A|$, since the longest chains from $\varnothing$ to A correspond to a permutation of the elements of $A = \{a_1, \ldots, a_k\}$: $\varnothing \subset \{a_1\} \subset \{a_1, a_2\} \subset \cdots \subset A$. In particular, the height of $\mathcal{B}(S)$ is equal to n, and the chains of length n correspond precisely to the $n!$ permutations of $\{1, 2, \ldots, n\}$.

We now ask how large an antichain in $\mathcal{B}(S)$ can be. It is clear that $\binom{S}{k} = \{A \subseteq S : |A| = k\}$ is an antichain for all k (since no k-set can be contained in another k-set), and therefore $|\binom{S}{k}| = \binom{n}{k}$. We know that the largest binomial coefficients occur in the middle, whence $\max_{0 \leq k \leq n} \binom{n}{k} = \binom{n}{\lfloor n/2 \rfloor}$ (see Exercise 1.8). That there are no larger antichains is the content of the following theorem of Sperner.

Theorem 11.3 (Sperner). *In $\mathcal{B}(S)$, $|S| = n$, one has*

$$\max\left(|\mathcal{A}| : \mathcal{A} \text{ an antichain}\right) = \binom{n}{\lfloor n/2 \rfloor}.$$

Proof. There are many proofs of this fundamental theorem. The following is the briefest and most elegant.

Let $|\mathcal{A}| = m$, and let f_i be the number of sets $A \in \mathcal{A}$ with $|A| = i$, that is, $\sum_{i=0}^{n} f_i = m$. We consider the $n!$ maximal chains in $\mathcal{B}(S)$ from $\varnothing$ to $\{1, \ldots, n\}$. Since $\mathcal{A}$ is an antichain, such a maximal chain meets the antichain $\mathcal{A}$ in at most one set. The number of chains that go through an $A \in \mathcal{A}$ is $|A|!(n - |A|)!$, since we may first go in all possible ways from $\varnothing$ to A and then from A to $S = \{1, \ldots, n\}$ (product rule). We thereby obtain

$$n! \geq \sum_{A \in \mathcal{A}} |A|! \, (n - |A|)! = \sum_{i=0}^{n} f_i \, i!(n-i)!,$$

or

(11.2)
$$1 \geq \sum_{i=0}^{n} f_i \frac{1}{\binom{n}{i}}.$$

If we replace $\binom{n}{i}$ with the maximum $\binom{n}{\lfloor n/2 \rfloor}$, then (11.2) yields

$$1 \geq \sum_{i=0}^{n} f_i \frac{1}{\binom{n}{\lfloor n/2 \rfloor}} = \frac{1}{\binom{n}{\lfloor n/2 \rfloor}} \sum_{i=0}^{n} f_i = \frac{1}{\binom{n}{\lfloor n/2 \rfloor}} m ,$$

and therefore as desired, $m \leq \binom{n}{\lfloor n/2 \rfloor}$. Since $\binom{S}{\lfloor n/2 \rfloor}$ is an antichain, the proof is complete. □

Moreover, it can easily be shown that the only maximal antichains in $\mathcal{B}(S)$ are precisely the families $\binom{S}{\frac{n}{2}}$ for n even and $\binom{S}{\frac{n-1}{2}}$, $\binom{S}{\frac{n+1}{2}}$ for n odd (see Exercise 11.30).

Let us now pose the following problem: We would like to decompose $\mathcal{B}(S)$ into the smallest possible number of disjoint chains. Clearly, we need at least $\binom{n}{\lfloor n/2 \rfloor}$ chains, since no two sets of an antichain can appear in a chain. However, is there a decomposition into $\binom{n}{\lfloor n/2 \rfloor}$ chains? Indeed there is, and the relevant general theorem of Dilworth tells us that this applies to arbitrary orders.

Theorem 11.4 (Dilworth). *Let* $(P, \leq)$ *be a finite order. Then*

$$\min \left(|\mathcal{C}| : \mathcal{C} \text{ a chain decomposition} \right) = \max \left(|A| : A \text{ an antichain} \right).$$

Proof. Let $\mathcal{C} = \{C_1, C_2, \ldots, C_t\}$ be a decomposition of P into t disjoint chains. Since the elements of an antichain A appear in different C_i's, we have $|\mathcal{C}| \geq |A|$, and hence

$$\min |\mathcal{C}| \geq \max |A| .$$

To prove the equality, we employ induction on $n = |P|$. For $n = 1$ there is nothing to prove. Let $\mathcal{A}(P)$ be the family of antichains of P. We assume inductively that $\min |\mathcal{C}| = \max |A|$ for all orders Q with $|Q| < |P|$. Let $m = \max \left(|A| : A \in \mathcal{A}(P) \right)$. We first assume that there exists an antichain A with $|A| = m$ that contains neither all maximal elements nor all minimal elements of P. We define P^+ and P^- by

$$P^+ = \{p \in P : p \geq a \text{ for some } a \in A\} ,$$
$$P^- = \{p \in P : p \leq a \text{ for some } a \in A\} .$$

The assumption on A implies $P^+ \neq P$, $P^- \neq P$ and $P = P^+ \cup P^-$, $A = P^+ \cap P^-$. (Why?) By induction, each of P^+ and P^- can be decomposed into m chains. Now paste these chains together in the elements of A to obtain a chain decomposition of P into m chains.

In the other case, each antichain A with $|A| = m$ contains either all maximal or all minimal elements. There can therefore be at most two such antichains: one comprising all maximal elements, the other all minimal elements. Let a be a maximal element and b a minimal element with $b \leq a$.

By induction, we can decompose $P \setminus \{a, b\}$ into $m - 1$ chains and then we can add on $b \leq a$. $\qquad \square$

Dilworth's theorem, with its statements about maxima and minima, is very similar to earlier theorems of this type, and that is no coincidence. Theorem 8.5 on maximum matchings in bipartite graphs is an immediate consequence of Theorem 11.4. To show this, we interpret $G = (S + T, E)$ as an order on $V = S + T$ by orienting the edges of S toward T, that is, we set $u < v$ if $u \in S$, $v \in T$, and $uv \in E$. A disjoint chain decomposition corresponds to a set of nonincident edges together with the remaining vertices. We clearly obtain a minimal decomposition including as many edges as possible, that is, taking a maximum matching M. We thereby obtain $|\mathcal{C}| = |M| + (|V| - 2|M|) = |V| - |M|$; hence

$$\min |\mathcal{C}| = |V| - \max \left(|M| : M \text{ a matching} \right).$$

An antichain corresponds to a set U of vertices that are not joined by any edges. The set complement $V \setminus U$ is therefore a vertex cover, and conversely, if D is a vertex cover, then $V \setminus D$ is an antichain. A maximal antichain A thus corresponds to a minimal vertex cover $D = V \setminus A$, and we conclude that

$$\max |A| = |V| - \min \left(|D| : D \text{ a vertex cover} \right).$$

From Dilworth's theorem we immediately obtain $\max |M| = \min |D|$.

We may also interpret families of sets $\mathcal{F} \subseteq \mathcal{B}(S)$ as a generalization of graphs. We call S the *vertices* and the sets $A \in \mathcal{F}$ the (hyper-) *edges*. Then $(S, \mathcal{F})$ is called a **hypergraph**. Sets A that appear multiple times in $\mathcal{F}$ are called *multiple edges*. Graphs, therefore, are the special case $\mathcal{F} \subseteq \binom{S}{2}$. Many notions and theorems of graph theory can be immediately extended to hypergraphs. For example, the *degree* $d(u)$, $u \in S$, is the number of $A \in \mathcal{F}$ with $u \in A$. If we write down the usual incidence matrix $M = (m_{ij})$, $S = \{u_1, \ldots, u_n\}$, $\mathcal{F} = \{A_1, \ldots, A_q\}$ with

$$m_{ij} = \begin{cases} 1 & \text{if } u_i \in A_j, \\ 0 & \text{otherwise,} \end{cases}$$

we obtain by double counting

$$\sum_{u \in S} d(u) = \sum_{j=1}^{q} |A_j| \, .$$

The exercises contain a number of additional examples of how graph-theoretic notions can be successfully applied to hypergraphs, that is, to families of sets. Of particular interest is the case $\mathcal{F} \subseteq \binom{S}{k}$. We then call the hypergraph $(S, \mathcal{F})$ **k-uniform**. Graphs are therefore precisely the 2-uniform hypergraphs.

Let us consider an example that is perhaps all too familiar: lotteries. The numbers involved are $S = \{1, 2, \ldots, 45\}$, and a lottery ticket consists of a number of wagers, each of which involves the selection of six numbers from S, that is, a family $\mathcal{F} \subseteq \binom{S}{6}$. Thus $(S, \mathcal{F})$ is a 6-uniform hypergraph. If X is the chosen 6-set, then a ticket wins a payout if $|X \cap A| \geq 3$ for at least one $A \in \mathcal{F}$. To *ensure* a full win of all six numbers, we must choose $\mathcal{F} = \binom{S}{6}$ with $|\mathcal{F}| = \binom{45}{6} = 8{,}145{,}060$. That won't do, of course (and that is why the state makes a profit on each drawing, even if there is a full winner, since it pays out only about 30% of the revenue from ticket sales in prize money).

In any event, we are dealing with a minimization problem: To be certain of winning at least something, we must construct a 6-uniform hypergraph $(S, \mathcal{F})$ such that for *every* $X \in \binom{S}{6}$ there exists an $A \in \mathcal{F}$ with $|X \cap A| \geq 3$. What is $\min |\mathcal{F}|$? No one knows, and even for considerably smaller numbers than 45, for example 15, the problem remains open.

Let us simplify the situation. A 6-uniform hypergraph $(S, \mathcal{F})$ is clearly a successful one if every triple of numbers appears in an $A \in \mathcal{F}$. We say that $\mathcal{F}$ *covers* all $\binom{S}{3}$. In general, we have the following problem: Let $n \geq k \geq t$ be given, $|S| = n$. The **covering number** $C(n, k, t)$ is the *minimal* size of a hypergraph $\mathcal{F} \subseteq \binom{S}{k}$ such that every t-set appears in *at least one* $A \in \mathcal{F}$. Analogously, the **packing number** $P(n, k, t)$ is the *maximal* size of $\mathcal{F} \subseteq \binom{S}{k}$ such that every t-set appears in *at most one* $A \in \mathcal{F}$.

Clearly, we have $C(n, k, 1) = \lceil \frac{n}{k} \rceil$, $P(n, k, 1) = \lfloor \frac{n}{k} \rfloor$. Let us consider the first case of interest: $k = 3$, $t = 2$. Let $\mathcal{F} \subseteq \binom{S}{3}$ be a covering set of minimal size $|\mathcal{F}| = C = C(n, 3, 2)$. We consider the pairs (X, A), $X \in \binom{S}{2}$, $A \in \mathcal{F}$, $X \subseteq A$. By double counting we obtain $\binom{n}{2} \leq 3C$, whence $C \geq \left\lceil \frac{n(n-1)}{3 \cdot 2} \right\rceil$. For $n = 4$ this yields $C(4, 3, 2) \geq 2$. However, two triples, e.g., $\{1, 2, 3\}$, $\{1, 2, 4\}$, do not suffice, since $\{3, 4\}$ is not covered.

Theorem 11.5. $C(n, 3, 2) \geq \left\lceil \frac{n}{3} \left\lceil \frac{n-1}{2} \right\rceil \right\rceil$, $P(n, 3, 2) \leq \left\lfloor \frac{n}{3} \left\lfloor \frac{n-1}{2} \right\rfloor \right\rfloor$.

Proof. Let $A_1, \ldots, A_C$, where $C = C(n, 3, 2)$ is the set of triples in $\mathcal{F}$, and let S be the underlying set. For $u \in S$ let r_u be the number of triples in $\mathcal{F}$ that contain u. Since every pair $\{u, v\}$ must be contained in a triple A_i, and every A_j contributes at most two pairs $\{u, x\}$, it follows that $r_u \geq \frac{n-1}{2}$, and thus $r_u \geq \left\lceil \frac{n-1}{2} \right\rceil$. Now we have $3C = \sum_{u \in S} r_u$, and therefore $C = \frac{1}{3} \sum_{u \in S} r_u \geq \frac{n}{3} \left\lceil \frac{n-1}{2} \right\rceil$, whence $C \geq \left\lceil \frac{n}{3} \left\lceil \frac{n-1}{2} \right\rceil \right\rceil$. The proof for $P(n, 3, 2)$ proceeds analogously. $\square$

We obtain the best case possible (without rounding) for $C(n, 3, 2) = \frac{n(n-1)}{6}$. If that is satisfied, then every pair must be in *precisely* one triple, and so $P(n, 3, 2) = \frac{n(n-1)}{6}$ must hold as well. When is this possible? First we have the *arithmetic* condition that $\frac{n(n-1)}{6}$ must be an integer. That is, n must be of the form $n = 6m + 1$ or $n = 6m + 3$, which is satisfied only for $n = 3, 7, 9, 13, 15, \ldots$. We consider the first interesting case: $n = 7$. The following family $\mathcal{F}$ with $|\mathcal{F}| = 7 = \frac{7 \cdot 6}{6}$ satisfies the condition:

$$\mathcal{F} = \{124, 235, 346, 457, 561, 672, 713\} .$$

The reader may well have observed that $\mathcal{F}$ has a special structure. The first triple is $1, 2, 4$ and we obtain the remaining ones by adding 1 to the elements cyclically, or in mathematical terminology, modulo 7.

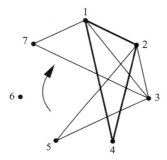

Generally, we shall expect that the extremal families will possess a particular internal structure, and this (algebraic) structure is the topic of the next chapter.

Exercises for Chapter 11

11.1 Verify the usual rules for $(\mathcal{B}(n), +, \cdot)$ such as commutativity, associativity, and distributivity and show furthermore that $x + \bar{x} = 1$, $x + x = x$, $x \cdot \bar{x} = 0$, $x \cdot x = x$, $x + xy = x(x + y) = x$.

11.2 Are the following statements consistent or inconsistent? (a) $A_1 = \{(x \to y) \to z, (\neg x \vee y) \to (y \wedge z), z \to (\neg x \to y)\}$, (b) $A_1 \cup \{\neg(y \to z)\}$, (c) $A_1 \cup \{y \to z\}$.

11.3 Suppose the Boolean functions $f, g \in \mathcal{B}(n)$ are monotonic, $s = x_1 \wedge \cdots \wedge x_n$, $t = x_1 \vee \cdots \vee x_n$. Show that: (a) $s \leq f \vee g \Rightarrow s \leq f$ or $s \leq g$, (b) $f \wedge g \leq t \Rightarrow f \leq t$ or $g \leq t$.

▷ **11.4** Show that CNF follows from DNF by applying de Morgan's laws.

11.5 Show that every $f \in \mathcal{B}(n)$ can be written in the form

$$f(x_1, \ldots, x_n) = x_1 f(1, x_2, \ldots, x_n) + \bar{x}_1 f(0, x_2, \ldots, x_n)$$

and apply this principle to $f = x_1 \bar{x}_2 + x_2 x_3 + x_2 \bar{x}_3 x_4$ to obtain the DNF.

▷ **11.6** Let $f : \mathcal{B}(3) \to \{0, 1\}$ be defined by $f(0, 0, 0) = f(0, 1, 1) = 0$ and 1 otherwise. Determine the DNF and CNF and find a simpler representation with three summands in which the variables appear five times in all.

11.7 Interpret all 16 Boolean functions f with two variables as logical expressions and as operations on sets.

11.8 Show that $\Omega = \{\oplus, \cdot\}$ is no longer a basis if the constants 0 and 1 are not given.

11.9 With each order P we associate a graph $G(P)$ as follows: The vertices of $G(P)$ are the elements of P, and $xy \in K(G(P))$ if and only if $x < y$ or $y < x$ in P. The graph $G(P)$ is called the *comparability graph* of P. Think about what Dilworth's theorem means for $G(P)$.

▷ **11.10** Prove the following analogue of Theorem 11.4: Let P be an order. Then the number of elements in a longest chain is equal to the minimum number of antichains into which P can be decomposed.

▷ **11.11** Let $\mathcal{H} = (S, \mathcal{F})$ be a hypergraph. A *circuit* in $\mathcal{H}$ consists of a sequence

$$u_1, F_1, u_2, F_2, \ldots, F_k, u_1$$

with $u_i \in S, F_j \in \mathcal{F}$ such that successive members are incident. The hypergraph $\mathcal{H}$ is said to be *connected* if every pair u, v is connected by a sequence $u, F_1, u_2, \ldots, F_k, v$. Show that $\mathcal{H}$ is connected and circuit-free if and only if we have $\sum_{F \in \mathcal{F}} (|F| - 1) = |S| - 1$. Hint: Consider the bipartite graph $G = (S + \mathcal{F}, E)$ with $uF \in E \Leftrightarrow u \in F$.

11.12 Suppose that every pair of edges of a hypergraph $\mathcal{H} = (S, \mathcal{F})$ have an element in common and $\mathcal{H}$ has no multiple edges. Show that $|\mathcal{F}| \le 2^{n-1}$, $n = |S|$. For which $\mathcal{F}$ is there equality?

▷ **11.13** Prove the ring-sum expansion normal form (RSE): Every Boolean function f can be expressed uniquely in the form $f(\boldsymbol{x}) = \sum_I \oplus a_I \boldsymbol{x}_I$, where for $I = \{i_1 < \cdots < i_k\} \subseteq \{1, \ldots, n\}$, we have $\boldsymbol{x}_I = x_{i_1} x_{i_2} \cdots x_{i_k}$ and $a_I \in \{0, 1\}$. Here $\sum \oplus$ denotes summation with $\oplus$.

11.14 Determine DNF, CNF, and RSE for the following Boolean functions: (a) $x_1 x_2 + x_3$, (b) $x_1 + \cdots + x_n$, (c) $x_1 x_2 \cdots x_n$.

▷ **11.15** Let $f \ne g \in \mathcal{B}(n)$ be given by their RSE representations. How can one find an input $\boldsymbol{a}$ with $f(\boldsymbol{a}) \ne g(\boldsymbol{a})$ without testing all inputs?

▷ **11.16** A problem from chemistry: If a chemical reaction produces a white precipitate, the test tube contains sodium or ammonia. If there is no sodium in the test tube, then it contains iron. If iron is present and there is a white precipitate, then no ammonia is present. What must be present if the reaction produces a white precipitate?

11.17 Show that the Sheffer stroke $\uparrow$ given by

x	y	$x \uparrow y$
0	0	1
1	0	1
0	1	1
1	1	0

together with the constants $0, 1$ forms a one-element basis. Among the 16 functions in $\mathcal{B}(2)$ there is another one-element basis. What is it?

▷ **11.18** Show that the basis $\Omega = \{+, \cdot\}$ realizes precisely the monotone functions.

11.19 Show that the SOPE representation of $x_1 \oplus \cdots \oplus x_n$ is equal to the DNF.

▷ **11.20** Construct a Karnaugh diagram for four variables x, y, z, w; that is, neighboring fields again differ in precisely one variable. Determine a SOPE representation for the function

$$g(x, y, z, w) = xy\bar{z}\,\bar{w} + xy\bar{z}w + x\bar{y}\,\bar{z}\,\bar{w} + xyzw + \bar{x}yz\,\bar{w} + \bar{x}\,\bar{y}\,\bar{z}\,w + \bar{x}\,\bar{y}\,z\,\bar{w}$$

with the help of the Karnaugh diagram. Hint: Three summands suffice.

11.21 Determine $C_\Omega^*(f)$ for each of the following functions in terms of the basis $\Omega = \{+, \cdot, {}^-, \oplus\}$:

(a) $f(x_1, \ldots, x_n, y_1, \ldots, y_n) = \begin{cases} 1 & x_i + y_i = 1 \text{ for all } i, \\ 0 & \text{otherwise;} \end{cases}$

(b) $f(x_1, \ldots, x_n, y_1, \ldots, y_n) = \begin{cases} 1 & x_i \neq y_i \text{ for all } i, \\ 0 & \text{otherwise.} \end{cases}$

11.22 A logical net has *fan-in* (*fan-out*) s if every gate has at most s entrances (exits). Let Ω be a basis and $C_s(f)$ the complexity for fan-out s. Show that $C_{s+1}(f) \leq C_s(f)$, $s \geq 1$.

▷ **11.23** Let Ω be a basis with fan-in r. We call $L_\Omega(f) = C_1(f)$ (fan-out 1) the *formula length* of $f \in \mathcal{B}(n)$. Show that $\log_r((r-1)L_\Omega(f) + 1) \leq D_\Omega(f)$, where $D_\Omega(f)$ is the depth of f.

11.24 Construct a logical net with fan-out 1 for $f(x_1, \ldots, x_n) = x_1 x_2 \cdots x_n$, $n = 2^k$, that assumes simultaneously the minimum for $L_\Omega(f)$ and for $D_\Omega(f)$. Conclude that equality can hold in the previous exercise.

▷ **11.25** Let $N(n)$ be the number of Boolean functions $f \in \mathcal{B}(n)$ that depend essentially on every variable. Prove that $\sum_{j=0}^{n} N(j)\binom{n}{j} = 2^{2^n}$ with $N(0) = 2$ and determine $N(n)$ for $n \leq 4$. Calculate $\lim_{n \to \infty} N(n)/2^{2^n}$.

11.26 Let $f_1, \ldots, f_m$ be Boolean functions with n variables such that $f_i \neq f_j, \bar{f}_j$ for all $i \neq j$ and $C_\Omega^*(f_j) \geq 1$ for all j, with Ω any basis. Show that $C_\Omega^*(f_1, \ldots, f_m) \geq m - 1 + \min_j C_\Omega^*(f_j)$.

▷ **11.27** The function $f_T^{(n)}(x_0, \ldots, x_{n-1}) = (g_0, \ldots, g_{2^n-1})$ is called a binary position transformation. One has $g_i(x_0, \ldots, x_{n-1}) = 1 \Leftrightarrow \sum_{j=0}^{n-1} x_j 2^j = i$, where $\sum$ is the usual addition of numbers. Show that $2^n + n - 2 \leq C_{\Omega_0}^*(f_T^{(n)}) \leq 2^n + n2^{\lceil n/2 \rceil} - 2$ for $\Omega_0 = \{+, \cdot, {}^-\}$. Hint: One has $g_i(x_0, \ldots, x_{n-1}) = m_c(x_0, \ldots, x_{n-1}) = x_0^{c_0} \cdots x_{n-1}^{c_{n-1}}$ (minterm function), where c is the binary representation of i.

11.28 Determine good lower and upper bounds for $C_\Omega^*(f)$ and $D_\Omega^*(f)$ from

$$f : \{0, 1\}^{(n+1)b} \to \{0, 1\} \text{ with } f(\boldsymbol{x}_1, \ldots, \boldsymbol{x}_n, \boldsymbol{y}) = \begin{cases} 1 & \boldsymbol{x}_i = \boldsymbol{y} \text{ for some } i, \\ 0 & \text{otherwise,} \end{cases}$$

over the basis $\Omega = \{+, \cdot, {}^-, \oplus\}$, where $\boldsymbol{x}_i, \boldsymbol{y} \in \{0, 1\}^b$.

11.29 A committee of three persons has the following rule: If a majority is "in favor," a lamp is lighted. Construct a logical net for this situation.

▷ **11.30** Show that the antichains of maximal size in $\mathcal{B}(S)$, $|S| = n$, are the families $\binom{S}{\lfloor n/2 \rfloor}$ and $\binom{S}{\lceil n/2 \rceil}$.

▷ **11.31** A lattice $(P, \leq)$ is said to be *distributive* if the laws $x \wedge (y \vee z) = (x \wedge y) \vee (x \wedge z)$ and $x \vee (y \wedge z) = (x \vee y) \wedge (x \vee z)$ hold. For example, the Boolean lattice is distributive. Show that in a lattice, the first law implies the second, and conversely.

11.32 Suppose the 3-uniform hypergraph $\mathcal{H} = (S, \mathcal{F})$ satisfies $|\mathcal{F}| = |S| - 1$. Prove that $\mathcal{H}$ contains a circuit.

11.33 Let $\mathcal{H} = (S, \mathcal{F})$ be a 3-uniform hypergraph without multiple edges, $|S| = n \geq 6$, such that every pair of edges has a nonempty intersection. Show that $|\mathcal{F}| \leq \binom{n-1}{2}$. Construct a hypergraph in which equality holds.

▷ **11.34** Let $(S, \mathcal{F})$ be a hypergraph with $|S| = n, |\mathcal{F}| = t$. Suppose $|F| < n$ for all $F \in \mathcal{F}$, and whenever $u \notin F$, we have $d(u) \geq |F|$. Show that one therefore has $t \geq n$. Hint: Show first that from $t < n$, it would follow that $\frac{d(u)}{t - d(u)} > \frac{|F|}{n - |F|}$ for every pair (u, F) with $u \notin F$.

11.35 Conclude the following assertion from the previous exercise: Let $\mathcal{H} = (S, \mathcal{F})$ be a hypergraph with $S \notin \mathcal{F}$, in which every pair of elements from S lie on exactly one edge. Then $|S| \leq |\mathcal{F}|$. Can there ever be equality?

11.36 Let S be a set of n points in the plane, not all lying on a single line. What can one say about the set $\mathcal{G}$ of straight lines determined by S? Hint: Use the previous exercise.

11.37 Let P be an order. A *filter* F is a set F with the property $x \in F$, $y \geq x \Rightarrow y \in F$, inclusive of $F = \emptyset$. Determine the number of filters for $P = \mathcal{B}(2)$ and $P = \mathcal{B}(3)$.

▷ **11.38** Let P be an order. Show that there are precisely as many filters in P as there are antichains.

11.39 Let P be an order and $\mathcal{A}$ the set of maximal antichains A (that is, A is maximal if $A \cup \{x\}$ is not an antichain for every $x \notin A$). For $A, B \in \mathcal{A}$ set $A \leq B$ if for each $x \in A$ there is a $y \in B$ with $x \leq y$. Show that $\leq$ defines a lattice $\mathcal{A}_{\leq}$ on $\mathcal{A}$. What does $\mathcal{A}_{\leq}$ look like for $P = \mathcal{B}(n)$. Hint: For $A, B \in \mathcal{A}$ consider the maximal and minimal elements in $A \cup B$.

11.40 In a lottery, 3 numbers out of n are drawn, and each wager contains 4 numbers. A winning wager is one containing all three of the numbers drawn. How many wagers must one make to be certain of winning? Construct an optimal set of lottery tickets for $n = 5, 6, 7$.

Chapter 12

Modular Arithmetic

12.1. Calculating with Congruences

The use of congruences is one of the most important techniques for dealing with number-theoretic questions. Although we shall assume a certain familiarity with the concept, we shall nevertheless review all the important ideas.

The object of study in this section is the set of integers $\mathbb{Z}$. The fundamental theorem of number theory, on which the entire subject rests, is the statement of the uniqueness of prime decomposition: Every natural number $n \geq 2$ can be decomposed *uniquely* into a product $n = p_1^{k_1} \cdots p_t^{k_t}$, where the p_i are prime numbers. In particular, we may conclude that if p is a prime number with $p \mid mn$ and if p and m are relatively prime, that is, $p \nmid m$, then $p \mid n$.

Let $x, y \in \mathbb{Z}$ and m a positive integer. We say that x is **congruent** to y modulo m, with the notation

$$x \equiv y \;(\text{mod } m),$$

if $x - y$ is divisible by m.

It should be immediately apparent that $\equiv$ is an equivalence relation for every integer m, and we always have $x \equiv r \;(\text{mod } m)$ when r is the *remainder* of the division $x \div m$. That is, if $x = qm + r$, $0 \leq r \leq m - 1$, then $x - r = qm$, and therefore $x \equiv r \;(\text{mod } m)$. Every integer x is therefore congruent to one of the integers $0, 1, \ldots, m - 1$. Since the integers $0, 1, \ldots, m - 1$ are mutually incongruent, it follows that there are precisely m congruence classes modulo m.

We can imagine modular calculation as taking place cyclically along a circle with circumference m:

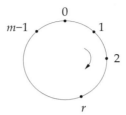

To determine the congruence class of x, we take x steps of unit length around the circle, starting at zero. After every m steps we again find ourselves at zero. If $x = qm + r$, $0 \le r \le m - 1$, we make q complete circumnavigations of the circle, and then continue until we end up at r.

For $x \in \mathbb{Z}$ we let $[x]$ denote the **congruence class**, or **residue class**, modulo m. Every integer $y \in [x]$, that is, $y \equiv x \pmod{m}$, is called a *representative* of $[x]$. If we select a single representative from each residue class, we say that we have a (complete) *residue system*. We know already that $0, 1, \ldots, m-1$ forms a residue system, which we call the *standard system* modulo m. However, we could equally well take $-3, -2, -1, 0, 1, 2, \ldots, m-4$ or any other set of m incongruent integers. Thus for $0 \le r \le m - 1$ the set $[r]$ consists of all integers that yield remainder r when divided by m. For example, for $m = 2$, the set $[0]$ is the set of even integers, and $[1]$ the set of odd integers.

Now that we know what congruences are, it is time to ask what they might be good for. First of all, they simplify the solution of problems about divisibility of integers. Thus without performing any division, we can say at once that the integer 4173905263 (in decimal notation) is divisible by 11. We shall soon see why this is the case.

We can calculate with congruence classes by setting

$$[x] + [y] = [x + y],$$
$$[x]\,[y] = [xy].$$

The definitions of sum and product are well defined, in that they do not depend on the choice of representatives x, y, since $x \equiv x'$, $y \equiv y'$ always yields $x + y \equiv x' + y'$, $xy \equiv x'y'$ (check this). The congruence classes therefore form a ring $\mathbb{Z}_m$, called the **residue class ring** modulo m, with additive identity $[0]$ and unit $[1]$.

Taking $m = 5$, for example, we obtain the following addition and multiplication tables for $\mathbb{Z}_5$:

+	0	1	2	3	4
0	0	1	2	3	4
1	1	2	3	4	0
2	2	3	4	0	1
3	3	4	0	1	2
4	4	0	1	2	3

·	1	2	3	4
1	1	2	3	4
2	2	4	1	3
3	3	1	4	2
4	4	3	2	1

We can now solve the question given above about divisibility by 11 effortlessly. Let $n = a_s a_{s-1} \ldots a_0$ be a number written in decimal notation. Thus $n = \sum_{i=0}^{s} a_i 10^i$, $0 \leq a_i \leq 9$. Since $10 \equiv -1 \pmod{11}$, it follows that $10^2 \equiv (-1)^2 = 1 \pmod{11}$, $10^3 = 10^2 \cdot 10 \equiv 1 \cdot (-1) = -1 \pmod{11}$, and in general, $10^{2i} \equiv 1 \pmod{11}$, $10^{2i+1} \equiv -1 \pmod{11}$. We thereby obtain $n = \sum_{i=0}^{s} a_i 10^i \equiv a_0 - a_1 + a_2 - \cdots \pmod{11}$. An integer n is therefore divisible by 11, that is, $n \equiv 0 \pmod{11}$, if and only if the alternating sum of the digits $a_0 - a_1 + a_2 - \cdots$ is divisible by 11. In our example, we obtain $3 - 6 + 2 - 5 + 0 - 9 + 3 - 7 + 1 - 4 = -22 \equiv 0 \pmod{11}$, and the divisibility is proven.

When is $\mathbb{Z}_m$ a field? That is, when does every element $[x] \neq [0]$ possess a multiplicative inverse $[y]$ with $[x][y] = [1]$? If $m = m_1 m_2$ is a composite number, then we have $[m_1][m_2] = [m] = [0]$, and therefore $[m_1]$ cannot possess a multiplicative inverse $[y]$, since that would yield $[m_2] = [1][m_2] = [y][m_1][m_2] = [0]$. However, in the case of a prime number p, we indeed have a field. Let $1 \leq r \leq p - 1$. Consider all multiples kr, $0 \leq k \leq p - 1$. From $p \mid kr - k'r = (k - k')r$ it follows that $p \mid k - k'$ and therefore $k = k'$, since p and r are relatively prime. Therefore, the integers kr, $0 \leq k \leq p - 1$, also form a system of representatives modulo p, which means that there is precisely one integer s, $1 \leq s \leq p - 1$, such that $sr \equiv 1 \pmod{p}$, which means that $[s]$ is an inverse to $[r]$.

Let us verify this for $m = 5$ and $r = 2$. The multiples are $0 \cdot 2 \equiv 0$, $1 \cdot 2 \equiv 2$, $2 \cdot 2 \equiv 4$, $3 \cdot 2 \equiv 1$, $4 \cdot 2 \equiv 3$, modulo 5, and therefore $[3]$ is the inverse of $[2]$.

One of the classical results of number theory is the theorem of Fermat, which provides an important method for testing a number for primality. Let p be a prime number, and n an integer relatively prime to p. Then $n^{p-1} \equiv 1 \pmod{p}$. For example, we have $4^2 \equiv 1 \pmod{3}$, or $20576^{42} \equiv 1 \pmod{43}$; that is, the enormous number 20576^{42} has remainder 1 on division by 43, and we obtain this result without performing any division, which given the size of the number, we couldn't do anyhow.

Theorem 12.1. *Let p be a prime number and n an integer relatively prime to p. Then $n^{p-1} \equiv 1 \pmod{p}$.*

Proof. By calculating with congruences, the proof is reduced to a couple of lines. We consider the residues $1, 2, \ldots, p-1$. We know that $1 \cdot n, 2 \cdot n, \ldots, (p-1)n$ again runs through all the nonzero residues, and so for the product $u = 1 \cdot 2 \cdots (p-1)$, we have $u \equiv (1 \cdot n)(2 \cdot n) \cdots (p-1)n \equiv u \cdot n^{p-1}$ modulo p. Since u possesses a multiplicative inverse v, it follows that $1 \equiv vu \equiv (vu)n^{p-1} \equiv n^{p-1}$ modulo p, and we are done. □

There is a lovely combinatorial proof of Fermat's theorem that also involves "cyclic" calculation modulo p.

Consider all necklaces with p beads each colored with one of n different colors. How many patterns of necklaces are there? We have known the answer for a long time. Pólya's theorem (Theorem 4.2) applied to the cyclic group C_p yields the number

$$\frac{1}{p}\sum_{d|p}\varphi(d)n^{p/d} = \frac{1}{p}\left(n^p + (p-1)n\right) = \frac{n^p - n}{p} + n\,.$$

Thus $\frac{n^p - n}{p}$ is an integer, and since n and p are relatively prime, it follows that $p \mid n^{p-1} - 1$.

12.2. Finite Fields

We know from the previous section that $\mathbb{Z}_p$ is a field with p elements whenever p is a prime number. In the previous example we saw that the fields $\mathbb{Z}_p$ exhibit a great deal of structure that is relevant to the solution of combinatorial problems. That is reason enough to ask whether there are any more finite fields. For the reader with only a modest interest in abstract algebra, it suffices to say that finite fields with q elements exist for all prime powers $q = p^n$, and up to isomorphism, precisely one for each such q. The field with $q = p^n$ elements is called the **Galois field** GF(q). In particular, for a prime number p, GF(p) = $\mathbb{Z}_p$.

Let K be a finite field. We let $K[x]$ denote the ring of polynomials $f(x) = a_0 + a_1 x + a_2 x^2 + \cdots$ with coefficients in K. We can add and multiply polynomials, and we have the zero element $f(x) = 0$ and the multiplicative unit $f(x) = 1$. Let us consider the example of $\mathbb{Z}_5[x]$. We choose the standard representatives $0, 1, 2, 3, 4$ and abuse notation by leaving off the brackets in the notation for the class $[i]$. Then for $f(x) = 2 + 3x + x^3$, $g(x) = 1 + 4x + x^2 + 4x^3$ we have $f(x) + g(x) = 3 + 2x + x^2$ and $f(x)g(x) = 2 + x + 4x^2 + 2x^3 + x^4 + x^5 + 4x^6$. For example, the coefficient of x^3 in $f(x)g(x)$ is $2 \cdot 4 + 3 \cdot 1 + 0 \cdot 4 + 1 \cdot 1 \equiv 3 + 3 + 0 + 1 = 7 \equiv 2 \pmod{5}$.

We now proceed just as we did in the ring of integers $\mathbb{Z}$. We call two polynomials $g(x), h(x)$ *congruent modulo* $f(x)$, with the notation

$$g(x) \equiv h(x) \,(\mathrm{mod}\; f(x)),$$

if $f(x)$ is a divisor of $g(x) - h(x)$, that is, if $g(x) - h(x) = q(x)f(x)$ for some polynomial $q(x) \in K[x]$. In complete analogy to the case $\mathbb{Z}$, we see that $\equiv$ is again an equivalence relation, and again we have for $g(x) \equiv g'(x)$, $h(x) \equiv h'(x)$ the relations $g(x) + h(x) \equiv g'(x) + h'(x) \,(\mathrm{mod}\; f(x))$ and $g(x)h(x) \equiv g'(x)h'(x) \,(\mathrm{mod}\; f(x))$. With the definition of addition and multiplication

$$[g(x)] + [h(x)] = [g(x) + h(x)],$$
$$[g(x)]\,[h(x)] = [g(x)h(x)],$$

$K[x] \,(\mathrm{mod}\; f(x))$ is a ring.

We have a standard system of representatives in $K[x] \,(\mathrm{mod}\; f(x))$. We proceed as in $\mathbb{Z}$. For $g(x) \in K[x]$, with polynomial division we obtain

$$g(x) = q(x)f(x) + r(x) \quad \text{with } \deg r(x) < \deg f(x).$$

We have $f(x) \mid g(x) - r(x)$, that is, $g(x) \equiv r(x) \,(\mathrm{mod}\; f(x))$. Thus every polynomial $g(x)$ is congruent to its remainder modulo $f(x)$, and since $\deg r(x) < \deg f(x)$, all of these residue polynomials $r(x)$ are mutually incongruent modulo $f(x)$.

Therefore, if $\deg f(x) = n$, the polynomials $r(x) = a_0 + a_1 x + \cdots + a_{n-1} x^{n-1}$, $a_i \in K$, form a complete system of representatives, and that is our *standard system of representatives*. The number of residue classes is $|K|^n$, since we can choose each of the a_i independently.

The role of the prime numbers in $\mathbb{Z}$ is now assumed by the *irreducible polynomials*. A polynomial $f(x) \in K[x]$ is said to be *irreducible* if $f(x) = g(x)h(x)$ implies $\deg g(x) = 0$ or $\deg h(x) = 0$, that is, $g(x)$ or $h(x)$ is a constant. A polynomial that is not irreducible is said to be *reducible*. For example, the polynomial $x^2 + 1$ is reducible over $\mathbb{Z}_5$, since $x^2 + 1 = (x + 2)(x + 3)$. On the other hand, it is easily checked that the polynomial $x^2 + 2$ is irreducible over $\mathbb{Z}_5$. The proof that $K[x] \,(\mathrm{mod}\; f(x))$ is a field if and only if $f(x)$ is irreducible carries over word for word from the proof in the case of $\mathbb{Z}_p$.

Let us consider a small example. Take the field $K = \mathbb{Z}_2$. The polynomial $x^2 + x + 1$ is irreducible over $\mathbb{Z}_2$, since if we had $x^2 + x + 1 = (x + a)(x + b)$, we would infer $x^2 + x + 1 = x^2 + (a + b)x + ab$, whence $a + b = 1$, $ab = 1$. But $ab = 1$ implies $a = b = 1$, whence $a + b = 0$, which is a contradiction. The standard system of representatives is $0, 1, x, x + 1$, and we obtain the following addition and multiplication tables for the four-element field $\mathbb{Z}_2[x] \,(\mathrm{mod}\; x^2 + x + 1)$:

+	0	1	x	$x+1$
0	0	1	x	$x+1$
1	1	0	$x+1$	x
x	x	$x+1$	0	1
$x+1$	$x+1$	x	1	0

$\cdot$	1	x	$x+1$
1	1	x	$x+1$
x	x	$x+1$	1
$x+1$	$x+1$	1	x

For example, $x(x+1) = x^2 + x \equiv 1 \pmod{x^2 + x + 1}$, and so $[x][x+1] = [1]$.

Let us summarize our findings thus far. Let $K = \mathbb{Z}_p$ and $f(x)$ an irreducible polynomial of degree n. The field $\mathbb{Z}_p[x] \pmod{f(x)}$ contains precisely p^n elements. It is not terribly difficult to show that for every n there exists an irreducible polynomial of degree n over $\mathbb{Z}_p$ (actually to construct such a polynomial is not always easy; see the bibliography for Part 3). Therefore, for every prime power p^n there exists a finite field with p^n elements.

Conversely, one can easily show that a finite field with q elements exists only for prime powers $q = p^n$ and that up to isomorphism there is only one field $\mathrm{GF}(q)$ with q elements. We shall omit the proof (see the bibliography) and instead content ourselves with an illustrative example.

Over $\mathbb{Z}_3$, the polynomials $f(x) = x^2 + x + 2$ and $g(x) = x^2 + 1$ are irreducible. In both cases we obtain the standard representatives $a_0 + a_1 x$, $a_i \in \mathbb{Z}_3$, which we consider first modulo $f(x)$ and then modulo $g(x)$. The reader may easily check that the following mapping is an isomorphism of $\mathrm{GF}(9)$:

$$0 \mapsto 0,$$
$$1 \mapsto 1,$$
$$2 \mapsto 2,$$
$$x \mapsto x + 1,$$
$$x + 1 \mapsto x + 2,$$
$$x + 2 \mapsto x,$$
$$2x \mapsto 2x + 2,$$
$$2x + 1 \mapsto 2x,$$
$$2x + 2 \mapsto 2x + 1 \ .$$

Consider, for example, $2x + 2$, $x + 1$ in $\mathbb{Z}_3[x] \pmod{x^2 + x + 2}$. Their product is $(2x+2)(x+1) = 2x^2 + 4x + 2 = 2x^2 + x + 2 = 2(x^2 + x + 2) + (2x+1)$, and so we have $(2x+2)(x+1) \equiv (2x+1) \pmod{x^2 + x + 2}$. The corresponding elements $2x + 1$, $x + 2$ have, on account of $(2x + 1)(x + 2) = 2x^2 + 5x + 2 =$

$2x^2 + 2x + 2 = 2(x^2 + 1) + 2x$, the product $2x$ in $\mathbb{Z}_3[x]$ (mod $x^2 + 1$), and again, $2x$ corresponds to $2x + 1$.

12.3. Latin Squares

In this and the following section we will apply modular arithmetic and finite fields to some significant combinatorial problems.

Suppose a tire company wishes to test five types of tire, A, B, C, D, E. On five successive days, five automobiles, each with a different type of tire, are tested. A *test protocol* is set up:

Auto \ Day	1	2	3	4	5
1	A	B	C	D	E
2	B	C	D	E	A
3	C	D	E	A	B
4	D	E	A	B	C
5	E	A	B	C	D

This protocol pairs up each car with each tire type. Now, it is possible that the five drivers, $\alpha, \beta, \gamma, \delta, \varepsilon$, influence the performance of the tires. It would therefore be advisable to set up the protocol in such a way that each driver tests each type of tire and each automobile exactly once. We therefore construct an additional protocol:

Auto \ Day	1	2	3	4	5
1	α	β	γ	δ	ε
2	ε	α	β	γ	δ
3	δ	ε	α	β	γ
4	γ	δ	ε	α	β
5	β	γ	δ	ε	α

One can easily check that in fact, each of the 25 tire–driver pairs appears exactly once.

Definition 12.2. Let A be an n-set. A **Latin square of order n** over A is a mapping $L : \{1, \ldots, n\} \times \{1, \ldots, n\} \to A$ such that $L(i, j) = L(i', j)$ implies $i = i'$, and analogously, $L(i, j) = L(i, j')$ implies $j = j'$. A Latin square is therefore an $n \times n$ matrix in which each element of A appears exactly once in every row and every column.

Two Latin squares L_1, L_2 over A are said to be **orthogonal** if for every $(a_1, a_2) \in A^2$, there is exactly one pair (i, j) with $L_1(i, j) = a_1$, $L_2(i, j) = a_2$.

Clearly, the definitions are independent of the set A. In our example we can first use Latin letters, and then Greek. The notions of Latin square and orthogonality remain the same. Thus our introductory example leads to a pair of orthogonal Latin squares of order 5. Now we can choose yet another parameter, say the road surface or weather conditions, and ask whether we can establish an additional protocol that is orthogonal to the other two.

This brings us to our main problem: Let $N(n)$ be the maximum number of mutually orthogonal Latin squares of order n; how large is $N(n)$?

It is here that our algebraic structures come into play. Consider a group of n elements, such as $\mathbb{Z}_n$, with addition as the binary operation. Number the rows and columns of the square with $0, 1, \ldots, n-1$. Then the addition table yields a Latin square if we set $L(i,j) = i + j \pmod{n}$, since $i + j \equiv i' + j \pmod{n}$ implies $i \equiv i' \pmod{n}$, and $i + j \equiv i + j' \pmod{n}$ implies $j \equiv j' \pmod{n}$. For $n = 6$, for example, we obtain the Latin square

$$
\begin{array}{c|cccccc}
 & 0 & 1 & 2 & 3 & 4 & 5 \\
\hline
0 & 0 & 1 & 2 & 3 & 4 & 5 \\
1 & 1 & 2 & 3 & 4 & 5 & 0 \\
2 & 2 & 3 & 4 & 5 & 0 & 1 \\
3 & 3 & 4 & 5 & 0 & 1 & 2 \\
4 & 4 & 5 & 0 & 1 & 2 & 3 \\
5 & 5 & 0 & 1 & 2 & 3 & 4 \\
\end{array}
$$

Now let $q = p^m$ be a prime power. We know, then, that there exists a finite field with q elements: $\mathrm{GF}(q) = \{a_0 = 0, a_1, a_2, \ldots, a_{q-1}\}$. We number the rows and columns with $a_0, a_1, \ldots, a_{q-1}$ and define for $h = 1, \ldots, q-1$ the Latin square L_h by

$$
L_h(a_i, a_j) = a_h a_i + a_j \quad (i, j = 0, 1, \ldots, q-1).
$$

Every L_h is a Latin square, since from $a_h a_i + a_j = a_h a_i' + a_j$ we obtain $a_h a_i = a_h a_i'$ by canceling, and since $a_h \neq 0$, we have $a_i = a_i'$. Likewise, from $a_h a_i + a_j = a_h a_i + a_j'$ we at once obtain $a_j = a_j'$. Now let L_h, L_k be two such Latin squares and $(a_r, a_s) \in \mathrm{GF}(q) \times \mathrm{GF}(q)$. The system of equalities

$$
a_h x + y = a_r,
$$
$$
a_k x + y = a_s,
$$

has a unique solution in x and y since $a_h \neq a_k$, and therefore there exist i, j such that

$$
L_h(a_i, a_j) = a_h a_i + a_j = a_r,
$$
$$
L_k(a_i, a_j) = a_k a_i + a_j = a_s,
$$

that is, L_h and L_k are orthogonal. We have thereby obtained $q-1$ mutually orthogonal Latin squares.

For $q = 5$ with $\mathrm{GF}(5) = \mathbb{Z}_5 = \{0, 1, 2, 3, 4\}$ this yields

$L_1(i,j) = i+j$	$L_2(i,j) = 2i+j$	$L_3(i,j) = 3i+j$	$L_4(i,j) = 4i+j$
0 1 2 3 4	0 1 2 3 4	0 1 2 3 4	0 1 2 3 4
1 2 3 4 0	2 3 4 0 1	3 4 0 1 2	4 0 1 2 3
2 3 4 0 1	4 0 1 2 3	1 2 3 4 0	3 4 0 1 2
3 4 0 1 2	1 2 3 4 0	4 0 1 2 3	2 3 4 0 1
4 0 1 2 3	3 4 0 1 2	2 3 4 0 1	1 2 3 4 0

In our tire problem we may therefore introduce at least two additional parameters, that is, $N(5) \geq 4$. Can we do better? We cannot, as the following theorem reveals.

Theorem 12.3. *For $n \geq 2$ we have $N(n) \leq n-1$, and $N(n) = n-1$ for a prime power $n = p^m$.*

Proof. Let $L_1, \ldots, L_t$ be orthogonal Latin squares of order n over A, where we take $A = \{1, \ldots, n\}$. We permute the elements in L_1 so that in the first row they appear in the order $1, 2, \ldots, n$. We proceed in the same manner with the remaining L_i. These permutations preserve the orthogonality. (Clear?) The new Latin squares, which we again denote by $L_1, \ldots, L_t$, therefore satisfy

$$L_i(1,1) = 1, \quad L_i(1,2) = 2, \quad \ldots, \quad L_i(1,n) = n \quad \text{for } i = 1, \ldots, t.$$

We now consider the place $(2,1)$. For all i we must have $L_i(2,1) \neq 1$, since 1 already appears in the first column. And furthermore, on account of the orthogonality we have $L_i(2,1) \neq L_j(2,1)$ for $i \neq j$, since the pairs (h,h), $1 \leq h \leq n$, are already accounted for in the first row. It follows that $t \leq n-1$, and we have already seen that we have equality for prime powers. $\qquad\square$

And what can we say about lower bounds for $N(n)$? In this regard we have the following result.

Theorem 12.4. *Let $n = n_1 n_2$. Then $N(n_1 n_2) \geq \min(N(n_1), N(n_2))$.*

Proof. Let $k = \min(N(n_1), N(n_2))$, and let $L_1, \ldots, L_k$ be orthogonal Latin squares of order n_1 over A_1, and $L'_1, \ldots, L'_k$ orthogonal Latin squares of order n_2 over A_2, $|A_1| = n_1$, $|A_2| = n_2$. We set $A = A_1 \times A_2$, and define L^*_h, $h = 1, \ldots, k$, on A by

$$L^*_h((i,i'),(j,j')) = (L_h(i,j), L'_h(i',j')).$$

It is immediately seen that the L^*_h are Latin squares. Let us prove the orthogonality. Let $((r,r'),(s,s')) \in A^2$, $1 \leq h \neq \ell \leq k$. From the orthogonality of the L_h we have $(i,j) \in A_1^2$ with $L_h(i,j) = r$, $L_\ell(i,j) = s$, and analogously, $(i', j') \in A_2^2$ with $L_h'(i', j') = r'$, $L_\ell'(i', j') = s'$. From this it follows that

$$\left(L_h^*((i, i'), (j, j')), L_\ell^*((i, i'), (j, j')) \right)$$
$$= \left((L_h(i, j), L_h'(i', j')), (L_\ell(i, j), L_\ell'(i', j')) \right)$$
$$= ((r, r'), (s, s')).$$

$\square$

From Theorem 12.4 we derive the following corollary.

Corollary 12.5. *Let $n = p_1^{k_1} \cdots p_t^{k_t}$ be the prime decomposition of n. Then $N(n) \geq \min_{1 \leq i \leq t}(p_i^{k_i} - 1)$. In particular, we have $N(n) \geq 2$ for all $n \not\equiv 2 \pmod 4$.*

Proof. The first assertion follows with $N(p_i^{k_i}) = p_i^{k_i} - 1$ by multiple applications of Theorem 12.4. For $n \not\equiv 2 \pmod 4$ we have $p_i^{k_i} \geq 3$ for all prime-power divisors of n, and we may apply the theorem. $\square$

For the first values of $N(n)$ our results yield the following table:

n	2	3	4	5	6	7	8	9	10	11	12	13	14
$N(n)$	1	2	3	4	≥ 1	6	7	8	≥ 1	10	≥ 2	12	≥ 1

The first nonobvious case is $n = 6$. Euler, who conjectured that $N(n) = 1$ for all $n \equiv 2 \pmod 4$, publicized this case in 1782 in the form of a clever puzzle: Thirty-six officers, of six different ranks in each of six different regiments, are to be arranged in a 6×6 formation in such a way that each row and column should contain precisely one officer of each regiment and of each rank. Is this possible? Of course, this would mean that $N(6) \geq 2$. Over a hundred years later, an exceedingly tedious listing of all the possibilities showed that it cannot be done, and so $N(6) = 1$.

But this is the only case in addition to $N(2) = 1$ in which Euler's conjecture is correct. In 1960, Bose, Shrikhande, and Parker showed that $N(n) \geq 2$ for *all* $n \neq 2, 6$. However, except for prime powers, not a single value of $N(n)$ is known. For example, today it is known that $N(10) \geq 2$, $N(12) \geq 5$, but no one has found three orthogonal Latin squares of order 10 or six of order 12; nor does one know whether such squares in fact exist.

12.4. Combinatorial Designs

Let us return to our lottery problem from the end of Section 11.4. We had set ourselves the following problem: Does there always exist for $n = 6m + 1$ or $n = 6m + 3$ a set $\mathcal{B}$ of triples from an underlying set S, $|S| = n$, such that every pair of elements of S appears in precisely one triple from $\mathcal{B}$?

We would like to formulate the problem in full generality: We are given natural numbers $v \geq k \geq t \geq 1$ and $\lambda \geq 1$. A **t-design** $(S, \mathcal{B})$ with *parameters* v, k, λ consists of a family $\mathcal{B}$ of subsets of S such that the following hold:

(i) $|S| = v$,
(ii) $|B| = k$ for all $B \in \mathcal{B}$,
(iii) each t-subset of S is contained in exactly λ sets from $\mathcal{B}$.

The sets from $\mathcal{B}$ are called the **blocks** of the design, where the blocks are not necessarily distinct. In the case $\lambda = 1$ a t-design is also called a **Steiner system** $S_t(v, k)$. Our lottery problem thus deals with the Steiner systems $S_2(v, 3)$.

The cases $k = v$, $k = v - 1$, $t = k$, $t = 1$ can be handled easily (see the exercises), so in the following we will always assume

$$2 \leq t < k \leq v - 2.$$

Example 12.6. The Steiner system $S_2(7, 3)$ that we constructed in Section 11.4 can be represented pictorially as follows:

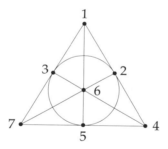

Here we interpret each "line" as a block (including the circle). This 2-design is unique up to a renumbering. It is called the **Fano plane**.

Example 12.7. Let S be the edge set of the complete graph K_5; therefore $|S| = 10$. As blocks $\mathcal{B}$ we take all edge sets of the form

Altogether, there are $5 \cdot 3 = 15$ sets of the first type, 10 of the second type, and 5 of the third, whence $|\mathcal{B}| = 30$. It is easily checked that $(S, \mathcal{B})$ is a 3-design with parameters $v = 10$, $k = 4$, $\lambda = 1$.

Example 12.8. Let $S = \{1, 2, 3, 4\} \times \{1, 2, 3, 4\}$. For every pair $(i, j) \in S$ we define a block $B_{ij} = \{(k, \ell) \in S : k = i$ or $\ell = j$, but $(k, \ell) \neq (i, j)\}$. Hence $|B_{ij}| = 6$. We can represent the B_{ij}'s in a checkerboard pattern; for

example, $B_{2,3}$ is

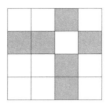

It is easily seen that $(S, \mathcal{B})$ represents a 2-design with $v = 16$, $k = 6$, $\lambda = 2$.

We may now state the main problem of design theory: Given v, k, λ, t, when does there exist a t-design with these parameters?

We have seen a couple of examples of designs for various parameters. But these were more or less special constructions. What really interests us are conditions that *exclude* the existence of a design and general methods for *constructing* designs explicitly for certain parameters.

First we have the following arithmetic conditions that must be satisfied by the parameters v, k, λ.

Theorem 12.9. *Let* $(S, \mathcal{B})$ *be a* t*-design with parameters* v, k, λ*. Then*

a. $|\mathcal{B}| = \lambda \binom{v}{t} / \binom{k}{t}$;

b. *every* i*-subset of* S *is contained in exactly* $\lambda_i = \lambda \binom{v-i}{t-i} / \binom{k-i}{t-i}$ *blocks,*
   $i = 1, 2, \ldots, t$. *In particular, a* t*-design is also an* i*-design for all* $i \le t$.

Proof. The first assertion follows at once by double counting of the pairs (T, B), $|T| = t$, $B \in \mathcal{B}$, $T \subseteq B$. The second statement is proved by induction on i with $\lambda_0 = |\mathcal{B}|$. □

Since $\lambda_i = \lambda \binom{v-i}{t-i} / \binom{k-i}{t-i} = \lambda \frac{(v-i)\cdots(v-t+1)}{(k-i)\cdots(k-t+1)}$ must be an integer, we obtain the following as a necessary condition for the existence of a t-design with parameters v, k, λ:

$$(12.1) \quad \lambda(v-i)\cdots(v-t+1) \equiv 0 \ (\mathrm{mod}\,(k-i)\cdots(k-t+1)), \quad i = 0, \ldots, t.$$

Take, for example, $t = 3$, $k = 4$, $\lambda = 1$. For what values of v are 3-designs possible? Condition (12.1) requires

$$v(v-1)(v-2) \equiv 0 \ (\mathrm{mod}\ 24),$$
$$(v-1)(v-2) \equiv 0 \ (\mathrm{mod}\ 6),$$
$$v - 2 \equiv 0 \ (\mathrm{mod}\ 2).$$

First of all, v must be even. From $(v-1)(v-2) \equiv 0 \ (\mathrm{mod}\ 6)$ it follows that $v \equiv 2, 4 \ (\mathrm{mod}\ 6)$, since $v \equiv 0 \ (\mathrm{mod}\ 6)$ is impossible, and these values also satisfy the first congruence $v(v-1)(v-2) \equiv 0 \ (\mathrm{mod}\ 24)$. The necessary conditions therefore give us $v \equiv 2, 4 \ (\mathrm{mod}\ 6)$, and thus with $v \ge k+2 = 6$ the initial values $v = 8, 10, 14, 16, \ldots$. The reader should construct an analogous design for $v = 8$ (the cube graph Q_3 should be of help). We have already realized $v = 10$ in Example 12.7 above.

In the following, we will restrict attention to the case $t = 2$, which by Theorem 12.9b is the simplest. A 2-design is frequently called a **block plan**. The necessary conditions in this case are

$$(12.2) \qquad \begin{aligned} \lambda v(v-1) &\equiv 0 \ (\mathrm{mod}\ k(k-1)), \\ \lambda(v-1) &\equiv 0 \ (\mathrm{mod}\ (k-1)) \,. \end{aligned}$$

A famous theorem of Wilson states that these necessary conditions are also sufficient in an asymptotic sense: If k, λ are given, then there exists v_0 such that for every $v \geq v_0$ satisfying (12.2), there in fact exists a block plan with parameters v, k, λ.

We move on to the construction methods, where we assume $\lambda = 1$ in addition to $t = 2$. Thus we are considering Steiner systems with parameters v, k. We set $b = |\mathcal{B}|$ and let $r = \lambda_1$ be the number of blocks that contain a fixed element. From Theorem 12.9 we have $bk(k-1) = v(v-1)$, $r(k-1) = v - 1$, and therefore

$$(12.3) \qquad \begin{aligned} b\,k &= vr, \\ r(k-1) &= v - 1, \end{aligned}$$

with $2 \leq r < b$ on account of $k < v$.

We see that in a Steiner system, on account of $\lambda = 1$, every pair of blocks have at most one element in common. We obtain an especially interesting case when every pair of blocks have intersection *exactly* one element, namely the classical (finite) projective planes. We interpret the set S as *points*, the blocks as *lines*, and say that the point u lies on the line B if $u \in B$.

Definition 12.10. A (finite) **projective plane** is a pair $(S, \mathcal{B})$ such that the following axioms are satisfied:

(i) 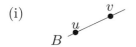 For $u \neq v \in S$ there exists precisely one line $B \in \mathcal{B}$ with $\{u, v\} \subseteq B$.

(ii) 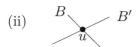 For $B \neq B' \in \mathcal{B}$ there exists exactly one point $u \in B \cap B'$.

(iii)  There exist four points of which no three lie on the same line.

Axiom (iii) is a nondegeneracy condition in order to exclude the following trivial structure:

Theorem 12.11. *Projective planes are block plans with $k = r$ and $\lambda = 1$. If we set $k = n + 1$, then $v = b = n^2 + n + 1$, $k = r = n + 1$, $\lambda = 1$.*

Proof. Let $B \neq B' \in \mathcal{B}$ be chosen arbitrarily. It is easily established that there exists a point $u \notin B \cup B'$ (otherwise, the trivial structure would result). Consider u and B. Every line through u has exactly one intersection with B (see the figure)

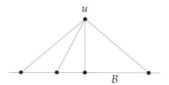

and conversely, every $v \in B$ determines a line via $\{u, v\}$. For the number r_u of lines through u we therefore have $r_u = |B|$. Applying the same considerations to u and B' yields $|B| = r_u = |B'|$. Since B, B' were arbitrary, all lines must contain the same number k of points, and we have $r_u = r = k$ for all points u, and on account of $bk = vr$, we have $b = v$ as well.

Setting $k = n+1$, we have $n \geq 2$ on account of $k \geq 3$ (since, for example, the two vertical lines in axiom (iii) intersect in a third point). Consider a point u. Every point not equal to u appears exactly once on a line through u (see the figures), and each of these lines contains $k - 1 = n$ points not equal to u. Hence $v = 1 + (n + 1)n = n^2 + n + 1$. $\qquad \square$

The integer n is called the *order* of the projective plane. One of the most famous problems in all of combinatorics asks, for what natural numbers $n \geq 2$ does there exist a projective plane of order n?

We already know *one* projective plane, the smallest of order 2. This is of course our Fano plane. The following theorem affirms the existence of projective planes for prime-power orders.

Theorem 12.12. *Let $q = p^m$ be a prime power. Then there exists a projective plane of order q.*

Proof. We begin with the Galois field $K = \mathrm{GF}(q)$ and consider a three-dimensional vector space over K. The "points" S are the one-dimensional subspaces, and the "lines" $\mathcal{B}$ are the two-dimensional subspaces. Every pair of one-dimensional subspaces are clearly contained in precisely one two-dimensional subspace (axiom (i)). Let $B \neq B'$. By the dimension equation for subspaces, it follows that

$$3 = \dim(B + B') = \dim(B) + \dim(B') - \dim(B \cap B'),$$

whence $\dim(B \cap B') = 2 + 2 - 3 = 1$, that is, B and B' intersect in one point (axiom (ii)). Finally, the four one-dimensional subspaces generated by the vectors $(1, 0, 0)$, $(0, 1, 0)$, $(0, 0, 1)$, $(1, 1, 1)$ satisfy axiom (iii). The system $(S, \mathcal{B})$ is therefore a projective plane. The order can be easily calculated. Altogether, there are $q^3 - 1$ nonzero vectors. Every one-dimensional subspace contains exactly $q - 1$ nonzero vectors, obtained by multiplying a nonzero vector by the nonzero elements of K, and we obtain $|S| = \frac{q^3 - 1}{q - 1} = q^2 + q + 1$, and therefore the order is q. $\qquad\square$

There are projective planes other than those we have mentioned, but to date, no one has succeeded in constructing one of order not a prime power. This is a familiar situation. We saw the same phenomenon with orthogonal Latin squares: there exist the maximal number $n - 1$ when n is a prime power; otherwise, nothing is known except for the case $n = 6$. This connection is not coincidental, for we have the following theorem, whose proof we shall reserve for the exercises.

Theorem 12.13. *There exists a projective plane of order $n \geq 2$ if and only if $N(n) = n - 1$, that is, if there exist $n - 1$ orthogonal Latin squares of order n.*

This takes care of the case $n = 6$: there is no projective plane of order 6. In 1988, an enormous computational effort established the nonexistence for $n = 10$. This implies $N(10) \leq 8$ for Latin squares. For the existence of a projective plane of order 12 one needs 11 orthogonal Latin squares; to date, one knows only that $N(12) \geq 5$.

To close, we return at last to our lottery problem, that is, the Steiner systems of triples $S_2(v, 3)$, $v \equiv 1, 3 \pmod 6$. We already know the smallest such system—our Fano plane $S_2(7, 3)$. How should one go about constructing such systems in general? Here modular arithmetic lends a helping hand.

Let us write out the blocks of the Fano plane modulo 7:

$$124, \quad 235, \quad 346, \quad 450, \quad 561, \quad 602, \quad 013.$$

If we set $B_0 = \{1, 2, 4\}$, the other triples are of the form $B_i = B_0 + i \pmod 7$, $i = 1, \ldots, 6$. That is, we add i to all the elements of B_0 modulo 7. We say that the Fano plane arises by *development of the initial block* B_0 modulo 7.

This looks quite promising for the general case. But how does this development work? Let us try another initial block: $C_0 = \{1, 2, 5\}$. By development we obtain

$$125, \quad 236, \quad 340, \quad 451, \quad 562, \quad 603, \quad 014.$$

This time, we do not obtain a Steiner system $S_2(7,3)$, since $\{1,5\}$ appears in C_0, but also in C_3. What is the reason for this? Let us examine the six differences in C_0 (mod 7), namely

$$1-2 \equiv 6, \quad 1-5 \equiv 3, \quad 2-1 \equiv 1, \quad 2-5 \equiv 4, \quad 5-1 \equiv 4, \quad 5-2 \equiv 3.$$

Since we always add the same number in the development, the (multi)set of differences remains the same. And now we see the reason for the duplication of $\{1,5\}$. Since the difference 3 appears twice, $1-5 \equiv 3$, $5-2 \equiv 3$, it follows that $\{5,2\} + i$ always has the difference 3, and thus we must find an i such that $5 + i \equiv 1$ (mod 7) in order to duplicate $1,5$, and this succeeds for the choice $i = 3$.

For the original initial block $B_0 = \{1,2,4\}$, on the other hand, all the differences are distinct:

$$1-2 \equiv 6, \quad 1-4 \equiv 4, \quad 2-1 \equiv 1, \quad 2-4 \equiv 5, \quad 4-1 \equiv 3, \quad 4-2 \equiv 2,$$

and therefore the development works. We would like to record this at once as a general result: Let $(G,+)$ be an abelian group with v elements. A family of blocks $B_i = \{p_{i1}, \ldots, p_{ik}\} \subseteq G, i \in I$, is called a (v,k,λ) **family of differences** if in the multiset

$$\Delta B_i = \{p_{ij} - p_{ij'} : 1 \leq j \neq j' \leq k\} \quad (i \in I)$$

of differences within the individual blocks, every nonzero element of G appears exactly λ times.

For example, let $G = \mathbb{Z}_9$, $B_1 = \{0,1,2,5\}$, $B_2 = \{0,2,3,5\}$. The differences are $\Delta B_1 = \{\pm 1, \pm 2, \pm 4, \pm 1, \pm 4, \pm 3\}, \Delta B_2 = \{\pm 2, \pm 3, \pm 4, \pm 1, \pm 3, \pm 2\}$. Each difference $\pm 1, \pm 2, \pm 3, \pm 4$ appears exactly three times, and therefore $\{B_1, B_2\}$ is a $(9,4,3)$ family of differences.

Theorem 12.14. *Let $(G,+)$ be an abelian group with v elements, $\overline{\mathcal{B}} = \{B_i : i \in I\} \subseteq \binom{G}{k}$, and $\mathcal{B} = \{B_i + g : i \in I, \ g \in G\}$. Then $(G,\mathcal{B})$ is a (v,k,λ)-block plan if and only if $\overline{\mathcal{B}}$ is a (v,k,λ)-family of differences.*

Proof. Let $B_i = \{p_{i1}, \ldots, p_{ik}\}, \Delta B_i$ the multiset of differences from B_i, and $\mathcal{B}_i = \{B_i + g : g \in G\}$. For $x \neq y \in G$ we have $x, y \in B_i + g$ if and only if there exist j, ℓ with $p_{ij} + g = x$, $p_{i\ell} + g = y$ if and only if $x - y = p_{ij} - p_{i\ell} \in \Delta B_i$. That is, the number of blocks in $\mathcal{B}_i$ that contain x, y is equal to the number of times $x - y \neq 0$ appears as a difference in ΔB_i. Summation over i now yields the result. $\qquad\square$

Our example $G = \mathbb{Z}_9$ from above thus yields, by development of B_1, B_2, a $(9,4,3)$ block plan.

We are interested primarily in the case $\lambda = 1$, that is, in Steiner systems. Thus we must construct initial blocks B_i such that every nonzero element appears *exactly once* as a difference. Let us consider as an example $G = \mathbb{Z}_{41}$, $B_1 = \{1, 10, 16, 18, 37\}$, $B_2 = \{5, 8, 9, 21, 39\}$. Then ΔB_i contains 20 differences, and one easily checks that every nonzero element appears as a difference (and therefore appears exactly once). The result is that by development we have obtained a $(41, 5)$ Steiner system.

With our methods we may finally carry out the construction of Steiner systems of triples $S_2(v, 3)$ for *all* $v \equiv 1, 3 \pmod 6$. We will discuss the case $v \equiv 3 \pmod 6$. The other possibility, $v \equiv 1 \pmod 6$, is treated similarly.

Let $v = 6m + 3$. We consider the additive group $G = \mathbb{Z}_{2m+1} \times \mathbb{Z}_3$. That is, the elements of G are all pairs (i, j), $i \in \mathbb{Z}_{2m+1}$, $j \in \mathbb{Z}_3$, with addition carried out coordinatewise. The group G thus has $6m + 3$ elements, and we must now find $b = \frac{v(v-1)}{6} = (2m + 1)(3m + 1)$ triples.

As initial blocks we take

$$B_0 = \{(0, 0), (0, 1), (0, 2)\},$$
$$B_h = \{(0, 0), (h, 1), (-h, 1)\} \text{ for } h = 1, \ldots, m,$$

and set

$$\mathcal{B}_0 = \text{the family of } \textit{distinct} \text{ blocks of the form } \{B_0 + g : g \in G\},$$
$$\mathcal{B}_h = \{B_h + g : g \in G\}, \ h = 1, \ldots, m,$$
$$\mathcal{B} = \mathcal{B}_0 \cup \bigcup_{h=1}^{m} \mathcal{B}_h.$$

Claim: $(G, \mathcal{B})$ is a Steiner system $S_2(v, 3)$.

Consider $\mathcal{B}_0$. Each triple $\{(i, j), (i, j + 1), (i, j + 2)\}$ is produced exactly three times by development of B_0, namely, $B_0 + (i, j)$, $B_0 + (i, j + 1)$, $B_0 + (i, j + 2)$. We therefore obtain

$$\mathcal{B}_0 = \{\{(i, 0), (i, 1), (i, 2)\} : i = 0, 1, \ldots, 2m\}, \quad |\mathcal{B}_0| = 2m + 1.$$

One sees at once that the triples in the $\mathcal{B}_h$'s are all distinct: $|\mathcal{B}_h| = 6m + 3$, $h = 1, \ldots, m$. Altogether, $\mathcal{B}$ therefore contains $2m + 1 + (6m + 3)m = (2m + 1)(3m + 1)$ triples; the number is correct. Now we must check that every pair $\{(x, y), (x', y')\}$ appears exactly once. For $h = 1, \ldots, m$ we have

$$\Delta B_h = \{\pm(h, 1), \pm(-h, 1), \pm(2h, 0)\};$$

that is, the $6m$ differences in $\bigcup \Delta B_h$ yield each element $g \in G$, $g \neq (0, 0)$, exactly once, with the exception of the elements $(0, 1), (0, 2)$, and those are precisely the two nonzero elements in B_0. For $\{x, y\}, \{x', y'\}$ with $x \neq x'$ it thus follows, as in the proof of Theorem 12.14, that they lie in exactly one block of $\bigcup_{h=1}^{m} \mathcal{B}_h$, and for $x = x'$ they lie, of course, in exactly one block of $\mathcal{B}_0$. Thus everything to be shown has been established.

Now we can answer, finally, our modified 45-number lottery. From $b = \frac{45 \cdot 44}{6} = 330$ we see that there are 330 triples in $\{1, 2, \ldots, 45\}$ that cover every pair, and 330 is the smallest number of triples with this property.

Exercises for Chapter 12

12.1 Prove the validity of the following tests for divisibility by 3 and 9: An integer $n = a_k a_{k-1} \ldots a_0$ in decimal notation is divisible by 3 if and only if the sum of digits $\sum_{i=0}^{k} a_i$ is divisible by 3. An integer $n = a_k a_{k-1} \ldots a_0$ in decimal notation is divisible by 9 if and only if the sum of digits $\sum_{i=0}^{k} a_i$ is divisible by 9.

12.2 Show in detail that $x \equiv x'$, $y \equiv y'$ (mod m) implies $x + y \equiv x' + y'$ and $xx' \equiv yy'$ (mod m).

▷ **12.3** Determine the remainders when 3^{15} and 15^{83} are divided by 13.

12.4 Solve the system of equations $x + 2y = 4, 4x + 3y = 4$ in $\mathbb{Z}_7$ and in $\mathbb{Z}_5$.

12.5 Prove Euler's generalization of Fermat's theorem: Let $\varphi(m)$ be the number of prime residue classes $[r]$ modulo m, that is, $\gcd(r, m) = 1$. Then $n^{\varphi(m)} \equiv 1$ (mod m) for every n such that $\gcd(n, m) = 1$.

▷ **12.6** Using the digits $1, 2, \ldots, 9$, write down a collection one- or two-digit numbers such that each digit appears exactly once and the sum is equal to 100. Example: 9, 37, 16, 28, 5, 4, except that here the sum is only 99.

12.7 Determine the remainder when $x^{81} + x^{49} + x^{25} + x^9 + x$ is divided by $x^3 - x$.

▷ **12.8** Find an irreducible polynomial $f(x)$ of degree 2 over GF(5) and use it to construct the Galois field GF(25).

12.9 Decompose $x^4 + 1$ into irreducible factors over $\mathbb{Z}_3$.

▷ **12.10** Take all four jacks, queens, kings, and aces from a regular deck of playing cards and arrange them in a 4×4 square in such a way that no two of the same suit or value appear in the same row, column, or diagonal. Can this be accomplished in such a way that the colors red and black alternate to form a checkerboard pattern?

12.11 The following problem comes from Sir Ronald Fisher: In a set of sixteen people, four are English, four are Scottish, four are Irish, and four Welsh. There are four each of the ages 35, 45, 55, and 65. Four are lawyers, four are doctors, four are soldiers, and four are clergymen. Four are single, four are married, four widowed, and four divorced. Finally, four are conservatives, four socialists, four liberals, and four fascists. No two of the same kind in one category are the same in another category. Three of the fascists are a single English lawyer of 65, a married Scottish soldier of 55, and a widowed Irish doctor of 45. Furthermore, the Irish socialist is 35, the Scottish conservative is 45, and the English clergyman is 55. What can you say about the Welsh lawyer?

▷ **12.12** How many rooks can be placed on an $n \times n \times n$ three-dimensional chessboard in such a way that no rook is attacking any other rook? How many on a d-dimensional $n \times \cdots \times n$ board?

12.13 Construct a pair of orthogonal Latin squares of order 12.

12.14 Show that permuting the rows, columns, or symbols preserves the orthogonality of Latin squares.

12.15 Determine the number of Latin squares of orders $1, 2, 3, 4$ on a given set of symbols. Show that up to permutation of the rows, columns, and symbols there is exactly one Latin square for $n \le 3$ and two for $n = 4$.

▷ **12.16** Let $(G, \cdot)$ be an abelian group of odd order. Define $(G, \pitchfork)$ by $x \pitchfork y = xy^{-1}$. Show that $L(x, y) = x \pitchfork y$ is a Latin square and is orthogonal to $L'(x, y) = xy$.

12.17 Let $\mathrm{PG}(q)$ denote the projective plane over $\mathrm{GF}(q)$. Three noncollinear points form a triangle. Show that the number of triangles is $\frac{1}{6} q^3 (q+1)(q^2 + q + 1)$.

▷ **12.18** Let $v = 2^n - 1$ and $S = \{0,1\}^n \setminus \mathbf{0}$. We add $\boldsymbol{x} + \boldsymbol{y}$ as usual. Three words $\boldsymbol{x}, \boldsymbol{y}, \boldsymbol{z}$ in S form a block if $\boldsymbol{x} + \boldsymbol{y} + \boldsymbol{z} = \mathbf{0}$. Show that we obtain a Steiner system $S_2(v, 3)$.

12.19 In Section 11.4 we defined the packing and covering numbers $P(n, k, t)$ and $C(n, k, t)$. Prove the following generalization of Theorem 11.5:

(a) $P(n, k, t) \le \left\lfloor \frac{n}{k} \left\lfloor \frac{n-1}{k-1} \left\lfloor \cdots \left\lfloor \frac{n-t+1}{k-t+1} \right\rfloor \cdots \right\rfloor \right\rfloor \right\rfloor$,

(b) $C(n, k, t) \ge \left\lceil \frac{n}{k} \left\lceil \frac{n-1}{k-1} \left\lceil \cdots \left\lceil \frac{n-t+1}{k-t+1} \right\rceil \cdots \right\rceil \right\rceil \right\rceil$.

▷ **12.20** Let p be a prime number. Show that $(a + b)^p \equiv a^p + b^p \pmod{p}$ holds for all $a, b \in \mathbb{Z}$. Hint: Use the binomial theorem or Fermat's theorem.

12.21 Prove Wilson's theorem: $(p - 1)! \equiv -1 \pmod{p}$ if and only if p is prime.

▷ **12.22** Let p be a prime number. Prove the following: (a) If $p \equiv 3 \pmod 4$, then there exists no $n \in \mathbb{Z}$ such that $p \mid n^2 + 1$. (b) If $p \equiv 1 \pmod 4$, then such n exist. Hint: Fermat's theorem and the previous exercise.

12.23 The integers 407 and 370 (in decimal notation) have the property that $407 = 4^3 + 0^3 + 7^3$ and $370 = 3^3 + 7^3 + 0^3$. Find all positive integers $\ne 1$ with this cubic property.

12.24 Let an integer be given in decimal notation, such as, for example, 145. Construct the sequence $145 \to 1^2 + 4^2 + 5^2 = 42 \to 4^2 + 2^2 = 20 \to 2^2 + 0^2 = 4 \to 4^2 = 16 \to 1^2 + 6^2 = 37 \to 3^2 + 7^2 = 58 \to 5^2 + 8^2 = 89 \to 8^2 + 9^2 = 145 \to 42 \to 20 \to \cdots$. Show that for every natural number, this sequence ends either in 1 or, as in our example, in a circuit that contains 145.

▷ **12.25** We would like to place n queens on an $n \times n$ chessboard in such a way that no two queens are attacking each other. Show that the following construction works for $n \equiv \pm 1 \pmod 6$, $n \ge 5$: The ith queen is placed on square $(i, 2i) \pmod n$, $0 \le i \le n - 1$.

12.26 We are given N matches. In the first move, one removes $a_1 \leq 1$ matches, $N \equiv a_1 \pmod 2$. The remaining pile now has $N - a_1$ matches. Now in the second move one removes $a_2 \leq 2$ matches with $N - a_1 \equiv a_2 \pmod 3$, in the third move $a_3 \leq 3$ with $N - a_1 - a_2 \equiv a_3 \pmod 4$, and so on. The game ends when at the ℓth move all the remaining matches are removed. Show that one always has $a_\ell = \ell$ for the final number.

▷ **12.27** Consider the function

$$f(m,n) = \frac{n-1}{2}[|B^2 - 1| - (B^2 - 1)] + 2$$

with $B = m(n+1) - (n!+1), m, n \in \mathbb{N}$. Show that $f(m,n)$ is always a prime number and that every odd prime is produced exactly once. Hint: Use Exercise 12.21.

▷ **12.28** Let A denote the sum of the digits of 4444^{4444} in decimal notation, and let B denote the sum of the digits of A. What is the sum of the digits of B?

12.29 What are the last three digits in 7^{9999}? Hint: Begin with $7^4 = 2401$ and consider $7^{4k} = (2400 + 1)^k$ using the binomial theorem.

12.30 For which prime numbers p is the polynomial $x^2 + 1$ irreducible over $\mathbb{Z}_p$?

12.31 Determine all numbers m for which $x^2 + mx + 2$ is irreducible over $\mathbb{Z}_{11}$.

▷ **12.32** Let $(G, \cdot)$ be an abelian group, $|G| = n$. Show that $a^n = 1$ for all $a \in G$. The smallest $d \geq 1$ such that $a^d = 1$ is called the order $\operatorname{ord}(a)$ of a. Show that $\operatorname{ord}(a) \mid n$ is always true. Hint: Use the proof of Fermat's theorem for the first part, and the Euclidean algorithm for the second.

12.33 Let $(G, \cdot)$ be a group of even order. Show that G contains an element of order 2.

▷ **12.34** Let q be an odd prime power. An element $a \in \mathrm{GF}(q)$, $a \neq 0$, is called a square if $a = b^2$ for some b. Let Q be the set of squares. Show that $|Q| = \frac{q-1}{2}$. For what q is -1 in Q?

12.35 Show that in $\mathrm{GF}(2^m)$, every element is a square.

12.36 Let $L : \mathbb{Z}_{2n} \times \mathbb{Z}_{2n} \to \mathbb{Z}_{2n}$ be the Latin square of order $2n$ with $L(i,j) \equiv i + j \pmod{2n}$. Show that there is no Latin square orthogonal to L.

▷ **12.37** A *magic square* Q of order n contains all numbers $1, \ldots, n^2$ in such a way that all rows, all columns, and both diagonals have the same sum. The square Q is called *half magic* if all row and column sums are the same. Think about how one could create a half-magic square from a pair of orthogonal Latin squares. What conditions must a pair of orthogonal Latin squares satisfy if a magic square is to be obtained? Construct magic squares for $n = 4, 5$. Hint: The two squares determine the numbers modulo 100.

12.38 Nine prisoners are divided into three groups of three such that the individuals in each group are linked by leg irons into a chain. For example, in the arrangement P_1–P_2–P_3, P_4–P_5–P_6, P_7–P_8–P_9, prisoner P_1 is chained to prisoner P_2, prisoner P_2 is chained to P_3 and so on. Devise a protocol for six days that provides for each pair of prisoners to be chained together exactly once.

▷ **12.39** Show that there exists a 1-design with parameters v, k, λ if and only if $k \mid \lambda v$.

12.40 Let $k = v-1$. Then there exists a t-design with parameters v, $k = v-1$, λ if and only if $v-t \mid \lambda$, and the only possible design is $\frac{\lambda}{v-t}\binom{S}{v-1}$, that is, all $(v-1)$-sets appear $\frac{\lambda}{v-t}$ times.

▷ **12.41** Let $D(v, k, \lambda)$ be a 2-design with b blocks, and $M = (m_{ij})$ the $v \times b$ incidence matrix, where $1 < k < v-1$, that is,

$$m_{ij} = \begin{cases} 1, & u_i \in B_j, \\ 0 & u_i \notin B_j. \end{cases}$$

Show that

$$MM^T = \begin{pmatrix} r & \cdots & \lambda \\ \vdots & \ddots & \vdots \\ \lambda & \cdots & r \end{pmatrix}$$

and conclude that $v \le b$. Hint: Calculate the determinant of MM^T.

12.42 In a Steiner system of triples $S_2(v, 3)$, construct the following graph G: The vertices are the blocks, and two blocks B, B' are joined by an edge if $B \cap B' \ne \varnothing$. Show that G possesses a Hamiltonian circuit.

12.43 We know that $P(n, 3, 2) \le \lfloor \frac{n}{3} \lfloor \frac{n-1}{2} \rfloor \rfloor$ holds for the packing number. Solve the following problems: (a) Show that $P(n, 3, 2) \le \lfloor \frac{n}{3} \lfloor \frac{n-1}{2} \rfloor \rfloor - 1$ for $n \equiv 5 \pmod 6$. Hint: Set $n = 6k + 5$ and determine the number of blocks. An integer n is said to be "good" if $P(n, 3, 2) = \lfloor \frac{n}{3} \lfloor \frac{n-1}{2} \rfloor \rfloor$. (b) Show that if the odd number n is good, then so is $n - 1$. (c) Determine $P(n, 3, 2)$ for $n \le 10$.

12.44 Let D be a projective plane of order n. Remove from D a block B and all of the points of B. Show that the resulting structure D' is a block plan with $v = n^2$, $k = n$, $\lambda = 1$, called an *affine plane* of order n.

▷ **12.45** Show that the blocks $\mathcal{B}$ of an affine plane of order n can be decomposed into $n + 1$ classes $\mathcal{B}_i$ such that two blocks have nonempty intersection if and only if they are in different classes.

12.46 Show that up to permutation, the Fano plane is the only projective plane of order 2. Show the same for the affine plane of order 3.

12.47 Prove Theorem 12.13.

▷ **12.48** Let q be a prime power of the form $4n + 3$. Prove that the nonzero squares in $\mathrm{GF}(q)$ form a set of differences with parameters $(4n + 3, 2n + 1, n)$.

12.49 Determine the nonzero squares in $\mathbb{Z}_{19}$ and use them to construct a block plan with parameters $(19, 9, 4)$.

▷ **12.50** In a projective plane P of order q, construct the bipartite graph $G = (S + B, E)$ with S the points of P, $\mathcal{B}$ the blocks of P, and $pB \in E \Leftrightarrow p \in B$. Show that the graph G is $(q + 1)$-regular and its shortest circuits have length 6. Show further that G with these conditions—being $(q + 1)$-regular and with length of the shortest circuit equal to 6—has the minimum number of vertices. Hint: Use Exercise 6.37.

Chapter 13

Coding

13.1. Statement of the Problem

A particularly interesting application of our algebraic structures involves the secure transmission of information. Before we get into any details, we shall first consider how information is actually transmitted. First of all, how do we formulate messages for transmittal? The usual forms are the spoken and written word. If we wish to input information into a computer, we use a keyboard, and the transmittal takes place by way of the mechanisms built into the hardware and electric current. The telephone, telegraph, radio, and so on are all means of communication.

If we wish to send a message, say, PUT THE PAPERS ON THE SUNDIAL, we **encode** the message according to whatever system we have chosen. For example, we can write the message in Morse code, using dots and dashes; or, what comes to the same thing, as a sequence of zeros and ones. Of course, we will probably wish to choose a system that offers reasonable efficiency. This first step of information transmittal is called **source encoding**.

We now send the encoded word over a "channel" to the designated recipient. And this is where the actual problem begins. We shall concentrate on two aspects: One is that the channel can be "noisy," due, for example, to cosmic or atmospheric disturbances (think of the information beamed down from satellites). In this case, we need a method that is able to **detect** and **correct** errors in transmission. The second aspect relates to data security. How do we protect data from unauthorized eavesdroppers? Such an eavesdropper can do more than just intercept and read a message; a malevolent interceptor can alter the message and then send it on to the intended recipient. How can we tell whether a message truly comes from the purported sender? These problems belong to the subject of **cryptography**, about which we shall have more to say in the next chapter.

The solutions to both aspects of the problem are in principle the same: the message is subjected to security measures—it is again encoded in such a way that the receiver can decode it correctly (the error-detection and -correction aspect) and that it is secure from the attack of an interceptor (cryptography). We call this second step **channel encoding**.

We therefore specify the following model:

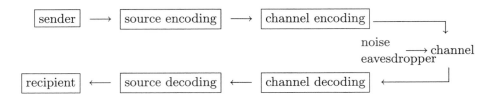

Both aspects of information transmittal are nice illustrations of the discrete methods that we have been discussing.

13.2. Source Encoding

In the following, we will restrict attention to $0, 1$ source encodings. That is, we are given a set X, e.g., $X = \{A, B, C, \ldots, Z, \sqcup\}$, where the symbol $\sqcup$ represents a space, and a map that takes each element of X to a $(0, 1)$-word. Let $A = \{0, 1\}$. As usual, $\boldsymbol{w} \in A^n$ is called a $(0, 1)$-*word of length* n, where we also allow the *empty word* $(\ ) \in A^0$. Finally, we set $A^* = \bigcup_{n \geq 0} A^n$. We thus seek a mapping $c : X \to A^*$ and call each such mapping a **source encoding** of X (using the code alphabet $\{0, 1\}$). The images $c(x)$, $x \in X$, are called the **codewords** of X under the encoding c, and $C = c(X)$ is called the (source) **code**.

It makes no difference whether we choose $A = \{0, 1, 2\}$ or some other alphabet. The techniques are similar. The only requirement is $|A| \geq 2$.

What properties do we require of a coding c? One thing is clear: c must be *injective*. Otherwise, a message cannot be uniquely decoded.

Consider the following example: $X = \{\alpha, \beta, \gamma\}$, $c(\alpha) = 0$, $c(\beta) = 1$, $c(\gamma) = 00$. Then c is injective. Suppose now that a recipient obtains 00. Then it is unclear whether $\alpha\alpha$ or γ was sent. The reason is clear: the codeword 0 is also the beginning, that is, a *prefix*, of another codeword, namely 00. We must also exclude such a possibility.

Definition 13.1. The set $C \subseteq A^*$ is called a **prefix code** if no word of C is the prefix to another word in C.

Clearly, a prefix code is also injective. Prefix codes allow for *unique* encryption. Again let us take $X = \{\alpha, \beta, \gamma\}$ and this time choose the prefix code $c(\alpha) = 00$, $c(\beta) = 1$, $c(\gamma) = 01$. Suppose a recipient receives the message 0001010010100. She begins decoding from left to right, moving rightward until the first valid codeword results, in our case 00. Since $C = c(X)$ is a prefix code, she knows that no other symbol from X is encoded with this beginning. Now she crosses out the 00 and looks for the next codeword, in our case 01. Again, the prefix property ensures uniqueness. Continuing in this manner, she finally obtains the unique text $\alpha\gamma\gamma\alpha\beta\gamma\alpha$.

Now we come to the question of finding the most *efficient* encoding. In our example, we could as well have come up with the prefix coding $c(\alpha) = 00$, $c(\beta) = 101$, $c(\gamma) = 1100$. The words would then have been longer, and encoding and decoding more time-consuming. One would certainly prefer the first encoding. It accomplishes the same thing at less cost.

If $\boldsymbol{w}$ is a word in A^n, then as usual, we let $\ell(\boldsymbol{w}) = n$ denote the *length* of the word $\boldsymbol{w}$. We therefore seek a prefix code C, with $|C| = |X|$, such that the sum $\sum_{\boldsymbol{w} \in C} \ell(\boldsymbol{w})$ is minimal. An important consideration is that the source letters generally have different frequencies of occurrence. For example, in English, the letter E is by far the most frequent (about 12.6%), followed by T at 9.1% and A at 8.2%. In contrast, J, K, Q, X, and Z each have frequencies under 1%, with Q and Z bringing up the rear at less than one-tenth of one percent each. Thus an efficient code would use its shortest codewords for the letters E, T, and A, and the longest for Q and Z.

This brings us to the fundamental problem of source encoding: Given a set $X = \{x_1, \ldots, x_n\}$ and a probability distribution $p_i = p(x_i)$, $i = 1, \ldots, n$, construct a prefix code $c : X \to A^*$, $\boldsymbol{w}_i = c(x_i)$, such that

$$(13.1) \qquad \overline{L}(C) := \sum_{i=1}^{n} p_i \ell(\boldsymbol{w}_i)$$

is minimal. The sum $\overline{L}(C)$ is called the *average length* of codewords in $C = \{\boldsymbol{w}_1, \ldots, \boldsymbol{w}_n\}$.

This problem should sound familiar. We posed a similar problem in Section 9.2 when we investigated the average length of rooted trees with n leaves. And this is no coincidence. These two problems are one and the same!

We represent the $(0, 1)$-words of A^* as a tree:

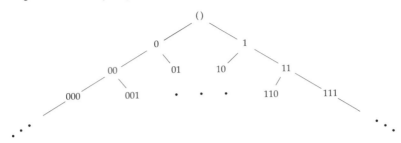

The root is the empty word ( ). If a zero appears, we move to the left, and for a one we move to the right. We may associate with $C \subseteq A^*$, $|C| = n$, the binary tree T whose leaves are the words of C. The prefix property means that no word of C appears in the tree T *before* another word of C. Therefore, T indeed has n leaves, $T \in \mathcal{T}(n, 2)$. Conversely, for each $T \in \mathcal{T}(n, 2)$ there is precisely one code $C \subseteq A^*$, $|C| = n$, whose words exactly correspond to the leaves.

Prefix codes $C \subseteq A^*$ and binary trees are therefore in bijective association, and with this association, the fundamental problem (13.1) is none other than the minimization problem from Section 9.2, since $\overline{L}(C) = \overline{L}(T)$ holds for the code and its associated code tree T.

Our code $c(X) \to A^*$, $c(\alpha) = 00$, $c(\beta) = 1$, $c(\gamma) = 01$ corresponds, for example, to the tree

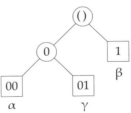

It should now be clear what we meant when we said that the size $|A| = q$ of the code alphabet was of no further importance. Instead of binary trees we would have q-ary trees $T \in \mathcal{T}(n, q)$.

Our main theorem, Theorem 9.6 from Section 9.2, can therefore be carried over word for word. It determines the optimal average length up to an error of size less than 1. And Huffman's algorithm gives the exact result. Thus we have successfully solved the problem of source encoding.

13.3. Error Detection and Correction

We now turn our attention to the first aspect of channel encoding. We are sending a $(0, 1)$-word $\boldsymbol{w} \in A^*$ over a channel. Normally, we expect 0 to be received as 0, and 1 as 1. Sometimes, however, channel disturbances appear, so that a 0 is received as a 1 or vice versa. Our task is to eliminate the effects of such errors.

The usual method for accomplishing this goal is the following: We decompose the chain of zeros and ones (or the symbols of any other alphabet) into blocks of a fixed length k. At the end of each block, r *check symbols* are appended that are determined in some way from the k *information symbols*. The entire word of length $n = k + r$ is the (channel) **codeword**, and our task consists in choosing the check symbols in such a way that transmission errors can be corrected or at least detected. We shall clarify all of this with two examples.

Example 13.2. We take $r = 2k$ and repeat the k symbols twice. For example, if the given word is 1011, then we encode 1011 1011 1011. We call this code a (double) **repetition code**. If we know that in transmission at most one error has occurred, say we have received the word 1011 1001 1011, we can correct the error at once. We need only look for the location(s) at which the digits disagree. In our case, there is disagreement in the third digit, and the single error must be the zero. The repetition code thus successfully corrects an error. However, it is not completely satisfying, since the **information rate** $\frac{k}{n} = \frac{1}{3}$ is very small, with two check symbols for each information symbol.

Example 13.3. In this example we take $r = 1$ and to the k information symbols we append their sum modulo 2. For example, we would encode $1011 \rightarrow 10111$ and $1001 \rightarrow 10010$. Therefore, each codeword always contains an even number of ones, for which reason it is called a **parity code**, and the last symbol a *parity check*. In this case the information rate is high, namely $\frac{k}{k+1}$, but the parity code can in general only detect an error (when the number of ones is odd), but not correct it. If, for example, we received 10110011, we would know from the odd number (5) of ones that an error had occurred, though we couldn't say where. The parity code thus detects a single error. Finally, we note that a parity code is used in the scanners in supermarket checkout lines.

We see, then, the fundamental problem of channel encoding: Construct a code with maximal information rate and minimal probability that the received word will be incorrectly decoded.

Clearly, these two goals pull in opposite directions. The higher the information rate, the greater will be the number of errors. However, the problem has a theoretical solution, for we have the following theorem of Shannon: Suppose we are given $\varepsilon > 0$ and $0 < I < K$, where K is a quantity, the *capacity* of the channel, that depends only on the probability distribution of an erroneous transmission. Then there exists a code with information rate greater than I and probability of error in decoding less than ε. For $I \geq K$ an error probability of less than ε is no longer guaranteed.

Are we finished? Alas, no. Shannon's theorem is purely an existence theorem: there *exist* such codes (the proof uses probability theory), but in general, we do not know how actually to construct them. Thus we are still faced with the task of finding good codes.

In our discussion we came to the following situation: We are given an **alphabet** A, usually $A = \{0, 1\}$. However, we assume in general $|A| = q \geq 2$. The data to be sent over the channel are words $\boldsymbol{w} \in A^k$. A **block code** C of length n over the alphabet A is a subset $C \subseteq A^n$ with $n \geq k$. The words of C are the **codewords**. If $A = \{0, 1\}$, then C is a *binary* code.

For each message $\boldsymbol{w}$, the encoder chooses a codeword $\boldsymbol{x} \in C$. The channel receives $\boldsymbol{x}$ and outputs $\boldsymbol{y} \in A^n$, which on account of noise in the channel is not necessarily in C. The task of the decoder is therefore to choose a codeword $\boldsymbol{x}' \in C$ for $\boldsymbol{y}$ and then to decode $\boldsymbol{x}'$ to a message $\boldsymbol{w}' \in A^k$. We thus have the following picture:

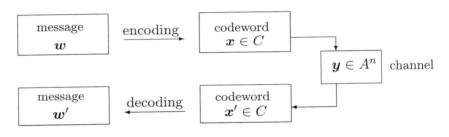

Two problems arise: What codeword $\boldsymbol{x}' \in C$ should the decoder choose if she receives $\boldsymbol{y} \in A^n$, and how can one encode and decode efficiently? In this section we will discuss the first question, and in the next section turn our attention to the practice of encoding and decoding.

The decoder proceeds by the greedy method. She works according to the rule that "fewer" errors are more probable than "more," and chooses for $\boldsymbol{y} \in A^n$ a codeword $\boldsymbol{x}' \in C$ that differs from $\boldsymbol{y}$ in the fewest possible places. If she has several codewords from which to choose, she chooses one of them. In particular, in the case $\boldsymbol{y} \in C$ she assumes that $\boldsymbol{y}$ was in fact sent and sets $\boldsymbol{x}' = \boldsymbol{y}$.

This leads us in a natural way to the **Hamming distance** between words, which we discussed already for $A = \{0, 1\}$ in Section 11.1.

Let $\boldsymbol{a} = (a_1, \ldots, a_n)$, $\boldsymbol{b} = (b_1, \ldots, b_n) \in A^n$. Then the *Hamming distance* is given by $\Delta(\boldsymbol{a}, \boldsymbol{b}) = |\{i : a_i \neq b_i\}|$. As in the binary case, Δ is again a metric on A^n, and therefore satisfies in particular the triangle inequality $\Delta(\boldsymbol{a}, \boldsymbol{b}) \leq \Delta(\boldsymbol{a}, \boldsymbol{c}) + \Delta(\boldsymbol{c}, \boldsymbol{b})$. We let $B_t(\boldsymbol{a}) = \{\boldsymbol{x} \in A^n : \Delta(\boldsymbol{a}, \boldsymbol{x}) \leq t\}$ denote the set of all words with distance less than or equal to t from $\boldsymbol{a}$ and give $B_t(\boldsymbol{a})$ the natural designation of the **ball** around $\boldsymbol{a}$ of **radius** t.

Suppose that we know that in the channel at most t errors can occur. That is, the output word $\boldsymbol{y} \in A^n$ differs from $\boldsymbol{x} \in C$ in at most t places. Thus we have $\boldsymbol{y} \in B_t(\boldsymbol{x})$. When does the decoder correctly fix errors? According to the general rule, she looks for the *closest* codeword, and she will choose the correct word $\boldsymbol{x}$ precisely when among all codewords, $\boldsymbol{x}$ is the *unique* closest one. Thus if the code C satisfies the condition $B_t(\boldsymbol{a}) \cap B_t(\boldsymbol{b}) = \varnothing$ for all $\boldsymbol{a} \neq \boldsymbol{b} \in C$, or equivalently, $\Delta(\boldsymbol{a}, \boldsymbol{b}) \geq 2t+1$, then the decoder always chooses the correct output word. However, this condition is also necessary. Namely, we have $\boldsymbol{y} \in B_t(\boldsymbol{x}) \cap B_t(\boldsymbol{z})$, and if $\Delta(\boldsymbol{z}, \boldsymbol{y}) \leq \Delta(\boldsymbol{x}, \boldsymbol{y}) \leq t$, then she may choose the incorrect codeword $\boldsymbol{z}$.

We proceed in the same way with the notion of detecting t errors. Suppose again that at most t can occur. If $\boldsymbol{x}$ is input into the channel and $\boldsymbol{y} \in B_t(\boldsymbol{x})$ with $\boldsymbol{y} \neq \boldsymbol{x}$ is received, then the decoder discovers that $\boldsymbol{y}$ is not the input word except when $\boldsymbol{y}$ itself is the codeword, since she then sets $\boldsymbol{x}' = \boldsymbol{y}$. Thus no codeword other than $\boldsymbol{x}$ may be in $B_t(\boldsymbol{x})$; that is, we must have $B_t(\boldsymbol{a}) \cap C = \{\boldsymbol{a}\}$ for all $\boldsymbol{a} \in C$, or equivalently, $\Delta(\boldsymbol{a}, \boldsymbol{b}) \geq t+1$ for all $\boldsymbol{a} \neq \boldsymbol{b} \in C$.

To sum up, a code $C \subseteq A^n$ is called **t-error-correcting** if $\Delta(\boldsymbol{a}, \boldsymbol{b}) \geq 2t+1$ for all $\boldsymbol{a} \neq \boldsymbol{b} \in C$, and **$t$-error-detecting** if $\Delta(\boldsymbol{a}, \boldsymbol{b}) \geq t+1$ for all $\boldsymbol{a} \neq \boldsymbol{b} \in C$.

We may now perform further analysis on our initial example. For the repetition code we have $\Delta(\boldsymbol{a}, \boldsymbol{b}) \geq 3$; the code detects two errors and corrects one error. For the parity code we have $\Delta(\boldsymbol{a}, \boldsymbol{b}) \geq 2$; the code detects one error but cannot correct it.

Therefore, we shall now concentrate on the following problem: Let $C \subseteq A^n$. The **distance** from C is $d(C) = \min_{\boldsymbol{a} \neq \boldsymbol{b}} \Delta(\boldsymbol{a}, \boldsymbol{b})$. The problem is to find a code $C \subseteq A^n$ for which

(i) $d(C)$ is as large as possible (i.e., good correction);

(ii) $|C|$ is as large as possible (i.e., large information rate).

Again the two goals are in opposition. The larger the code C, the smaller, of course, the distance. We have therefore formulated the following extremal problem: Given n, d, and A, denote by $\mathcal{C}(n, d; A)$ the set of all codes $C \subseteq A^n$ with distance greater than or equal to d, that is, $\mathcal{C}(n, d; A) = \{C \subseteq A^n : d(C) \geq d\}$. Determine $M(n, d; A) = \max(|C| : C \in \mathcal{C}(n, d; A)\}$.

Of course, the alphabet A plays no role, only the size $|A| = q \geq 2$ has any significance. We may therefore also write $\mathcal{C}(n, d; q)$ and $M(n, d; q)$. A first upper bound, called the **Hamming bound**, follows at once from our discussion of t-error-correcting.

Theorem 13.4. *Let $d = 2t + 1$ be odd, $|A| = q$. Then*

$$M(n, d; q) \leq \frac{q^n}{\sum_{i=0}^{t} \binom{n}{i}(q-1)^i} \cdot$$

Proof. Let $C \in \mathcal{C}(n, 2t + 1; q)$. Then $\Delta(\boldsymbol{a}, \boldsymbol{b}) \geq 2t + 1$, and therefore $B_t(\boldsymbol{a}) \cap B_t(\boldsymbol{b}) = \varnothing$ for $\boldsymbol{a} \neq \boldsymbol{b} \in C$. An arbitrary word $\boldsymbol{a} \in A^n$ has precisely $\binom{n}{i}(q-1)^i$ words with distance i, since we can choose the i error locations in $\binom{n}{i}$ ways and then for each of these have $q - 1$ possibilities for the error. We thereby obtain $|B_t(\boldsymbol{a})| = \sum_{i=0}^{t} \binom{n}{i}(q-1)^i$, and from the disjointness of the balls, we have

$$\left| \bigcup_{\boldsymbol{a} \in C} B_t(\boldsymbol{a}) \right| = |C| \sum_{i=0}^{t} \binom{n}{i}(q-1)^i \leq |A^n| = q^n.$$

$\square$

From this theorem we immediately derive a most attractive problem. A t-error-correcting code has the property that all balls $B_t(\boldsymbol{a})$, $\boldsymbol{a} \in C$, are disjoint. An arbitrary word $\boldsymbol{w} \in A^n$ therefore lies in at most one such ball. If now every word lies in *precisely* one ball, that is, when the balls $B_t(\boldsymbol{a})$ completely cover the space A^n, then C is said to be **t-perfect**. From Theorem 13.4 we therefore conclude that

$$C \subseteq A^n \text{ is } t\text{-perfect} \iff d(C) \geq 2t + 1 \text{ and } |C| = q^n \bigg/ \sum_{i=0}^{t} \binom{n}{i}(q-1)^i.$$

Let $A = \{0, 1\}$, $n = 2t + 1$, and let $C = \{00 \ldots 0, 11 \ldots 1\}$ be the binary repetition code. Then $d(C) = n = 2t + 1$, and since $\sum_{i=0}^{t} \binom{n}{i} = \sum_{i=0}^{\frac{n-1}{2}} \binom{n}{i} = 2^{n-1}$, we also have $|C| = 2 = 2^n / \sum_{i=0}^{t} \binom{n}{i}$. The repetition code is therefore t-perfect for every t.

To be sure, this example is not particularly exciting, but here is a nicer one, for which we summon our Fano plane. Let $A = \{0, 1\}$, $C \subseteq A^n$. As usual, we interpret words $\boldsymbol{a} = (a_1, \ldots, a_n)$ as subsets $U_{\boldsymbol{a}} = \{i : a_i = 1\} \subseteq S = \{1, \ldots, n\}$. The distance $\Delta(\boldsymbol{a}, \boldsymbol{b})$ then corresponds to the size of the symmetric difference $|U_{\boldsymbol{a}} \oplus U_{\boldsymbol{b}}|$. A code $C \in \mathcal{C}(n, d; 2)$ is thus nothing other than a family of sets $\mathcal{U} \subseteq \mathcal{B}(S)$ with $|U \oplus V| \geq d$ for all $U \neq V \in \mathcal{U}$. In particular, a t-perfect code thus corresponds to a family of sets $\mathcal{U} \subseteq \mathcal{B}(S)$, $|S| = n$, with $|U \oplus V| \geq 2t + 1$ for all $U \neq V \in \mathcal{U}$ and $|\mathcal{U}| = 2^n / \sum_{i=0}^{t} \binom{n}{i}$.

How do we go about constructing such "t-perfect" families of sets? Designs are certainly a good start. Let us try $n = 7$, $t = 1$. We must find $\mathcal{U} \subseteq \mathcal{B}(S)$ with $|U \oplus V| \geq 3$ and $|\mathcal{U}| = \frac{2^7}{1+7} = 2^4 = 16$. We take $\varnothing$, the seven triples of the Fano plane, and then the eight complements of these sets. Hence $|\mathcal{U}| = 16$. It is easily checked that the condition $|U \oplus V| \geq 3$ is satisfied for all $U \neq V$. Translated into $(0, 1)$-words, we obtain the following 1-perfect code $C \subseteq \{0, 1\}^7$:

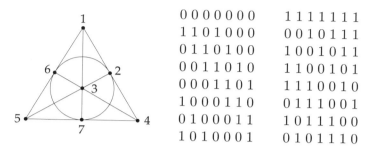

0 0 0 0 0 0 0	1 1 1 1 1 1 1
1 1 0 1 0 0 0	0 0 1 0 1 1 1
0 1 1 0 1 0 0	1 0 0 1 0 1 1
0 0 1 1 0 1 0	1 1 0 0 1 0 1
0 0 0 1 1 0 1	1 1 1 0 0 1 0
1 0 0 0 1 1 0	0 1 1 1 0 0 1
0 1 0 0 0 1 1	1 0 1 1 1 0 0
1 0 1 0 0 0 1	0 1 0 1 1 1 0

Because of the complementation the second eight codewords arise by exchanging 0 and 1 in the first eight words. In the next section and in the exercises we will look more closely at perfect codes.

13.4. Linear Codes

We now would like to construct codes systematically and investigate their capacity for error-correction. Here linear algebra will be of help. Let $K = \mathrm{GF}(q)$ be the Galois field with q elements, q a prime power, and K^n the n-dimensional vector space over K.

Every subspace $C \subseteq K^n$ is called a **linear code** over K. The dimension k of the subspace C is called the **dimension** of C, and C, for short, an (n, k)-code. For example, as one easily checks (and later we shall prove it), the Fano code that we just constructed is a $(7, 4)$-code over $\mathrm{GF}(2)$.

With linear codes we can more conveniently express the distance $d(C)$. For $\boldsymbol{a} \in K^n$ we call $w(\boldsymbol{a}) = |\{i : a_i \neq 0\}|$ the **weight** of $\boldsymbol{a}$. Let $\boldsymbol{a}, \boldsymbol{b} \in K^n$. Since $a_i \neq b_i \iff a_i - b_i \neq 0$, we have $\Delta(\boldsymbol{a}, \boldsymbol{b}) = w(\boldsymbol{a} - \boldsymbol{b})$, and in particular, $\Delta(\boldsymbol{0}, \boldsymbol{c}) = w(\boldsymbol{c})$ for $\boldsymbol{c} \in C$. In a subspace C, however, $\boldsymbol{a}, \boldsymbol{b} \in C$ implies $\boldsymbol{a} - \boldsymbol{b}$ in C. We conclude that for a linear code C,

$$d(C) = \min_{\boldsymbol{a} \neq \boldsymbol{b} \in C} \Delta(\boldsymbol{a}, \boldsymbol{b}) = \min_{\boldsymbol{0} \neq \boldsymbol{c} \in C} w(\boldsymbol{c}).$$

We therefore do not have to calculate a distance, but need consider only the weights of the words.

To every linear code $C \subseteq K^n$ there exists a **dual code** $C^\perp$. Let C be an (n, k)-code. From linear algebra we know that the set of all vectors $\boldsymbol{a} \in K^n$ that are orthogonal to all vectors in C form a subspace $C^\perp$ of dimension $n - k$, that is,

$$C^\perp = \{\boldsymbol{a} \in C : \boldsymbol{a} \cdot \boldsymbol{c} = 0 \text{ for all } \boldsymbol{c} \in C\},$$

where $\boldsymbol{a} \cdot \boldsymbol{c} = a_1 c_1 + \cdots + a_n c_n$ is the usual scalar product. Furthermore, we have $(C^\perp)^\perp = C$. From the linearity of the product, it suffices to prove $\boldsymbol{a} \cdot \boldsymbol{g}_i = 0$ for a basis $\{\boldsymbol{g}_1, \ldots, \boldsymbol{g}_k\}$.

This gives us two useful descriptions of a linear code C. First, we choose a basis $\boldsymbol{g}_1, \ldots, \boldsymbol{g}_k$ of C and form the $k \times n$ matrix G with the $\boldsymbol{g}_i$ as rows. The matrix G is called a *generator matrix* of C. Every codeword $\boldsymbol{c}$ is determined uniquely by the coefficients w_i, $\boldsymbol{c} = \sum_{i=1}^{k} w_i \boldsymbol{g}_i$; that is, we have $\boldsymbol{c} = \boldsymbol{w} G$ as the vector–matrix product. Second, we can take a generator matrix H of the *dual* code $C^\perp$. Hence H is an $(n - k) \times n$ matrix, and we have

$$(13.2) \qquad\qquad\qquad \boldsymbol{c} \in C \iff H\boldsymbol{c} = \boldsymbol{0}.$$

Every such matrix H is called a *parity check matrix* of C.

The following theorem shows how one can construct a linear code of distance d.

Theorem 13.5. *Let C be an (n, k)-code over K, and H a parity check matrix. Then $d(C) \geq d$ if and only if every $d - 1$ columns in H are linearly independent.*

Proof. Let $\boldsymbol{u}_1, \ldots, \boldsymbol{u}_n$ be the columns of H; the $\boldsymbol{u}_i$ are therefore vectors in K^r, $r = n - k$. From (13.2) we wee that

$$\boldsymbol{c} \in C \iff c_1 \boldsymbol{u}_1 + \cdots + c_n \boldsymbol{u}_n = \boldsymbol{0}.$$

Suppose $w(\boldsymbol{c}) \leq d - 1$ and $c_{i_1}, \ldots, c_{i_\ell}$, $\ell \leq d - 1$, are the nonzero entries. Then we have $c_{i_1} \boldsymbol{u}_{i_1} + \cdots + c_{i_\ell} \boldsymbol{u}_{i_\ell} = \boldsymbol{0}$, that is, the $\ell \leq d - 1$ columns $\boldsymbol{u}_{i_1}, \ldots, \boldsymbol{u}_{i_\ell}$ are linearly dependent. Conversely, if $\boldsymbol{u}_{j_1}, \ldots, \boldsymbol{u}_{j_h}$ are linearly dependent with $h \leq d - 1$, then there exists a nontrivial linear combination $c_{j_1} \boldsymbol{u}_{j_1} + \cdots + c_{j_h} \boldsymbol{u}_{j_h} = \boldsymbol{0}$, that is, the vector $\boldsymbol{c}$ with coordinates c_{j_i} for $i = 1, \ldots, h$ and 0 otherwise in C and satisfies $w(\boldsymbol{c}) \leq d - 1$. $\square$

Example 13.6. With Theorem 13.5 we can immediately construct a class of 1-perfect codes, the so-called **Hamming codes**. We take $K = \mathrm{GF}(q)$, $r \geq 2$. For $d = 3$ we must find vectors $\boldsymbol{u}_1, \ldots, \boldsymbol{u}_n$ in K^r such that every pair of them are linearly independent. Since every vector generates $q - 1$ nonnull multiples, there are altogether $\frac{q^r - 1}{q - 1}$ such vectors. Setting $n = \frac{q^r - 1}{q - 1}$, $k = n - r$, we obtain from the theorem an (n, k)-code $C \subseteq K^n$ that satisfies $d(C) \geq 3$ and thus is 1-error-correcting. Finally, we have

$$|C| = q^k = \frac{q^n}{q^r} = \frac{q^n}{1 + \frac{q^r - 1}{q - 1}(q - 1)} = \frac{q^n}{1 + n(q - 1)},$$

and so C is in fact 1-perfect. For $q = 2$ this construction yields for $r = 2$ the binary repetition code, and for $r = 3$ the Fano code described above, which shows again that the Fano code is a $(7, 4)$-code.

Moreover, it follows from the second condition of the theorem that $d-1 \leq r = n - k$, since K^r can contain no more than r linearly independent vectors, and we therefore have the bound

(13.3) $$d(C) \leq n - k + 1 \quad \text{for an } (n, k)\text{-code } C.$$

We now turn to the other two questions from the previous section, namely how one can code and decode messages effectively. For linear codes we have the following procedure.

Let C be an (n, k)-code over $K = \mathrm{GF}(q)$. We choose a generator matrix G with basis $\boldsymbol{g}_1, \ldots, \boldsymbol{g}_k$. We know already that every vector $\boldsymbol{c} \in C$ can be written as a product $c = \boldsymbol{w}G$ with $\boldsymbol{w} \in K^k$. We now identify the *message* with the q^k vectors in K^k and encode using

$$\phi : \boldsymbol{w} \in K^k \to \boldsymbol{w}G \in C.$$

Of course, there are various encodings ϕ, corresponding to the various generator matrices G. A generator matrix is said to be *systematic* if $G = (E_k, G_1)$, where E_k is the $k \times k$ identity matrix and G_1 is a $k \times (n - k)$ matrix. In this case we have

$$\phi\boldsymbol{w} = \boldsymbol{w}G = (\boldsymbol{w}, \boldsymbol{w}G_1),$$

that is, the k information symbols appear in the first k places, and the $r = n - k$ check symbols are *linear functions* of the information symbols. Consider, for example, the $(n, n-1)$-code over $\mathrm{GF}(2)$ with generator matrix

$$G = \begin{pmatrix} 1 & & & 1 \\ & 1 & 0 & 1 \\ & 0 & \ddots & \vdots \\ & & 1 & 1 \end{pmatrix}.$$

We obtain the encoding $\phi\boldsymbol{w} = \left(w_1, \ldots, w_{n-1}, \sum_{i=1}^{n-1} w_i\right)$, that is, our parity code. The dual code $C^\perp$ is an $(n, 1)$-code with generator matrix $G = (1, 1, \ldots, 1)$, and the corresponding encoding is $\phi w = (w, w, \ldots, w)$. That is, $C^\perp$ is nothing other than the repetition code.

The decoding into the source code proceeds conversely by solving a system of equations. If we have received $\boldsymbol{c} \in C$, we determine $\boldsymbol{w}$ from $\boldsymbol{w}G = \boldsymbol{c}$. In the case of a systematic encoding there is nothing to do. The message consists simply of the first k symbols of the codeword.

For channel decoding we use the second description using a parity check matrix H. It is particularly easy to determine a parity check matrix when the code is given by a systematic generator matrix. In this case one sees at once that $H = (-G_1^T, E_{n-k})$ is a parity check matrix, where G_1^T is the matrix transpose of G_1. For example, $H = (1, 1, \ldots, 1)$ over $\mathrm{GF}(2)$ is the parity check matrix of the parity code, which of course can be seen directly on account of $Hc = 0 \iff c_1 + \cdots + c_n = 0$.

What is the parity check matrix of the Fano code? By construction of the Hamming code for $n = 7$, $r = 3$, we must write all linearly independent vectors as columns. Since all pairs of nonnull vectors over $GF(2)$ are linearly independent, *all* $2^3 - 1 = 7$ nonnull vectors appear as columns, and we obtain

$$H = \begin{pmatrix} 1 & 0 & 0 & 1 & 1 & 0 & 1 \\ 0 & 1 & 0 & 1 & 0 & 1 & 1 \\ 0 & 0 & 1 & 0 & 1 & 1 & 1 \end{pmatrix}.$$

The check equations for the codeword c in the Fano code are therefore

$$\begin{aligned} c_1 + \quad\quad\quad + c_4 + c_5 \quad\quad + c_7 &= 0, \\ c_2 + \quad\quad c_4 + \quad + c_6 + c_7 &= 0, \\ c_3 + \quad\quad\quad c_5 + c_6 + c_7 &= 0. \end{aligned}$$

How, then, does one actually go about decoding? Suppose that C corrects up to t errors. If $x \in C$ is sent and $y \in K^n$ received, then we have $\Delta(x, y) \leq t$ and $\Delta(c, y) > t$ for all $c \in C$, $c \neq x$. We can look in the list of codewords until we come upon the unique codeword x for which $\Delta(x, y) \leq t$. A better approach is the following. Let H be an $r \times n$ parity check matrix, $r = n - k$. For $a \in K^n$ we call

$$s(a) = Ha \in K^r$$

the *syndrome* of a. The function s is therefore a linear mapping from K^n to K^r with kernel C, since

$$s(a) = 0 \iff a \in C.$$

Hence if a, b are in the same residue class of C, that is, $b = a + c$, $c \in C$, then we have $s(b) = s(a)$, and conversely, from $s(a) = s(b)$ we immediately obtain $b - a \in C$, and therefore $b = a + c$.

The concept of syndromes leads to the following procedure. If $x \in C$ is sent and $y = x + e$ received, we call e the *error vector*. We have $s(y) = s(e)$ and weight $w(e) \leq t$. We then search in the list of possible syndromes for the unique error vector e with $w(e) \leq t$ and $s(e) = s(y)$ and decode y to $y - e$. The case is particularly simple for 1-error-correcting codes. Here the error vectors of minimal weight are of the form

$$e = (0, \ldots, \overset{\downarrow i}{a}, \ldots, 0),$$

whence

$$s(e) = a \cdot [\text{ith column of } H].$$

That is, if a received vector y has an error, then the error occurred in the ith place if and only if $s(y)$ is a multiple of the ith column of H.

Example 13.7. Consider the parity check matrix H of the Fano code. Suppose $x = 1\ 1\ 1\ 0\ 0\ 0\ 1$ is sent and $y = 1\ 1\ 0\ 0\ 0\ 0\ 1$ received. We calculate Hy,

$$
\begin{pmatrix} 1\ 0\ 0\ 1\ 1\ 0\ 1 \\ 0\ 1\ 0\ 1\ 0\ 1\ 1 \\ 0\ 0\ 1\ 0\ 1\ 1\ 1 \end{pmatrix}
\begin{pmatrix} 1 \\ 1 \\ 0 \\ 0 \\ 0 \\ 0 \\ 1 \end{pmatrix}
=
\begin{pmatrix} 0 \\ 0 \\ 1 \end{pmatrix},
$$

and conclude that the error occurred in the third position. Since the sum of two columns of H is always nonnull, it follows that $He \neq 0$ for all error vectors e with $w(e) = 2$. Thus the decoder *detects* two errors, but in general is unable to localize them, since, for example,

$$
\begin{pmatrix} 0 \\ 0 \\ 1 \end{pmatrix} + \begin{pmatrix} 1 \\ 1 \\ 1 \end{pmatrix} = \begin{pmatrix} 1 \\ 0 \\ 0 \end{pmatrix} + \begin{pmatrix} 0 \\ 1 \\ 0 \end{pmatrix}.
$$

Another important class of codes is that of **Reed–Solomon codes** C, which are used for error correction in compact discs. Let $\mathrm{GF}(q) = \{0, 1, a_1, \ldots, a_{q-2}\}$ be the Galois field of the prime power q. The parity check matrix H is the $(d-1) \times (q+1)$ matrix, $d \geq 3$, defined by

$$
H = \begin{pmatrix}
0 & 1 & 1 & 1 & \cdots & 1 \\
0 & 0 & 1 & a_1 & & a_{q-2} \\
0 & 0 & 1 & a_1^2 & \cdots & a_{q-2}^2 \\
\vdots & \vdots & \vdots & & & \\
0 & 0 & 1 & a_1^{d-3} & \cdots & a_{q-2}^{d-3} \\
1 & 0 & 1 & a_1^{d-2} & & a_{q-2}^{d-2}
\end{pmatrix}.
$$

One easily checks that every $d - 1$ columns of H are linearly independent (note that columns 3 through $q+1$ form a Vandermonde matrix). Thus by Theorem 13.5, C forms a $(q+1, q+2-d)$-code with $d(C) \geq d$. Since (13.3) implies the inequality $d(C) \leq q+1 - (q+2-d) + 1 = d$, we indeed have precisely $d(C) = d$. The information rate is $\frac{k}{n} = \frac{q+2-d}{q+1}$, which is close to 1.

By this point the reader may well have asked why we are interested at all in linear codes over $\mathrm{GF}(q)$ with $q > 2$. Since any data can be coded with $(0, 1)$-words, why don't we simply restrict our attention to $\mathrm{GF}(2)$? The following example provides a nice answer to this question.

Example 13.8. Consider $K = \mathrm{GF}(2^8)$ and $d = 11$. The Reed–Solomon code C is then a $(257, 247)$-code with $d(C) = 11$. We know from Section

12.2 that K is a vector space of dimension 8 over $\mathrm{GF}(2)$. That is, we can write every element $\boldsymbol{a} \in K$ as a $(0,1)$-word of length 8, and therefore every code vector $\boldsymbol{c} = (c_1, \ldots, c_{257}) \in C$ as a $(0,1)$-word of length $257 \cdot 8 = 2056$:

$$\boldsymbol{c} = \left(c_1^{(1)}, \ldots, c_1^{(8)}, \ldots, c_{257}^{(1)}, \ldots c_{257}^{(8)} \right) .$$

In this manner we obtain a new code C' over $\mathrm{GF}(2)$ of length 2056. We claim that C' can be corrected for up to 33 *sequential* errors (a so-called burst error, perhaps the result of a lightning strike). The proof is simple. The 33 errors occur in 5 successive 8-blocks, and since our original code C can correct these 5 errors, C' is able to correct all 33 errors. If one wished to construct a 33-error-correcting code directly over $\mathrm{GF}(2)$, one would obtain a significantly inferior information rate, as can be read off, for example, from the Hamming bound.

13.5. Cyclic Codes

In practice, the decoding of linear codes requires us to specify an error vector for each syndrome. For a $(40, 10)$-code over $\mathrm{GF}(2)$ there are already $2^{30} > 10^9$ syndromes to store. We must therefore look for linear codes exhibiting considerably more algebraic structure.

   A particularly important class of such codes is that of the cyclic codes. As usual, let $K = \mathrm{GF}(q)$. We write vectors $\boldsymbol{a} \in K^n$ in the form $\boldsymbol{a} = (a_{n-1}, a_{n-2}, \ldots, a_1, a_0)$. This has the advantage that we can identify $\boldsymbol{a}$ with the *polynomial* $a(x) = \sum_{i=0}^{n-1} a_i x^i$. The mapping $\phi : \boldsymbol{a} \to a(x)$ is an isomorphism from K^n to the vector space $K^n[x]$ of all polynomials of degree less than n.

Definition 13.9. A code $C \subseteq K^n$ is **cyclic** if C is linear and such that if $\boldsymbol{c} = (c_{n-1}, \ldots, c_0) \in C$, then $\hat{\boldsymbol{c}} = (c_{n-2}, \ldots, c_0, c_{n-1}) \in C$ as well.

   For example, over $\mathrm{GF}(2)$ the repetition code $\{(0, \ldots, 0), (1, \ldots, 1)\}$ is cyclic, as is the parity code in which all codewords have even weight. Likewise, we see that the $(7, 4)$ Fano code is cyclic. How can we tell whether a linear code is cyclc? Here is where the map $\boldsymbol{c} \longrightarrow c(x)$ comes into play.

Theorem 13.10. *Let $C \subseteq K^n$ be cyclic, $\dim C > 0$. Then there exists a monic polynomial $g(x)$ such that*

   (i) $g(x) \mid x^n - 1$,
   (ii) $\boldsymbol{c} \in C \iff g(x) | c(x)$.

*The polynomial $g(x)$ is called the **generator polynomial** of C.*

   Conversely, if $g(x)$ is a monic polynomial and $g(x) \mid x^n - 1$, then $C = \{\boldsymbol{c} \in K^n : g(x) \mid c(x)\}$ is a cyclic code.

Proof. Let $g(x) \in C(x)$ be the unique monic polynomial of minimal degree. For $c(x) = c_{n-1}x^{n-1} + \cdots + c_0 \in C(x)$ we have $xc(x) = c_{n-2}x^{n-1} + \cdots + c_0 x + c_{n-1}x^n$, and thus for the translated word $\hat{c}$ we have

$$\hat{c}(x) = xc(x) - c_{n-1}(x^n - 1). \tag{13.4}$$

If $g(x)$ has degree r, then we have $xg(x), \ldots, x^{n-1-r}g(x) \in C(x)$, and hence every product $a(x)g(x)$ with $\deg a(x) \le n - 1 - r$ is in $C(x)$ as well. If $c(x) \in C(x)$, then we know from the division algorithm for polynomials that

$$c(x) = a(x)g(x) + r(x), \quad \deg r(x) < \deg g(x).$$

Since $c(x), a(x)g(x) \in C(x)$, then $r(x) \in C(x)$ as well, and we conclude that $r(x) = 0$ on account of the minimality of $g(x)$, whence $g(x) \mid c(x)$.

Now let $c(x) = x^{n-1-r}g(x) \in C(x)$. It then follows from (13.4) that

$$\hat{c}(x) = x^{n-r}g(x) - (x^n - 1).$$

Since we now have, as we saw earlier, $g(x) \mid \hat{c}(x)$, we have as well $g(x) \mid x^n - 1$, as claimed.

Conversely, suppose $C = \{\boldsymbol{c} \in K^n : g(x)|c(x)\}$. Then C is clearly a subspace, and from (13.4) and $g(x) \mid c(x)$, $g(x) \mid x^n - 1$ we immediately obtain $g(x) \mid \hat{c}(x)$, and so $\hat{\boldsymbol{c}} \in C$. $\qquad\square$

For cyclic codes we can easily give a generator matrix and a parity check matrix. Let $g(x) = x^r + g_{r-1}x^{r-1} + \cdots + g_0$ be the generator polynomial. Then $\{g(x), xg(x), \ldots, x^{k-1}g(x)\}$ is a basis for $C(x)$, $k = n - r$. We thereby obtain the generator matrix

$$G = \begin{pmatrix} 1 & g_{r-1} & \cdots & g_0 & 0 & \cdots & 0 \\ 0 & 1 & g_{r-1} & \cdots & g_0 & \cdots & 0 \\ \vdots & & \ddots & & & & \\ 0 & & & 1 & g_{r-1} & \cdots & g_0 \end{pmatrix} = \begin{pmatrix} x^{k-1}g(x) \\ x^{k-2}g(x) \\ \vdots \\ g(x) \end{pmatrix}.$$

The dimension of C is therefore $n - \deg g(x)$.

If $g(x)$ is the generator polynomial of C, we call $h(x) = \frac{x^n - 1}{g(x)}$ the *check polynomial*. Let $h(x) = x^k + h_{k-1}x^{k-1} + \cdots + h_0$. Then by comparing coefficients in $g(x)h(x) = x^n - 1$, we obtain

$$g_i + h_{k-1}g_{i+1} + h_{k-2}g_{i+2} + \cdots + h_i g_k = 0 \quad (i = 0, 1, \ldots, r - 1),$$

and thus we conclude that

$$H = \begin{pmatrix} 0 & \cdots & \cdots & h_0 & \cdots & \cdots & h_{k-1} & 1 \\ 0 & \cdots & h_0 & h_1 & \cdots & h_{k-1} & 1 & 0 \\ \vdots & & & \cdots & & & & \\ h_0 & h_1 & \cdots & h_{k-1} & 1 & 0 & \cdots & 0 \end{pmatrix}$$

is the parity check matrix for C.

Example 13.11. In order to determine all binary cyclic codes of length 7, we must go through all the divisors of $x^7 - 1$. The decomposition of $x^7 - 1$ into irreducible factors is (note that $1 = -1$)

$$x^7 - 1 = (x + 1)(x^3 + x + 1)(x^3 + x^2 + 1).$$

To $g(x) = 1$ belongs the code $C = \mathrm{GF}(2)^7$. For $g(x) = x + 1$ we obtain

$$G = \begin{pmatrix} 1 & 1 & & & & \\ & 1 & 1 & & 0 & \\ & & & \ddots & & \\ & 0 & & & & \\ & & & & 1 & 1 \end{pmatrix},$$

which is the parity code. For $g(x) = x^3 + x + 1$ the check polynomial is $h(x) = (x + 1)(x^3 + x^2 + 1) = x^4 + x^2 + x + 1$, and we obtain the parity check matrix

$$H = \begin{pmatrix} 0 & 0 & 1 & 1 & 1 & 0 & 1 \\ 0 & 1 & 1 & 1 & 0 & 1 & 0 \\ 1 & 1 & 1 & 0 & 1 & 0 & 0 \end{pmatrix},$$

and we have the Fano code, as we already knew. The other possibilities are resolved with equal ease.

Encoding with cyclic codes is particularly simple. We define $\phi : K^k \longrightarrow C$ by $\phi \boldsymbol{a} = \boldsymbol{c}$, where $c(x) = a(x)g(x)$. Conversely, for decoding we set $\boldsymbol{c} \in C \longrightarrow \boldsymbol{a} \in K^k$ with $a(x) = \frac{c(x)}{g(x)}$. If the received word $v(x)$ is not divisible by $g(x)$, then an error has occurred, and $v(x) = c(x) + e(x)$, $c(x) \in C(x)$, and $e(x)$ are determined from

$$v(x)h(x) = c(x)h(x) + e(x)h(x) \equiv e(x)h(x) \pmod{x^n - 1}$$

using the Euclidean algorithm.

How does one find "good" cyclic codes C, that is, codes C with $d(C) \geq d$? From $g(x) \mid c(x)$ one knows that every root α of $g(x)$ is also a root of $c(x)$. We can therefore also describe C by specifying the roots (in a suitable field):

(13.5) $\boldsymbol{c} \in C \Longleftrightarrow c(\alpha_i) = 0, \quad i = 1, \ldots, r$.

This can be clarified with an example: Let $K = \mathrm{GF}(q)$ and let K^* be the multiplicative group of K. We know from Exercise 12.32 that $\beta^{q-1} = 1$ for all $\beta \in K^*$, that is, the β are precisely the roots of the polynomial $x^{q-1} - 1$, and we can write

(13.6) $$x^{q-1} - 1 = \prod_{\beta \in K^*} (x - \beta).$$

In the study of abstract algebra one learns that in every finite field K there are elements α with ord $(\alpha) = q - 1$. These are called the *primitive elements* of K. In other words, $K^* = \{1, \alpha, \alpha^2, \ldots, \alpha^{q-2}\}$.

So let α be a primitive element and

$$g(x) = (x - 1)(x - \alpha)(x - \alpha^2) \cdots (x - \alpha^{d-2}).$$

From (13.6) we know that $g(x)$ is a divisor of $x^{q-1} - 1$, and with $g(x)$ as generator polynomial we obtain a cyclic code C of length $q - 1$. From (13.5) we see that

$$\boldsymbol{c} = (c_{q-2}, \ldots, c_0) \in C \iff (\alpha^i) = 0 \quad (i = 0, \ldots, d - 2)$$

$$\iff \sum_{j=0}^{n-1} c_j \alpha^{ij} = 0 \quad (i = 0, \ldots, d - 2).$$

Therefore,

$$H = \begin{pmatrix} 1 & 1 & \cdots & 1 \\ \alpha^{q-2} & \alpha^{q-3} & & 1 \\ \vdots & \vdots & & \vdots \\ \alpha^{(q-2)(d-2)} & \alpha^{(q-3)(d-2)} & \cdots & 1 \end{pmatrix}$$

is a parity check matrix for C. This matrix should look familiar. It is (except for the first two columns) the parity check matrix of the Reed–Solomon code. We therefore have $d(C) \geq d$, and all Reed–Solomon codes are cyclic.

Exercises for Chapter 13

13.1 Consider the set $\{0, 1, \ldots, 99\}$ of the first hundred nonnegative integers. The usual decimal representation is clearly not a prefix code. Determine an optimal binary source code C on the assumption that all numbers are equally probable. How large is $\overline{L}(C)$?

▷ **13.2** Is there a prefix code over $\{0, 1\}$ with six codewords of lengths $1, 3, 3, 3, 3, 3$? Answer the same question for the lengths $2, 3, 3, 3, 3, 3$.

13.3 For every n construct a prefix code over $\{0, 1\}$ with codewords of lengths $1, 2, \ldots, n$. Show that exactly one digit in each such code is superfluous. Where is it located?

13.4 Determine $\overline{L}(C)$ for a prefix code over $\{0, 1\}$ with the probabilities $\frac{1}{64}(27, 9, 9, 9, 3, 3, 3, 1)$.

▷ **13.5** Construct a code $C \subseteq \{0, 1\}^6$ with $|C| = 5$ that corrects one error. Does this work also with $C \subseteq \{0, 1\}^5$, $|C| = 5$?

13.6 Show that for a 2-error-correcting code $C \subseteq \{0, 1\}^8$ one has $|C| \leq 4$. Does there exist such a code?

▷ **13.7** Let $M(n, d) = M(n, d; 2)$. Show that (a) $M(n, 2d - 1) = M(n + 1, 2d)$; (b) $M(n, d) \leq 2M(n - 1, d)$.

13.8 A code $C \subseteq GF(3)^6$ is given by the check matrix

$$H = \begin{pmatrix} 2 & 0 & 1 & 1 & 0 & 0 \\ 1 & 2 & 0 & 0 & 1 & 0 \\ 0 & 2 & 2 & 0 & 0 & 1 \end{pmatrix}.$$

Determine a generator matrix G and encode the messages 102, 101, 210, 122.

13.9 Consider the binary code C given by the check matrix

$$H = \begin{pmatrix} 1 & 1 & 0 & 0 & 1 & 0 & 0 \\ 0 & 0 & 1 & 1 & 0 & 1 & 0 \\ 1 & 1 & 1 & 1 & 0 & 0 & 1 \end{pmatrix}.$$

Is C 1-error-correcting? Write down a list of syndromes and decode the following words: 0110111, 0111000, 1101011, 1111111.

▷ **13.10** Let C denote the Fano code. Assume that every symbol is transmitted incorrectly with probability p. What is the probability that a received word will be decoded correctly?

13.11 Let $C \in \mathcal{C}(n, 2t; 2)$. Prove that there exists another code $C' \in \mathcal{C}(n, 2t; 2)$ in which all codewords have the same weight.

13.12 Show that in a binary linear code, either all codewords have even weight or half have odd weight and half even.

13.13 Let q be a prime power and $1 + n(q - 1) \mid q^n$. Show that $1 + n(q - 1) = q^r$.

13.14 Prove in detail that in the check matrix of the Reed–Solomon code, every $d - 1$ columns are linearly independent.

▷ **13.15** Show that the linear code over $GF(3)$ with generator matrix

$$\begin{pmatrix} 0 & 1 & 1 & 2 \\ 1 & 0 & 1 & 1 \end{pmatrix}$$

is 1-perfect.

13.16 Let C_1, C_2 be two cyclic codes over $GF(q)$ with generator polynomials $g_1(x)$, $g_2(x)$. Show that $C_1 \subseteq C_2 \iff g_2(x) \mid g_1(x)$.

▷ **13.17** Let C be a binary cyclic code that contains a codeword of odd weight. Prove that $(1, 1, \ldots, 1) \in C$.

13.18 A source emits ten different signals, two of them with probability 0.14 and eight with probability 0.09. Determine $\overline{L}(C)$ over the alphabet $\{0, 1, 2\}$, and the same for the alphabet $\{0, 1, 2, 3\}$.

▷ **13.19** Suppose a source sends n signals with a certain distribution $(p_1, \ldots, p_n)$. Show that the lengths $\ell_1, \ldots, \ell_n$ of the codewords in an optimal code over $\{0, 1\}$ always satisfy $\sum_{i=1}^n \ell_i \leq \frac{n^2 + n - 2}{2}$ and that equality holds for certain distributions. Hint: Analyze the Huffman algorithm.

13.20 Show in analogy to the previous exercise that one always has $\sum_{i=1}^n \ell_i \geq n \lg n$. Can equality hold in this case?

▷ **13.21** Let $\kappa : X \to A^*$ be a source encoding. We extend κ to $\kappa^* : X^* \to A^*$ via $\kappa^*(x_1 x_2 \ldots x_k) = \kappa(x_1)\kappa(x_2) \ldots \kappa(x_k)$. The code $C = \kappa(X)$ is said to be *uniquely decodable* if κ^* is injective. In other words, $\boldsymbol{v}_1 \ldots \boldsymbol{v}_s = \boldsymbol{w}_1 \ldots \boldsymbol{w}_t$ with $\boldsymbol{v}_i, \boldsymbol{w}_j \in C$ implies $s = t$ and $\boldsymbol{v}_i = \boldsymbol{w}_i$ for all i. Prove the following generalization of Kraft's inequality, Theorem 9.3: If $|A| = q$, $C = \{\boldsymbol{w}_1, \ldots, \boldsymbol{w}_n\}$ is uniquely decodable, then $\sum_{i=1}^{n} q^{-\ell(\boldsymbol{w}_i)} \leq 1$. Hint: Consider the number $N(k, \ell)$ of codewords $\boldsymbol{w}_{i_1} \ldots \boldsymbol{w}_{i_k} \in C^k$ with total length ℓ.

13.22 You are given $A = \{1, \ldots, n\}$ and orthogonal Latin squares $L_1, \ldots, L_t$. Construct the following code $C \subseteq A^{t+2}$. The codeword $\boldsymbol{c}_{ij}$ is $\boldsymbol{c}_{ij} = (i, j, L_1(i, j), \ldots, L_t(i, j))$; hence $|C| = n^2$. Show that $t + 1$ is the minimal distance.

▷ **13.23** The Hamming bound states that $M(n, 2t + 1; q) \leq q^n / \sum_{i=0}^{t} \binom{n}{i}(q - 1)^i$. Let q be a prime power. Show that conversely, $\sum_{i=0}^{2t-1} \binom{n-1}{i}(q - 1)^i < q^{n-k}$ implies the existence of a linear (n, k)-code C over $\mathrm{GF}(q)$ with $d(C) \geq 2t + 1$. Hint: Use Theorem 13.5 and construct successive vectors of $\mathrm{GF}(q)^{n-k}$ of which every $2t$ are linearly independent.

13.24 Let $D = S_t(v, k)$ be a Steiner system. Show that the blocks (written as $0, 1$ incidence vectors) yield a code C with $|C| = b$ and $d(C) \geq 2(k - t + 1)$.

▷ **13.25** A binary code C is called *self-dual* if $C = C^{\perp}$. Suppose the self-dual code C has a basis $\boldsymbol{g}_1, \ldots, \boldsymbol{g}_k$ with $w(\boldsymbol{g}_i) \equiv 0 \pmod{4}$ for all i. Show that then $w(\boldsymbol{c}) \equiv 0 \pmod{4}$ for all $\boldsymbol{c} \in C$.

13.26 Let H_r be the Hamming code with parameters $n = 2^r - 1$, $k = 2^r - 1 - r$, $d = 3$, and add a parity-check column to the end (so that all codewords have an even number of ones). Call the new code $\hat{H}_r$. Show that (a) the codewords in H_r of weight 3, considered as incidence vectors of the sets of ones, form a Steiner system $S_2(2^r - 1, 3)$; (b) the codewords in $\hat{H}_r$ of weight 4 form a Steiner system $S_3(2^r, 4)$.

13.27 Determine the number of codewords in H_r of weight 3 and the number of codewords in $\hat{H}_r$ of weight 4 from the previous exercise.

▷ **13.28** Let $C \in \mathcal{C}(n, d; 2)$, $|C| = M$. Show that $\binom{M}{2}d \leq \sum_{i,j} \Delta(\boldsymbol{c}_i, \boldsymbol{c}_j) \leq \frac{nM^2}{4}$, where the summation runs over all pairs $\boldsymbol{c}_i, \boldsymbol{c}_j \in C$. Hint: Consider the $M \times n$ matrix with the codewords as rows.

13.29 From the previous exercise with $M(n, d) = M(n, d; 2)$ conclude the following: (a) $M(n, d) \leq 2\lfloor \frac{d}{2d-n} \rfloor$, d even, $2d > n$; (b) $M(2d, d) \leq 4d$; (c) $M(n, d) \leq 2\lfloor \frac{d}{2d+1-n} \rfloor$, d odd, $2d + 1 > n$; (d) $M(2d + 1, d) \leq 4d + 4$.

▷ **13.30** Given k and $d \geq 2$, let $N(k, d; q)$ denote the smallest n such that there exists a linear (n, k)-code over $\mathrm{GF}(q)$. For example, (13.3) says that $N(k, d; q) \geq d + k - 1$. The following result improves on this. Show that $N(k, d; q) \geq d + N(k - 1, \lceil \frac{d}{q} \rceil; q)$ and conclude that $N(k, d; q) \geq d + \lceil \frac{d}{q} \rceil + \cdots + \lceil \frac{d}{q^{k-1}} \rceil$. Hint: Consider a generator matrix G with first row $(c_1, \ldots, c_d, 0, \ldots, 0)$, $c_i \neq 0$.

13.31 Determine the largest possible dimension k of a binary code C of length n with $d(C) = 1$, 2, and 3.

▷ **13.32** The following exercises deal in general with perfect codes. Let C be a t-perfect code, $C \subseteq \{0, 1\}^n$. Let $\hat{C}$ be derived from C by adding a parity-check column. Show that the codewords in C of weight $2t + 1$ form a Steiner system $S_{t+1}(n, 2t+1)$, and the codewords of $\hat{C}$ of weight $2t+2$ a Steiner system $S_{t+2}(n+1, 2t+2)$. Let h_{2t+1} be the number of codewords of weight $2t+1$ in C. Show that $h_{2t+1}\binom{2t+1}{t+1} = \binom{n}{t+1}$. Hint: Consider the balls $B_t(\boldsymbol{a})$ with $\boldsymbol{a} \in C$.

13.33 Conclude from the previous exercise that the existence of a t-perfect code $C \subseteq \{0,1\}^n$ implies $\binom{n-i}{t+1-i} \equiv 0 \pmod{\binom{2t+1-i}{t+1-i}}$ for $i = 0, \ldots, t+1$, and therefore in particular, $n + 1 \equiv 0 \pmod{t+1}$.

▷ **13.34** For 2-perfect codes $C \subseteq \{0,1\}^n$ one has from Theorem 13.4 that $|C|(1 + n + \binom{n}{2})) = 2^n$, whence $n^2 + n = 2^{r+1} - 2$ for some r. Show that this equation implies $(2n+1)^2 = 2^{r+3} - 7$. From number theory one knows that the equation $x^2 + 7 = 2^m$ has exactly the solutions $x = 1, 3, 5, 11, 181$. Discuss these cases and show that aside from the linear repetition code there is no other 2-perfect code.

13.35 Show that there is no 1-perfect code C with $n = 7$, $q = 6$. Hint: Show that every 5-tuple $(a_1, \ldots, a_5) \in A^5$, $|A| = 6$, appears exactly once in a codeword and from this it would follow that there exist two orthogonal Latin squares of order 6.

▷ **13.36** Another interesting class of codes is generated from Hadamard matrices. A *Hadamard matrix* of order n is an $n \times n$ matrix with entries ± 1 for which $HH^T = nE_n$ holds (E_n the identity matrix). Prove the following: (a) If there exists a Hadamard matrix of order n, then $n = 1, 2$ or $n \equiv 0 \pmod 4$. (b) If H is a Hadamard matrix, then so is H^T. (c) If H_n is a Hadamard matrix of order n, then $H_{2n} = \left(\begin{smallmatrix} H_n & H_n \\ H_n & -H_n \end{smallmatrix}\right)$ is a Hadamard matrix of order $2n$. (d) Conclude that there exist Hadamard matrices for all $n = 2^k$.

13.37 A Hadamard matrix with all ones in the first row and first column is said to be *normalized*. Show that the existence of a normalized Hadamard matrix of order $n = 4t \geq 8$ is equivalent to a block plan with parameters $v = b = 4t - 1$, $k = r = 2t - 1$, $\lambda = t - 1$. Hint: Replace the -1's by 0's and delete the first row and first column.

▷ **13.38** Let H be a normalized Hadamard matrix of order $n \geq 4$. Use it to construct codes A, B, C with $A \subseteq \{0,1\}^{n-1}$, $|A| = n$, $d(A) = \frac{n}{2}$; $B \subseteq \{0,1\}^{n-1}$, $|B| = 2n$, $d(B) = \frac{n}{2} - 1$; $C \subseteq \{0,1\}^n$, $|C| = 2n$, $d(C) = \frac{n}{2}$. What do B and C look like for $n = 8$?

13.39 Let C_1, C_2 be error-correcting codes of length n over $\{0,1\}$, $|C_1| = m_1$, $|C_2| = m_2$, $d(C_1) = d_1$, $d(C_2) = d_2$. The code $C_3 = C_1 * C_2$ is defined by $C_3 = \{(\boldsymbol{u}, \boldsymbol{u} + \boldsymbol{v}) : \boldsymbol{u} \in C_1, \boldsymbol{v} \in C_2\}$. Hence C_3 is a code of length $2n$ with $|C_3| = m_1 m_2$. Show that $d(C_3) = \min(2d_1, d_2)$.

▷ **13.40** For each $m \in \mathbb{N}$ and $0 \leq r \leq m$ we define the code $C(r, m)$ over $\{0, 1\}$ of length 2^m recursively. We have $C(0, m) = \{\boldsymbol{0}, \boldsymbol{1}\}$, $C(m, m) = $ the set of all $(0, 1)$-words of length 2^m, and $C(r + 1, m + 1) = C(r + 1, m) * C(r, m)$, where $*$ is defined as in the previous exercise. Example: $m = 1$, $C(0, 1) = \{00, 11\}$, $C(1, 2) = C(1, 1) * C(0, 1) = \{0000, 0011, 1010, 1001, 0101, 0110, 1111, 1100\}$. The codes $C(r, m)$ are called *Reed–Muller codes*. Prove that $|C(r, m)| = 2^a$ with $a = \sum_{i=0}^r \binom{m}{i}$, $d(C(r, m)) = 2^{m-r}$. The code $C(1, 5)$ was used by NASA in its spaceflights.

13.41 The Kronecker product $A \,\square\, B$ of two matrices is given by

$$\begin{pmatrix} a_{11}B & a_{12}B \dots \\ \vdots & \\ a_{m1}B & \dots \end{pmatrix}.$$

Show that the Reed–Muller code (see the previous exercise) $C(1, m)$ is equal to the Hadamard code C from Exercise 13.38, where the Hadamard matrix H is equal to the m-fold Kronecker product of $\left(\begin{smallmatrix} 1 & 1 \\ 1 & -1 \end{smallmatrix}\right)$.

▷ **13.42** Let C be a binary 3-perfect code of length $n = 7$. The necessary condition is $1 + n + \binom{n}{2} + \binom{n}{3} = 2^r$. Show that aside from the repetition code ($n = 7$, $r = 6$), only $n = 23$ with $r = 11$ is possible. Hint: Write the condition in the form $(n+1)(n^2 - n + 6) = 3 \cdot 2^{r+1}$ and consider the possible divisors of $n + 1$.

13.43 The next two exercises show that for $n = 23$ there actually exists a 3-perfect binary code, the so-called Golay code G_{23}. Let I be the icosahedral graph

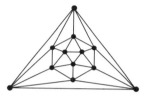

shown in the figure, A the adjacency matrix of I, and B the 12×12 matrix resulting from A by $0 \longleftrightarrow 1$. Show that $G = (E_{12}, B)$ is the generator matrix of a self-dual code G_{24} each of whose codewords has weight congruent to zero modulo 4. Hint: Show from the properties of I that every two rows of B are orthogonal (over GF(2)) and use Exercise 13.25.

▷ **13.44** Show that for all $\mathbf{0} \neq \mathbf{c} \in G_{24}$, one has $w(\mathbf{c}) \geq 8$. Conclude that the code G_{23} that results from eliminating a coordinate is 3-perfect. Incidentally, it is known that except for the repetition code, there are no t-perfect codes for $t \geq 4$.

13.45 Construct Steiner systems $S_4(23, 7)$ and $S_5(24, 8)$ from G_{23} and G_{24} .

13.46 Show that the dual code of a cyclic code is again cyclic.

Chapter 14

Cryptography

14.1. Cryptosystems

We turn now to the second aspect of channel encoding: security of data transmission. In order to send data securely, one *encodes*, or *encrypts*, a **text**, or **plaintext**, T as $c(T)$. Corresponding to c there is a mapping d, which we shall call *decoding*, or *decryption*, such that $d(c(T)) = T$. The encoded text $C = c(T)$ is called a **cryptogram** (or encrypted text or ciphertext). Our task is to develop a cryptogram that cannot be decoded by anyone other than the recipient. We are not concerned here with the issue of transmission errors. We shall assume that a message that is sent over a channel reaches the intended recipient without error.

In practice, c is a mapping that depends on a number of parameters k, the keys. Formally, then, a **cryptosystem** is a triple $(\mathcal{T}, \mathcal{C}, \mathcal{K})$, where $\mathcal{T}$ is the set of texts, $\mathcal{C}$ the cryptograms, and $\mathcal{K}$ the set of keys together with two mappings $c : \mathcal{T} \times \mathcal{K} \longrightarrow \mathcal{C}$, $d : \mathcal{C} \times \mathcal{K} \longrightarrow \mathcal{T}$, which for every pair (T, k) satisfy

$$d(c(T, k), k) = T.$$

We thus have the following situation:

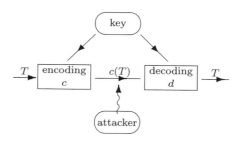

Secret codes have been employed in all eras of human history; their utility for the transmittal of military secrets or in general for the security of data is apparent. One of the oldest forms of encryption is traditionally attributed to Julius Caesar. Each letter of a text is transposed one place in the alphabet (mod 26). We could just as easily transpose by 4 or in general k letters or even use some combination of transpositions. The set of keys $\mathcal{K}$ in this case is $\{0, 1, \ldots, 25\}$.

For example, if we use alternately the keys 1 and 4, the text

$$T \;=\; \text{EAT \quad THIS \quad MESSAGE}$$

would be encoded as the cryptogram

$$c(T) \;=\; \text{FEU \quad XIMT \quad QFWTEHI}\,.$$

One can easily think up further variants of this method: permutations of the letters or other substitutions—more on this in the exercises.

A somewhat more refined method uses matrices. We represent the letters A–Z as the numbers 0–25, and the space, comma, and period as 26, 27, 28. We now calculate modulo the prime $p = 29$. Another possibility would use 0–9 for the usual base-10 digits, 10 for the space, and 11–36 for the letters, in which case we would work modulo 37. Things work better if the total number of symbols is a prime, since calculation in $\mathbb{Z}_p$ is particularly simple.

We now decompose our text T into blocks $\boldsymbol{x}_1, \ldots, \boldsymbol{x}_k$ of length n, select an invertible matrix A (over $\mathbb{Z}_p$), and encrypt $\boldsymbol{x}$ into $c(\boldsymbol{x}) = A\boldsymbol{x}$. Decryption takes place using the inverse matrix $\boldsymbol{y} \to d(\boldsymbol{y}) = A^{-1}\boldsymbol{y}$. With the 29-letter alphabet and block length $n = 3$ we might let, for example,

$$A = \begin{pmatrix} 2 & 0 & 21 \\ 5 & 4 & 1 \\ 3 & 3 & 7 \end{pmatrix}.$$

The message $\text{HELP},_{-}\,\text{POLICE}$ (where the underscore _ represents a space) would then be represented as

$$7\ 4\ 11 \mid 15\ 27\ 26 \mid 15\ 14\ 11 \mid 8\ 2\ 4,$$

and from

$$\begin{pmatrix} 2 & 0 & 21 \\ 5 & 4 & 1 \\ 3 & 3 & 7 \end{pmatrix} \begin{pmatrix} 7 & 15 & 15 & 8 \\ 4 & 27 & 14 & 2 \\ 11 & 26 & 11 & 4 \end{pmatrix} = \begin{pmatrix} 13 & 25 & 0 & 13 \\ 4 & 6 & 26 & 23 \\ 23 & 18 & 19 & 0 \end{pmatrix},$$

where the arithmetic is performed modulo 29, we obtain the cryptogram

$$13\ 4\ 23 \mid 25\ 6\ 18 \mid 0\ 26\ 19 \mid 13\ 23\ 0 = \text{N E X Z G S A \_ T N X A}\,.$$

All these encryption methods do not appear to be very secure (and indeed they are not), but what do we really mean by "security"?

To answer this question we should begin by placing ourselves in the position of the attacker and try to make realistic assumptions. The attacker knows how the encoding c works and has additional information, such as the frequency of letters or perhaps a general idea of the subject matter of the encrypted text. What he lacks is knowledge of the key. One may imagine three possible modes of attack (in order of increasing danger):

A. The attacker knows only the cryptogram $c(T)$ and attempts to determine the text from it.

B. A more realistic variant posits that the attacker is in possession of a fairly long text T *together with* the associated cryptogram $c(T)$. From this he wishes to determine the key.

C. Even more a danger to sender and receiver is the situation in which we assume that the attacker has gotten hold of a number of pairs $(T, c(T))$ *of his choice*.

Let us consider the substitution method in which the key is a permutation π of the 26 letters of the alphabet. That is, X is encoded to πX. With variant C our opponent has an easy time of it. He has simply to compare a piece of the text T with $c(T)$ in which all the letters appear. In case B the situation is somewhat different. If the attacker has intercepted only the meaningless message $A\,A\ldots A$ together with the cryptogram $\pi A\,\pi A\ldots\pi A$, it will be of no help to him, regardless of the length of the text. But even here (and with variant A as well) he will be able to crack the system using the frequency of the letters. Thus probability plays a role here, and this leads to the question of the meaning of perfect security.

Suppose we are given the cryptosystem $(\mathcal{T}, \mathcal{C}, \mathcal{K})$, where the set of texts $\mathcal{T} = \{T_1, T_2, \ldots, T_n\}$ is finite. We assume the probabilities $p_i = p(T_i)$ that T_i has been sent, where we always assume $p_i > 0$. Furthermore, let $p(k_j)$ be the probability that the key k_j was used. The following definition, which goes back to Shannon, is intuitively plausible.

Definition 14.1. The system $(\mathcal{T}, \mathcal{C}, \mathcal{K})$ has **perfect security** if for all $T \in \mathcal{T}$ and $C \in \mathcal{C}$, we have

(14.1) $$p(T \mid C) = p(T),$$

where $p(T \mid C)$ is the conditional probability.

In other words, knowledge of the cryptogram C tells absolutely nothing about the text T that was sent. The condition

$$p(C \mid T) = \frac{p(C \wedge T)}{p(T)} = \frac{p(T \mid C)p(C)}{p(T)}$$

immediately implies the equivalent condition

(14.2) $$p(C \mid T) = p(C) \quad \text{for all } T \in \mathcal{T},\ C \in \mathcal{C}.$$

Perfect security is of course a worthy goal toward which to strive. However, does such a perfectly secure system actually exist? The classic example, the **one-time pad**, goes back to the AT&T engineer Gilbert Vernam and U.S. Army Signal Corps captain Joseph Mauborgne, who patented their idea in the mid 1920s. It works as follows: Take the usual 26-letter alphabet A–Z represented as the set 0–25, and consider a text $T = x_1 x_2 \ldots x_n$. To encrypt T, we generate a *random* sequence $z = z_1 z_2 \ldots z_n$ from the set $\{0, 1, \ldots, 25\}$, where each number is selected independently with probability $\frac{1}{26}$. Then z is the key. The cryptogram C is now defined by

$$C = c(T, z) = y_1 y_2 \cdots y_n$$

with $y_i = x_i + z_i \pmod{26}$ for all i. Decoding is then accomplished by $d(y_i) = y_i - z_i$.

It is clear that all 26^n cryptograms C are equally probable, and therefore $p(C) = \frac{1}{26^n}$, as well as $p(C \mid T) = \frac{1}{26^n}$, since each individual letter y_i in C is generated from x_i with equal probability $\frac{1}{26}$. Thus the one-time pad offers perfect security. The name "one-time pad" derives from the fact that we create a new key sequence each time the system is used.

For all the comfort this system provides, it also suffers from enormous drawbacks. First of all, the key is as long as the text, and furthermore, the entire key must be sent under separate cover to the recipient, which in itself introduces a measure of insecurity. And on the mathematical side there is a difficulty as well: no one knows how actually to produce such a random sequence using a numerical algorithm. One is forced to use **pseudorandom sequences**, which exhibit to a greater or lesser extent the features that a truly random sequence is expected to possess, and it is unclear whether such a system guarantees the security of a random sequence. See the bibliography for pointers to the literature on the fascinating subject of pseudorandom-number generators. In spite of all these reservations, the one-time pad is said to have been used for communication between Washington and Moscow over the "red telephone."

14.2. Linear Shift Registers

We would like now to examine an example of such a number generator that illustrates very nicely some of the connections between discrete mathematics and abstract algebra.

We will work over the finite field $K = \mathrm{GF}(q)$. A **linear shift register** is a sequence of n registers $R_1, \ldots, R_n$ together with n elements $c_1, \ldots, c_n$

of K. The figure shows the operation of a shift register:

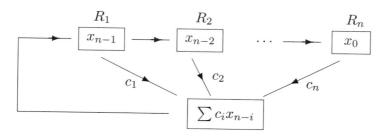

The registers are initialized to the value $x(0) = (x_{n-1}(0), \ldots, x_0(0))$, and so after the execution of the shift register we have the new vector

$$x(1) = \left(\sum_{i=1}^{n} c_i x_{n-i}(0), \; x_{n-1}(0), \; \ldots, \; x_1(0) \right),$$

and in general we have

$$(14.3) \qquad x(t+1) = \left(\sum_{i=1}^{n} c_i x_{n-i}(t), \; x_{n-1}(t), \; \ldots, \; x_1(t) \right).$$

We always assume $c_n \neq 0$, since otherwise, the last register is unnecessary. We shall now explain why we choose the indices $x_{n-1}, \ldots, x_0$ from left to right.

We begin with an arbitrary input vector $s = (s_{n-1}, \ldots, s_0)$. Then a single application yields a new element s_n in the first register; the other registers are shifted, and we obtain a new element s_{n+1} in R_1, and so on. The shift register defined by $c_1, \ldots, c_n$ determines together with the input vector s the **register sequence** $(s_0, s_1, s_2, \ldots)$, and it is this sequence in which we are interested.

This should all seem somewhat familiar. A linear shift register with n registers is simply a *recurrence* of length n with the constant coefficients $c_1, \ldots, c_n$, such as we discussed in detail in Chapter 3. By (14.3) we have

$$(14.4) \qquad s_{i+n} = c_1 s_{i+n-1} + c_2 s_{i+n-2} + \cdots + c_n s_i \quad (i \geq 0),$$

and now it is clear why the indices were chosen as they were. Equation (14.4) corresponds to the recurrence equation (A1) in Theorem 3.1. If we bring everything over to one side, we obtain the polynomial $f(x) = x^n - c_1 x^{n-1} - \cdots - c_n$, which corresponds to the polynomial $q(x)$ in Theorem 3.1. The polynomial $f(x)$ is called the **characteristic polynomial** of the shift register. The condition $c_n \neq 0$ means that $f(0) \neq 0$. Using the equivalence of (A1) and (A2) in Theorem 3.1 (which of course holds as well for finite fields), we obtain the following: If $S(x) = \sum_{k \geq 0} s_k x^k$ is the generating function of the register sequence (s_k), then

$$(14.5) \qquad S(x) = \sum_{k \geq 0} s_k x^k = \frac{g(x)}{f(x)}, \quad \deg g < n,$$

where $g(x)$ is determined by the input sequence $s = (s_{n-1}, \ldots, s_0)$. We also obtain from the equivalence of (A1) and (A2) the converse statement: For every polynomial $g(x)$ with $\deg g < n$, the function $S(x) = \frac{g(x)}{f(x)}$ is the generating function of a register sequence, and for each of the 2^n possible input vectors there is associated exactly one polynomial $g(x)$.

All right, then, all this is nothing new. However, for *finite* fields, another phenomenon comes into play: All register sequences (s_k) are periodic! This can be seen as follows. Let A be the following $n \times n$ matrix over K:

$$A = \begin{pmatrix} c_1 & c_2 & \cdots & & c_n \\ 1 & 0 & \cdots & & 0 \\ \vdots & 1 & & & \vdots \\ \vdots & & \ddots & & \vdots \\ 0 & \cdots & & 1 & 0 \end{pmatrix}.$$

Then a shift register is described by $x \longrightarrow Ax$. The matrix A is invertible, since $\det A = (-1)^{n-1} c_n \neq 0$. If the input vector is $s = 0$, we obtain of course the null sequence. We therefore assume in the following that $s \neq 0$. The action of the shift register thus yields the vectors

$$s, As, A^2 s, \ldots, A^i s, \ldots .$$

If there are $q^n - 1$ nonnull vectors in K^n, then sooner or later there must be a pair $i < j$ such that $A^i s = A^j s$, that is, $A^t s = s$ with $t = j - i$. The sequence of vectors is therefore of the form

$$s, As, \ldots, A^{t-1} s, s, As, \ldots ,$$

that is, it repeats itself every t terms. And the same holds of course for the register sequence (s_k):

$$s_{k+t} = s_k \quad (k \geq 0) \text{ with } t \leq q^n - 1.$$

To summarize, for *every* input vector s, the register sequence $(s_0, s_1, s_2, \ldots)$ is periodic. If (s_k) has period t, then every multiple of t is also a period. What is of interest is therefore the *smallest* period p, which we denote by $\text{per}(s)$.

The smallest period can depend heavily on s, as shown by the following example. Let $K = \text{GF}(2)$, $n = 4$, $c_1 = c_2 = c_3 = 0$, $c_4 = 1$. Then for the input vectors $(1,1,1,1)$, $(1,0,1,0)$, $(1,1,1,0)$ we obtain

0 0 0 1	0 0 0 1	0 0 0 1
1 1 1 1	1 0 1 0	1 1 1 0
....	0 1 0 1	0 1 1 1
1 1 1 1	....	1 0 1 1
	1 0 1 0	1 1 0 1
period 1		....
	period 2	1 1 1 0

period 4

If one tries out all the other vectors, one sees that the period lengths are always at most 4, while the maximum of $2^4 - 1 = 15$ was possible in principle. What shift registers have the maximal possible period $q^n - 1$? One thing is clear. If $A^i s$ runs through all $q^n - 1$ nonnull vectors before $A^{q^n} s = s$ appears again, this must hold as well for *all* other inputs—indeed, they run through the same cycle.

Example 14.2. Again let $K = \mathrm{GF}(2)$ and let the two shift registers be given by $(c_1, c_2, c_3, c_4) = (1, 0, 0, 1)$ and $(1, 0, 1, 1)$. With the input $(1, 0, 0, 0)$ we obtain

1 0 0 1	1 0 1 1
1 0 0 0	1 0 0 0
1 1 0 0	1 1 0 0
1 1 1 0	1 1 1 0
1 1 1 1	0 1 1 1
0 1 1 1	0 0 1 1
$\vdots$	0 0 0 1
	....
0 0 0 1	1 0 0 0
....	
1 0 0 0	
period 15	period 6

In the first case this yields the maximum period $2^4 - 1$, while in the second case this cannot happen for any other input vector. Why this is so can be answered easily with the help of generating functions.

If the sequence (s_k) has period t, then for the generating function $S(x) = \sum_{k \geq 0} s_k x^k$ this clearly implies

$$S(x) = (s_0 + s_1 x + \cdots + s_{t-1} x^{t-1})(1 + x^t + x^{2t} + \cdots),$$

and therefore

(14.6) $$S(x) = \frac{s(x)}{1 - x^t}, \quad \deg s(x) < t,$$

and conversely, the sequence determined by $S(x) = \frac{s(x)}{1-x^t}$ has period t. From this simple idea we can immediately derive an important theorem of abstract algebra.

Theorem 14.3. *Let $f(x) \in K[x]$ be a nonzero polynomial of degree n. Then there exists $t \leq q^n - 1$ such that $f(x) \mid x^t - 1$. Conversely, if $f(x) \mid x^t - 1$, then the shift register determined by $f(x)$ has period t for every input vector s.*

Proof. Let $S(x) = \frac{1}{f(x)}$. Then we know that the sequence (s_k) has a certain period t. From (14.6) we thus have

(14.7) $$S(x) = \frac{1}{f(x)} = \frac{s(x)}{1 - x^t},$$

that is, $f(x)(-s(x)) = x^t - 1$ and so $f(x) \mid x^t - 1$.

Conversely, if $f(x)h(x) = x^t - 1$, then for an arbitrary input vector s we obtain

$$S(x) = \frac{g(x)}{f(x)} = \frac{g(x)h(x)}{x^t - 1} \quad \text{with } \deg(gh) < t,$$

and therefore (s_k) has period t by (14.6).   □

In analogy to how we defined the period $\mathrm{per}(s)$, we define the *exponent* of f as $\exp(f) = \min(t : f(x) \mid x^t - 1)$. The theorem then asserts that

(14.8) $\mathrm{per}(s) \leq \exp(f) \leq q^n - 1$ for all $s \in K^n \setminus \mathbf{0}, \quad n = \deg f$.

And another thing is at once clear. If f is an *irreducible* polynomial over $GF(q)$ and $p = \mathrm{per}(s)$, then from (14.6) and (14.2) we have

$$S(x) = \frac{g(x)}{f(x)} = \frac{s(x)}{1 - x^p},$$

that is, $f(x)s(x) = g(x)(1 - x^p)$. Since f is irreducible, $f(x)$ must divide either $g(x)$ or $x^p - 1$, and on account of $\deg g < n$ this can only be $x^p - 1$. This yields us the following corollary.

Corollary 14.4. *If $f(x)$ is an irreducible polynomial, then $\exp(f) = \mathrm{per}(s)$ for all $s \in K^n \setminus \mathbf{0}$.*

This last result can be used to show that a polynomial $f(x)$ is *not* irreducible. One simply needs to find two input vectors with different periods.

For example, let $K = \mathrm{GF}(2)$, $f(x) = x^4 + x^2 + 1$. Then $c_1 = c_3 = 1$, $c_2 = c_4 = 0$ (in $\mathrm{GF}(2)$ we have $1 = -1$). We calculate

$$
\begin{array}{cccc}
0 & 1 & 0 & 1
\end{array}
\qquad\qquad
\begin{array}{cccc}
0 & 1 & 0 & 1
\end{array}
$$

$$
\begin{array}{cccc}
0 & 0 & 0 & 1 \\
1 & 0 & 0 & 0 \\
0 & 1 & 0 & 0 \\
1 & 0 & 1 & 0 \\
0 & 1 & 0 & 1 \\
0 & 0 & 1 & 0 \\
\cdots\cdots\cdots \\
0 & 0 & 0 & 1
\end{array}
\qquad\qquad
\begin{array}{cccc}
0 & 1 & 1 & 0 \\
1 & 0 & 1 & 1 \\
1 & 1 & 0 & 1 \\
\cdots\cdots\cdots \\
0 & 1 & 1 & 0
\end{array}
$$

$$\text{period } 3$$

$$\text{period } 6$$

Therefore, $x^4 + x^2 + 1$ is not irreducible, and in fact we have $x^4 + x^2 + 1 = (x^2 + x + 1)^2$. From $x^6 + 1 = (x^4 + x^2 + 1)(x^2 + 1)$ we see with the help of (14.8) that $\mathrm{per}(s) \le 6 = \exp(f)$ for all s.

We can now easily establish our main result.

Theorem 14.5. *If R is a shift register over $\mathrm{GF}(q)$ with characteristic polynomial $f(x)$, then R has maximal period $q^n - 1$ if and only if $f(x)$ is irreducible with $\exp(f) = q^n - 1$.*

Proof. The implication $\Longleftarrow$ we already know. Conversely, if $\mathrm{per} = q^n - 1$, then we have $\exp(f) = q^n - 1$ in any case on account of (14.8). Suppose, to obtain a contradiction, we had $f(x) = h_1(x)h_2(x)$ with $\deg h_i(x) = n_i < n$. Then we consider

$$S(x) = \frac{h_1(x)}{f(x)} = \frac{1}{h_2(x)}.$$

The sequence (s_k) generated by $S(x) = \frac{h_1(x)}{f(x)}$ has period $q^n - 1$. On the other hand, we see from $S(x) = \frac{1}{h_2(x)}$ and (14.8) that the same sequence has period less than or equal to $q^{n_2} - 1 < q^n - 1$, a contradiction. $\qquad\square$

Remark 14.6. In algebra one learns that there always exist irreducible polynomials $f(x)$ with $\exp(f) = q^n - 1$ (they are called *primitive* polynomials) and that the number of them is $\varphi(q^n - 1)/n$, where φ is Euler's φ-function. Therefore, for every n there exist shift registers with maximal period $q^n - 1$. For example, for $q = 2$, $n = 3$ there are exactly $\varphi(7)/3 = 2$ primitive polynomials, namely $x^3 + x + 1$ and $x^3 + x^2 + 1$.

Let us take another look at the use of shift registers as one-time pads. Let $T = t_0 t_1 t_2 \ldots$ be a sequence with $t_i = \mathrm{GF}(q)$ and let $s_0 s_1 s_2 \ldots$ be a register sequence. We encode as usual $t_i \xrightarrow{c} t_i + s_i$ and decode $y_i \xrightarrow{d} y_i - s_i$. How secure is this system? An attacker under assumption C has available a text sample and the corresponding cryptogram of his choosing and hopes to extract the key $c_1, \ldots, c_n$ from this information. If he chooses $2n$ *sequential* symbols $s_k, \ldots, s_{k+2n-1}$ (which he knows on account of $s_i = y_i - t_i$), he then obtains $c_1, \ldots, c_n$ by solving the system of linear equations

$$
\begin{aligned}
s_{k+n} &= c_1 s_{k+n-1} &+ \cdots + c_n s_k, \\
s_{k+(n+1)} &= c_1 s_{k+n} &+ \cdots + c_n s_{k+1}, \\
&\ \ \vdots \\
s_{k+(2n-1)} &= c_1 s_{k+(2n-2)} &+ \cdots + c_n s_{k+n-1},
\end{aligned}
$$

and the system has been broken.

Linear shift registers are therefore not particularly secure, but they can be used to generate cyclic codes in an elegant way. Let us take $K = \mathrm{GF}(q)$ as before and let

$$
A = \begin{pmatrix}
c_1 & \cdots & & c_n \\
1 & & & 0 \\
& 1 & & \vdots \\
0 & & \ddots & 1 & 0
\end{pmatrix}
$$

be the matrix of the register and $\boldsymbol{s}$ a nonnull input vector. Now let $p = \mathrm{per}(\boldsymbol{s})$ be the maximal period and $\boldsymbol{s}, A\boldsymbol{s}, \ldots, A^{p-1}\boldsymbol{s}$ the sequence of vectors. Let the code $C \subseteq K^p$ be given by the parity check matrix

$$
H = (A^{p-1}\boldsymbol{s},\ A^{p-2}\boldsymbol{s},\ \ldots,\ A\boldsymbol{s},\ \boldsymbol{s}),
$$

that is,

$$
\boldsymbol{a} = (a_{p-1}, \ldots, a_0) \in C \iff \sum_{i=0}^{p-1} a_i(A^i \boldsymbol{s}) = \boldsymbol{0}.
$$

Multiplying this equation by A, we obtain

$$
a_{p-2}(A^{p-1}\boldsymbol{s}) + \cdots + a_0(A\boldsymbol{s}) + a_{p-1}\boldsymbol{s} = \boldsymbol{0},
$$

that is, $(a_{p-2}, \ldots, a_0, a_{p-1}) \in C$, and C is cyclic.

Example 14.7. Let us take a register over $\mathrm{GF}(2)$ with r registers and maximal period $2^r - 1$. Then H contains *all* $2^r - 1$ nonnull vectors as columns, and the result is our old friend the binary perfect Hamming code from Section 13.4. And thus we have also proved that all of these codes are cyclic.

14.3. Public-Key Cryptosystems

In the first section of this chapter we discussed the notions of cryptosystems and perfect security and remarked that no one knows how such a system might be realized in practice. We shall therefore take a more modest definition of "secure," namely "hard to crack." For this we will make use of the concepts of "easy" and "hard" from our discussion of the complexity of algorithmic problems from Section 8.5.

Our task is to find an encoding c such that knowledge of a cryptogram $c(T)$ does not lead (easily) to the original text. Such mappings go under the name **one-way functions**. Our conditions on such a function f will therefore be as follows:

(A) The calculation of $f(T)$ from T should be easy, that is, in polynomial time. In other words, f should be in the class P.

(B) The determination of $f^{-1}(C)$ from C should be difficult. That is, we require that no polynomial algorithm for f should be known.

Condition (B) is precisely our security requirement that an unauthorized third party be unable to crack the code, and (A) is a requirement of efficiency: we would like to be able to send a message relatively quickly. Of course, the recipient, and she only, should be able to decode the cryptogram efficiently. We therefore shall add a third condition to (A) and (B). We shall see shortly how that is done.

One of the first candidates for a one-way function was the **discrete logarithm**. We choose a large prime number p. By Fermat's theorem from Section 12.1 we know that

$$x^{p-1} \equiv 1 \ (\text{mod } p) \quad \text{for all } x = 1, \ldots, p-1.$$

Let $a \in \{1, \ldots, p-1\}$. If the numbers $a, a^2, \ldots, a^{p-1}$ are all incongruent modulo p, we call a a **primitive root** modulo p. It will be shown in Exercise 14.27 that every prime p has a primitive root and that the number of them is equal to the number of positive integers less than p that are relatively prime to $p-1$.

For example, $p = 11$ has the primitive roots $2, 6, 7, 8$, and we obtain, for example, for 7 the following sequence of powers 7^i modulo 11:

$$7^1 \equiv 7, \quad 7^2 \equiv 5, \quad 7^3 \equiv 2, \quad 7^4 \equiv 3, \quad 7^5 \equiv 10,$$
$$7^6 \equiv 4, \quad 7^7 \equiv 6, \quad 7^8 \equiv 9, \quad 7^9 \equiv 8, \quad 7^{10} \equiv 1.$$

We can now define the discrete logarithm. For the prime number p with primitive root a we define the *exponential function* modulo p by

$$y \equiv a^x \ (\text{mod } p).$$

The integer x, $1 \leq x \leq p-1$, is then called the *discrete logarithm* of y modulo p (to the base a).

We claim that $f(x) = a^x$ is a one-way function. Condition (A) is easy to check. If we wish to compute, for example, a^{39}, we set $a^{39} = a^{32} \cdot a^4 \cdot a^2 \cdot a$, and we obtain the separate powers a^{2^k} by successive squaring, for example, $a^{32} = ((((a^2)^2)^2)^2)^2$. Thus the total number of multiplications is bounded by $2\lceil \lg p \rceil$ (see Exercise 5.10), and we have established that f is in the class P. Note that we make the reductions of the individual multiplications modulo p using the Euclidean algorithm, which is also polynomial.

In contrast, there in no known polynomial algorithm for computing the discrete logarithm, and it is generally assumed that the complexity of this problem is equivalent to that of the factorization of integers.

In practice, as we know, the encoding depends on the number of keys k. The cryptogram is of the form $c(T, k)$. The keys, however, represent a problem, as we have seen in several examples: they are a security risk and they require the exchange of information. To deal with this problem, in 1976, Diffie and Hellman suggested the following method, called a **public-key cryptosystem**. The idea is attractively simple. Every user i has a *public* key k_i and a *secret* key g_i. The public key is available to everyone, together with the applicable encoding c and decoding d. For every text T, for each i one should have

$$(14.9) \qquad\qquad d\big(c(T, k_i), g_i\big) = T \,.$$

If a participant j now wishes to send a message T to recipient i, he uses i's key k_i to send

$$C = c(T, k_i),$$

and then user i decodes the encrypted message using (14.9). To guarantee efficiency and security, the system must satisfy the following conditions:

(A) From T and k_i, the computation of $C = c(T, k_i)$ is easy.
(B) Given the cryptogram C, the decoding $d(C)$ *without* knowledge of g_i is difficult.
(C) Given the cryptogram C *and* the key g_i, then computing $d(C, g_i)$ is easy.

Conditions (A) and (B) tell us that the encoding is a one-way function. A one-way function that also satisfies (C) (that is, enables efficient inversion with knowledge of the secret key) is called a **trapdoor function**.

We would now like to examine the most famous trapdoor system of all, the RSA cryptosystem, named for its developers Rivest, Shamir, and Adleman. Again one uses modular arithmetic. The encoding and decoding work in the following manner:

1. A user chooses two large prime numbers p and q and a pair of integers k, g, $1 \leq k, g \leq (p-1)(q-1)$, relatively prime to $(p-1)(q-1)$ with $kg \equiv 1 \pmod{(p-1)(q-1)}$.
2. The user makes known as public key the product $n = pq$ and k.

3. A text T is represented as a number in $\{0, \ldots, n-1\}$; if T is too large, it is decomposed into blocks.
4. Encoding takes place via

$$C \equiv T^k \pmod{n},$$

and decoding, using the secret key g, via

$$D \equiv C^g \pmod{n}.$$

Before we check conditions (A), (B), (C), we shall first look at a small example based on the work of Rivest, Shamir, and Adleman.

Let $p = 47$, $q = 59$, $n = 47 \cdot 59 = 2773$, $(p-1)(q-1) = 2668$. We choose $g = 157$ (a prime) and $k = 17$ with $17 \cdot 157 \equiv 1 \pmod{2668}$. If we wish to send a text, we encode our alphabet, for example, as space $= 00$, $A = 01$, $B = 02$, $\ldots$, $Z = 26$. We then form each pair of letter into a block of four digits. For example, the message

EAT THIS MESSAGE

becomes the text

0501 2000 2008 0919 0013 0519 1901 0705.

Since we have chosen $n = 2773$, all blocks of four digits represent integers less than n. The individual blocks T are now encrypted via $T^{17} \pmod{2773}$, and we end up with the cryptogram

2326 0317 1567 2244 0219 1493 1281 1692.

We would now like to show that the RSA system represents a trapdoor function. We have already seen that exponentiation $T^k \pmod{n}$ is polynomial. For g relatively prime to $(p-1)(q-1)$ we can take some prime $g > \max(p, q)$ and then compute k with the Euclidean algorithm. Therefore condition (A) is satisfied, and likewise (C), since decoding corresponds to an exponentiation. Before we discuss condition (B), we must show that $d(c(T)) = T$, that is, that

$$C \equiv T^k \pmod{n}$$

always implies

$$T \equiv C^g \pmod{n}.$$

In other words, we wish to show that

$$T^{kg} \equiv T \pmod{n}.$$

If $T \equiv 0 \pmod{p}$, then we also of course have $T^{kg} \equiv T \pmod{p}$. Therefore let T be relatively prime to p. By Fermat's theorem we have $T^{p-1} \equiv$

$1 \pmod{p}$. From $kg \equiv 1 \left(\bmod{\,(p-1)(q-1)}\right)$ we have $kg = t(p-1)(q-1)+1$ for some $t \in \mathbb{N}$ and hence

$$T^{kg} = T^{t(p-1)(q-1)}T = (T^{p-1})^{t(q-1)}T \equiv T \pmod{p}.$$

Analogously we conclude that $T^{kg} \equiv T \pmod{q}$. The two congruences mean that $p \mid T^{kg} - T$ and $q \mid T^{kg} - T$, and it follows that $pq \mid T^{kg} - T$, or equivalently, $T^{kg} \equiv T \pmod{n}$, as desired.

We now deal with condition (B). The integers n, k are public. If someone were able to factor n, then from $n = pq$ he would immediately obtain $(p-1)(q-1)$ and thereby g from $kg \equiv 1 \left(\bmod{\,(p-1)(q-1)}\right)$. Can one compute g *without* knowing p, q? From knowledge of $(p-1)(q-1)$ and $n = pq$ we can immediately determine p and q, and thereby factor n. Namely, we have

$$p + q = pq - (p-1)(q-1) + 1, \qquad (p-q)^2 = (p+q)^2 - 4pq,$$

and from $p + q$ and $p - q$ we immediately obtain p and q.

Thus everything depends on whether one can factor large integers efficiently, and at present, no such algorithm is known. Of course, one could say that $C \equiv T^k \pmod{n}$ is actually a problem of extracting kth roots modulo n and has nothing immediately to do with factorization. However, Rivest, Shamir, and Adleman conjecture that *every* efficient method of cracking their system would imply an efficient algorithm for factoring integers. This conjecture is unproven, so that we can say only this: according to our present knowledge, the RSA system is secure.

There is another important problem that can be solved with the RSA system. How does recipient i know whether a message actually comes from j? To solve this problem, we use encodings c and decodings d that in addition to $d\big(c(T)\big) = T$ also satisfy $c\big(d(T)\big) = T$. The RSA system of course satisfies this condition. If now i wishes to check whether j was really the sender, she simply chooses a random word x and sends $u = c(x, k_j)$ to j. He simply sends a "signature" $d(u, g_j)$ back to i, and she checks whether $c\big(d(u, g_j), k_j\big)$ is in fact u. If so, she can be sure, on account of property (B), that the message was actually sent by j.

14.4. Zero-Knowledge Protocols

Cryptography as we have been considering it is concerned with questions of data security and the problem of key exchange. In our Internet age, however, it is not just securing data that is important, but also managing the way in which communication takes place between two parties. We call such a manner of communication a **protocol**.

A particularly interesting protocol requires the following. One person, P (the prover; let us call him Pasha), wishes to convince another, V (the verifier, Victoria), that he possesses certain information (or as we also say, a secret) *without* revealing this information. And what is more, Victoria (or some unauthorized eavesdropper) should be incapable of reconstructing even a small part of the secret from the data of the protocol.

Consider the following example. In authentication systems, Pasha usually has to establish his identity by convincing the verification authority Victoria that he possesses a secret. This can be a password or a secured credit-card number. Then Victoria compares the password with a preset comparison password and thereby confirms that they are in accord. The risk is clear: the protocol can be intercepted by an attacker, or Victoria herself could abuse the system. How is one to proceed?

Before we look at an example, we would like to make precise the requirements of such a **zero-knowledge protocol**. A protocol is set up between Pasha and Victoria, and at the end, Victoria **accepts** that Pasha possesses the secret or else she **rejects**. The rules by which Victoria accepts or rejects (as well as the modalities of the protocol) are established beforehand in a two-way agreement. The protocol should satisfy the following conditions:

(A) Executability: If Pasha possesses the secret, then Victoria accepts with an arbitrarily high probability (close to 1).
(B) Correctness: If Pasha does not have the information (Pasha is bluffing), then Victoria accepts with only an arbitrarily low probability (near 0).
(C) Zero Knowledge: Victoria learns from the protocol nothing about the secret (she is no wiser than before).

In addition, the protocol should be efficient, that is, it should run in polynomial time.

The following amusing example should help to clarify these concepts. Consider the following diagram:

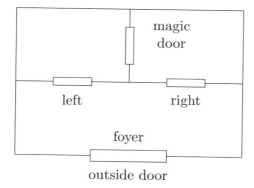

outside door

Pasha tells Victoria that he possesses the secret password to the magic door. The protocol is as follows:

1. Pasha enters the foyer and closes the outside door. He then tosses a coin to determine whether to take the left or right door, goes through the chosen door, and closes the door behind him.
2. Now Victoria enters the foyer, makes a random choice of a door, and calls out to Pasha to come through that door.
3. If Pasha happens to be behind that door, he simply walks through it. Otherwise, Pasha must know the secret password so that he can pass through the magic door and come out through the door chosen by Victoria.

This procedure is repeated n times, and if Pasha manages to come through the correct door every time, then Victoria is convinced and accepts. Otherwise, she rejects.

Let us check the conditions. If Pasha knows the password, then Victoria accepts with probability 1. Condition (B) concerns the probability that Pasha doesn't know the code but somehow happens to come through the correct door every time, the probability of which is 2^{-n}, which can be made arbitrarily small. The zero-knowledge property is obviously satisfied.

And now back to real life. The best known and most frequently implemented zero-knowledge protocol was suggested in 1985 by Fiat and Shamir. Again modular arithmetic is lurking in the background.

Let us recall the concepts that we will need. As usual, $\mathbb{Z}_n$ is the ring of residue classes modulo n. We call r a *prime* residue if $\gcd(r, n) = 1$. The prime residues form a group under multiplication, the *prime residue class group* $\mathbb{Z}_n^*$. For $n = 14$, for example, we have $\mathbb{Z}_{14}^* = \{1, 3, 5, 9, 11, 13\}$, and it is clear that if r is a prime residue, then so is $n - r$. An element $r \in \mathbb{Z}_n^*$ is called a **quadratic residue** if the equation $r \equiv x^2 \pmod{n}$ can be solved in $\mathbb{Z}_n^*$. Of course, $r \equiv x^2$ means that we also have $r \equiv (-x)^2 \pmod{n}$. Thus in $\mathbb{Z}_{14}^*$ the quadratic residues are $(\pm 1)^2 = 1$, $(\pm 3)^2 = 9$, $(\pm 5)^2 = 11$. If $n = p$ is a prime, then every equation $x^2 \equiv r \pmod{p}$ has at most two solutions ($\mathbb{Z}_p$ is a field!), but for composite n there can be more solutions. For example, in $\mathbb{Z}_8$ each of the prime residues $1, 3, 5, 7$ is a solution of $x^2 \equiv 1 \pmod 8$.

The security of the Fiat–Shamir protocols rests on the assumption that it is difficult (in the sense of complexity theory) to compute square roots in the group $\mathbb{Z}_n^*$ for large n. We now proceed as follows. Prover Pasha (or a reliable trust center) chooses two primes p and q and forms the product $n = pq$. Then $s \in \mathbb{Z}_n^*$ is chosen as well as $v \equiv s^2 \pmod{n}$. The integers n and v are made known to Victoria, while p, q, and s remain secret. Now Pasha proves his identity to Victoria by asserting that he knows a square root of v. This leads to the following protocol:

1. Pasha chooses a random $r \in \mathbb{Z}_n^*$ and sends $x \equiv r^2 \pmod{n}$ to Victoria.
2. Victoria chooses randomly a bit $b \in \{0, 1\}$ and sends it to Pasha.
3. Pasha sends $y \equiv rs^b$ back to Victoria.
4. Victoria checks whether $y^2 \equiv xv^b \pmod{n}$ is satisfied.

This exchange is repeated n times, and if each time $y^2 \equiv xv^b \pmod{n}$ holds, then Victoria accepts. Otherwise, she rejects.

Condition (A) is certainly satisfied. If Pasha does not know a square root of v (Pasha is bluffing), then Pasha can answer correctly to at most one of the possibilities $b = 0$ and $b = 1$. For if $y_0^2 \equiv x$ and $y_1^2 \equiv xv$ were satisfied, then $(\frac{y_1}{y_0})^2 \equiv v$ would be a square root of v, which Pasha, however, doesn't know. Thus Pasha can answer correctly at best with probability $\frac{1}{2}$. One can easily see that Pasha is always correct with probability $\frac{1}{2}$ (Exercise 14.29). After n repetitions the probability of a successful bluff is thus $\frac{1}{2^n}$. For the zero-knowledge property it remains only to note that Victoria acquires from $y^2 \equiv xv^b$ only the information that x is a quadratic residue, since r was chosen randomly. But she knows that already.

Exercises for Chapter 14

▷ **14.1** Why is the permutation method $X \to \pi X$ likely more secure than the Caesar system $X \to X + i$?

14.2 An affine cryptosystem with block length n consists of an invertible matrix A and a fixed vector $\boldsymbol{b} \in K^n$. Encryption takes place via $c(\boldsymbol{x}) = A\boldsymbol{x} + \boldsymbol{b}$. How much text T together with $c(T)$ does an attacker require under condition (C) in Section 14.1 to crack the code?

▷ **14.3** A text is decomposed into blocks of length n, and encryption is via $c(\boldsymbol{x}) = \pi\boldsymbol{x} + \boldsymbol{b}$, where π is a permutation of the coordinates and $\boldsymbol{b} \in K^n$ a fixed vector. Show that this system is an affine cryptosystem (as defined in the previous exercise).

14.4 An affine Caesar encryption is of the form $c : \mathbb{Z}_n \to \mathbb{Z}_n$, $c(x) \equiv ax+b \pmod{n}$ with $0 < a, b < n$. Show that c is a bijection if and only if a and n are relatively prime. How many different such systems are there?

14.5 Suppose the following message was encrypted with an affine Caesar system on 26 letters. Decrypt VBEDXSXIXKPXS.

▷ **14.6** Let per$(\boldsymbol{s})$ be the minimal period of a shift register with input $\boldsymbol{s}$. Show that t is a period if and only if per$(\boldsymbol{s}) \mid t$.

14.7 Determine all irreducible polynomials over GF(2) up to degree 5. Which of them are primitive?

▷ **14.8** Suppose the sequences (a_k) and (b_k) are periodic with minimal periods r and s. Is the sequence of the sums $(a_k + b_k)$ periodic? If yes, what can one say about the minimal period?

14.9 Suppose that a is not a primitive root of p. Show that then $y \equiv a^x \pmod{p}$ does not necessarily have a unique solution in x for a given y (or perhaps no solution at all).

14.10 Let a be a primitive root of p. Show that for the discrete logarithm one has $\log_a(xy) = \log_a x + \log_a y \pmod{p-1}$.

14.11 Let $n = pq$, where p and q are prime. How many solutions can an equation of the form $x^2 \equiv s \pmod{n}$ have in $\mathbb{Z}_n^*$?

$\triangleright$ **14.12** Let $(\mathcal{T}, \mathcal{C}, \mathcal{K})$ be a cryptosystem with perfect security. Show that $|\mathcal{K}| \geq |\mathcal{T}|$ must hold.

14.13 Let $(\mathcal{T}, \mathcal{C}, \mathcal{K})$ be given with $|\mathcal{T}| = |\mathcal{C}| = |\mathcal{K}|$. Show that perfect security is guaranteed if and only if for each pair (T, C) there is precisely one key k such that $c(T, k) = C$, and all keys are equally probable.

14.14 How many invertible $n \times n$ matrices are there over $\mathrm{GF}(q)$?

$\triangleright$ **14.15** Suppose that a shift register over $\mathrm{GF}(2)$ with n has maximal period $2^n - 1$, with register sequence (s_k). Show that every segment of $2^n - 1$ successive symbols contains 2^{n-1} ones and $2^{n-1} - 1$ zeros.

14.16 You are given a register as in the previous exercise with input vector $s_0 = \cdots = s_{n-2} = 0$, $s_{n-1} = 1$. We order $s_0, \ldots, s_{2^n-1}$ clockwise in a circle. Show that the set of 2^n words of n successive symbols contain every $0, 1$ word of length n exactly once. Such circular arrangements are called *de Bruijn words*.

$\triangleright$ **14.17** We construct de Bruijn words (see the previous exercise) using graphs. Let $\vec{G} = (V, E)$ be the following directed graph: $V = \{0, 1\}^{n-1}$ and $a_1 a_2 \ldots a_{n-1} \to b_1 b_2 \ldots b_{n-1}$ if $a_2 = b_1, a_3 = b_2, \ldots, a_{n-1} = b_{n-2}$. Example: For $n = 3$ we have $10 \to 00$, $10 \to 01$, $11 \to 10$, $11 \to 11$. Show that every vertex has in-degree 2 and out-degree 2. Construct from $\vec{G}$ a de Bruijn word. Hint: $\vec{G}$ is Eulerian.

14.18 Show that the coefficients $c_1, \ldots, c_n$ of a shift register cannot necessarily be obtained from $2n$ register numbers s_i if the s_i do not follow one another sequentially.

$\triangleright$ **14.19** Given the recurrence $s_{n+5} = s_{n+1} + s_n$ over $\mathrm{GF}(2)$, show that (s_k) can be realized by a linear shift register and that the possible minimal periods are $1, 3, 7, 21$.

14.20 Let $f_1(x), f_2(x)$ be the characteristic polynomials of two shift registers R_1, R_2. Show that every register sequence (s_k) of R_1 is also one of R_2 if and only if $f_1(x) \mid f_2(x)$.

$\triangleright$ **14.21** Show that all perfect Hamming codes over $\mathrm{GF}(q)$ are cyclic. Hint: Consider what $A^i \boldsymbol{s} = \lambda \boldsymbol{s}$ means, where A is the matrix of the register.

14.22 We are using the RSA system. The recipient announces her public key $k = 43$, $n = 77$. A message M is sent as $C = 5$ to the recipient and intercepted. What is M?

$\triangleright$ **14.23** The following public key system was suggested by ElGamal. All users know the same large prime number p and a primitive root a. User j chooses a natural number x_j randomly, and that becomes his secret key. As public key he gives

$y_j \equiv a^{x_j} \pmod{p}$. Suppose that user i wishes to send a message M, $1 \leq M \leq p-1$, to j. He proceeds as follows: (1) He chooses at random an integer k such that $1 \leq k \leq p-1$. (2) He calculates $K = y_j^k \pmod{p}$. (3) He sends the pair (C_1, C_2) with $C_1 \equiv a^k \pmod{p}$, $C_2 \equiv K \cdot M \pmod{p}$. Show that this system satisfies the requirements of a public key system.

14.24 Show that a polynomial algorithm for calculating the discrete logarithm would break the ElGamal system introduced in the previous exercise.

▷ **14.25** Consider an ElGamal system (see the previous two exercises) with $p = 71$ and primitive root 7. Suppose $y_j = 3$ and user i chooses the key $k = 2$. What does the cryptogram of $M = 30$ look like? Suppose that with a new k' it turns out that $M = 30$ is sent as $(2, C_2)$. What is C_2?

14.26 Show that the encryption of a message in the ElGamal system requires roughly $2 \log p$ multiplications modulo p.

▷ **14.27** Let p be a prime number. Show that the group $\mathbb{Z}_p^*$ is cyclic and that it contains exactly $\varphi(p-1)$ generators. Hint: Use $\sum_{d|p-1} \varphi(d) = p - 1$.

14.28 Compute the number of quadratic residues in $\mathbb{Z}_n^*$, $n = 2^k$ ($k \geq 3$). Hint: Consider r and $\frac{n}{2} - r$ for $r < \frac{n}{4}$.

14.29 Show that in the Fiat–Shamir protocol, an impostor P can always achieve that one of the two equations $y_0^2 \equiv x$ and $y_1^2 \equiv xv \pmod{n}$ is correct.

14.30 A famous unsolved problem in complexity theory is the graph-isomorphism problem GI. Given two graphs $G_1 = (V_1, E_1)$ and $G_2 = (V_2, E_2)$, are G_1 and G_2 isomorphic? The problem is in NP, but no polynomial algorithm is known. The following protocol is based on the supposed difficulty of GI. Pasha claims that he knows an isomorphism $\varphi : V_1 \to V_2$:

(1) He chooses a permutation π of V_1 at random, with all permutations equiprobable, and sends the graph $H = \pi G_1$ to Victoria.

(2) Victoria chooses at random $i \in \{1, 2\}$ and sends i to Pasha with the request that he specify an isomorphism between G_i and H.

(3) Pasha sets $\rho = \pi$ (if $i = 1$) or $\rho = \pi \circ \varphi^{-1}$ (if $i = 2$) and sends ρ to Victoria.

(4) Victoria verifies whether $\rho : G_i \to H$ is an isomorphism.

This protocol is repeated n times. If ρ is an isomorphism every time, then Victoria accepts Pasha's claim. Otherwise, she rejects it. Show that this is a zero-knowledge protocol.

Chapter 15

Linear Optimization

This last chapter exceeds somewhat the scope of an introductory text on discrete mathematics. It goes more deeply into underlying mathematical structures, and is therefore in our necessarily brief presentation more theoretical than the previous chapters. However, today linear optimization is such an important subject, with an enormous wealth of applications, above all (but not only) discrete problems, that every "discrete" mathematician should be familiar with its essential ideas and methods. In this chapter we shall only introduce the basic results, and the reader is referred to the bibliography for further reading.

15.1. Examples and Definitions

In Section 8.2 we brought our job-assignment problem with suitability coefficients (w_{ij}) into the following form: Given a matrix (w_{ij}), one seeks (x_{ij}) with $x_{ij} = 0$ or 1 such that

$$\sum_{i=1}^{n}\sum_{j=1}^{n} w_{ij}x_{ij}$$

is minimal, under the additional constraints

$$\sum_{j=1}^{n} x_{ij} = 1 \text{ for all } i, \qquad \sum_{i=1}^{n} x_{ij} = 1 \text{ for all } j.$$

The knapsack problem has a similar structure. Given w_i, g_i, we seek x_i, $1 \le i \le n$, with $x_i = 1$ or 0 (here $x_i = 1$ means that object i is packed, while

$x_i = 0$ means that it is not), such that

$$\sum_{i=1}^{n} w_i x_i$$

is maximal, under the constraints

$$\sum_{i=1}^{n} g_i x_i \leq G \,.$$

The supply and demand problem from Section 8.4 was also of this form, and the inequalities in the treatment of flows in networks also look quite similar. And of course, the traveling salesman problem can also be cast in this form. How such problems are handled generally is the content of this chapter.

We begin with a few definitions. We consider $m \times n$ matrices A over $\mathbb{R}$. However, all of our results will hold for $\mathbb{Q}$ as well. The set of all $m \times n$ matrices is $\mathbb{R}^{m \times n}$, and we will denote vectors by $\boldsymbol{a}, \boldsymbol{b}, \boldsymbol{c}, \dots$. If $\boldsymbol{a}, \boldsymbol{b} \in \mathbb{R}^n$, then we set, as usual, $\boldsymbol{a} \leq \boldsymbol{b}$ if $a_i \leq b_i$ for all coordinates. Vectors $\boldsymbol{a}$ are generally column vectors, and row vectors will be denoted by $\boldsymbol{a}^T$, just as A^T denotes the matrix transpose of A. In particular, $\boldsymbol{a} \geq \boldsymbol{0}$ means that all coordinates a_i are greater than or equal to 0.

Definition 15.1. Given $A \in \mathbb{R}^{m \times n}$, $\boldsymbol{b} \in \mathbb{R}^m$, $\boldsymbol{c} \in \mathbb{R}^n$, a **standard program** is the problem of finding $\boldsymbol{x} \in \mathbb{R}^n$ such that

$$
\begin{array}{lll}
(1) & A\boldsymbol{x} \leq \boldsymbol{b}, \; x \geq \boldsymbol{0}, & \\
(2) & \boldsymbol{c}^T \boldsymbol{x} = \max, & \text{standard maximum program}
\end{array}
$$

or

$$
\begin{array}{lll}
(1) & A\boldsymbol{x} \geq \boldsymbol{b}, \; x \geq \boldsymbol{0}, & \\
(2) & \boldsymbol{c}^T \boldsymbol{x} = \min. & \text{standard minimum program}
\end{array}
$$

An $\boldsymbol{x}^* \in \mathbb{R}^n$ satisfying (1) is called an **admissible solution**. If in addition, $\boldsymbol{x}^*$ satisfies (2), then $\boldsymbol{x}^*$ is called an **optimal solution**, and $\boldsymbol{c}^T \boldsymbol{x}^*$ the **value** of the linear program; $\boldsymbol{c}^T \boldsymbol{x}$ is called the **objective function** of the program.

Example 15.2. Ellie the orchardist produces two kinds of fruit juice: J_1 and J_2. In addition to fruit concentrate, sugar, and water, she uses two additives A_1, A_2. She would like to set up a production plan that takes into account the supply of the various ingredients and produces the maximum profit. The result is the following program:

Ingredient	J_1	J_2	Supply
Concentrate	0.4	0.25	30
Sugar	0.2	0.3	25
Water	0.25	0.15	100
A_1	0.15	0	10
A_2	0	0.3	20
Profit per liter	7	3	

Constraints

$$R_1: \quad 0.4x_1 \quad + \quad 0.25x_2 \quad \leq \quad 30$$
$$R_2: \quad 0.2x_1 \quad + \quad 0.3x_2 \quad \leq \quad 25$$
$$R_3: \quad 0.25x_1 \quad + \quad 0.15x_2 \quad \leq \quad 100$$
$$R_4: \quad 0.15x_1 \qquad\qquad\qquad \leq \quad 10$$
$$R_5: \qquad\qquad\qquad\quad 0.3x_2 \quad \leq \quad 20$$
$$Q: \quad 7x_1 \quad + \quad 3x_2 \quad = \text{max}$$

We thereby obtain a standard maximum program with objective function Q and constraints $R_1, \ldots, R_5$. The following figure shows the geometric structure of the linear program. The shaded part is the domain of admissible solutions. The function $Q(x_1, x_2)$ describes with $Q(x_1, x_2) = m$ a family of parallel lines, and the unique maximal value is attained at the vertex $\widetilde{x}$. The linear program therefore has the solution $\widetilde{x} = (66.6, 13.3)$, and the optimal profit is 506.6.

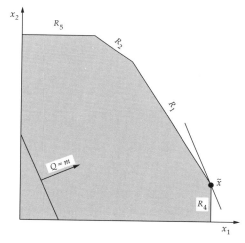

This example indicates the fundamental questions of linear optimization:

1. Given a standard program, when does there exist an admissible solution? When does there exist an optimal solution?
2. What is the structure of the set of admissible solutions? What is the structure of the set of optimal solutions? How does one find an optimal solution?

15.2. Duality

A fact of fundamental importance for both theory and practice is that for every standard program there exists a unique dual standard program.

Definition 15.3. Let

$$
\text{(I)} \qquad
\begin{aligned}
A\boldsymbol{x} &\leq \boldsymbol{b},\ \boldsymbol{x} \geq \boldsymbol{0}, \\
\boldsymbol{c}^T \boldsymbol{x} &= \max,
\end{aligned}
$$

be a standard maximum program. Program (I) is called the **primal program**. The standard minimum program

$$
\text{(I*)} \qquad
\begin{aligned}
A^T \boldsymbol{y} &\geq \boldsymbol{c},\ \boldsymbol{y} \geq \boldsymbol{0}, \\
\boldsymbol{b}^T \boldsymbol{y} &= \min,
\end{aligned}
$$

is called the **dual program** to (I). Conversely, if (I*) is given, then (I*) is called the primal program and (I) the dual program. Clearly, (I**) = (I), that is, the dual of the dual is the original program.

Theorem 15.4. *Given the programs* (I) *and* (I*), *and admissible solutions* $\boldsymbol{x}$ *of* (I) *and* $\boldsymbol{y}$ *of* (I*), *then*

$$
\boldsymbol{c}^T \boldsymbol{x} \leq \boldsymbol{y}^T A \boldsymbol{x} \leq \boldsymbol{b}^T \boldsymbol{y} \ .
$$

Proof. Since $\boldsymbol{x} \geq \boldsymbol{0}$, $\boldsymbol{y} \geq \boldsymbol{0}$, we have

$$
\boldsymbol{c}^T \boldsymbol{x} \leq (\boldsymbol{y}^T A)\boldsymbol{x} = \boldsymbol{y}^T (A\boldsymbol{x}) \leq \boldsymbol{y}^T \boldsymbol{b} = \boldsymbol{b}^T \boldsymbol{y} \ .
$$

$\square$

An immediate consequence is the following result.

Theorem 15.5. *Let* $\boldsymbol{x}, \boldsymbol{y}$ *be admissible solutions of* (I) *and* (I*). *If* $\boldsymbol{c}^T \boldsymbol{x} = \boldsymbol{b}^T \boldsymbol{y}$, *then* $\boldsymbol{x}$ *is an optimal solution of* (I) *and* $\boldsymbol{y}$ *an optimal solution of* (I*).

Example 15.6. Consider the usual matching problem on a bipartite graph $G = (S + T, E)$ without isolated vertices. Let A denote the $n \times q$ incidence matrix of the vertices and edges, and let the vectors $\boldsymbol{b} = \boldsymbol{1} \in \mathbb{R}^n$ and $\boldsymbol{c} = \boldsymbol{1} \in \mathbb{R}^q$ have coordinates all ones. If we interpret $\boldsymbol{x} \in \mathbb{R}^q$, $x_i = 1$ or 0, as usual as characteristic vector of an edge set X, then $A\boldsymbol{x} \leq \boldsymbol{1}$ means that X is incident with each vertex at most once and hence is a *matching*. If we interpret $\boldsymbol{y} \in \mathbb{R}^n$, $y_i = 1$ or 0, as a vertex set Y, then $A^T \boldsymbol{y} \geq \boldsymbol{1}$ requires that every edge be incident with at least one vertex from Y, and therefore Y is a *vertex cover*. The objective functions are $\boldsymbol{1}^T \boldsymbol{x} = |X|$, $\boldsymbol{1}^T \boldsymbol{y} = |Y|$. From Theorem 15.4 we therefore obtain our well-known inequality

$$
\max\,(|X| : X \text{ a matching }) \leq \min(|Y| : Y \text{ a vertex cover}).
$$

The fundamental theorem of linear optimization, which we shall prove in Section 15.3, is the converse of Theorem 15.5. If (I) and (I*) possess admissible solutions, then they also have optimal soltuions, and the value of (I) is equal to the value of (I*).

In our matching problem, $x = 0$ and $y = 1$ are clearly admissible solutions of $Ax \leq 1$ and $A^T y \geq 1$, and thus there are optimal solutions $\tilde{x}, \tilde{y}$. Does our Theorem 8.5, $\max(|M| : M$ a matching$) = \min(|D| : D$ a vertex cover$)$, follow from this? No, not immediately, since characteristic vectors are integral, with zeros and ones, but for the optimal solutions $\tilde{x}, \tilde{y}$ we can say at the outset only that the coordinates are rational numbers between zero and one. The question of when *integral* optimal solutions exist is of fundamental importance; we shall return to this question in the last section. In the matching case there are, in fact, always such solutions (as we shall see), and Theorem 8.5 is therefore a special case of the fundamental theorem of linear optimization.

To answer the question of when a given program possesses an admissible solution, we investigate general systems of equations and inequalities. The following three statements are so-called alternative theorems. The "either/or" is always meant exclusively. The significance resides in the fact that we get a *positive* condition under which a certain system is *not* solvable. For what follows, let $a^1, \ldots, a^n$ denote the column vectors of the matrix A, and $a_1^T, \ldots, a_m^T$ the row vectors. Let $r(A)$ denote the rank of the matrix and $\langle b_1, \ldots, b_k \rangle$ the subspace spanned by the vectors $b_1, \ldots, b_k$.

Theorem 15.7. *Exactly one of the following holds:*

(A) $Ax = b$ *is solvable,*

(B) $A^T y = 0, \quad b^T y = -1$ *is solvable.*

Proof. If (A) and (B) were both solvable with x, y, then we would have $0 = x^T A^T y = (Ax)^T y = b^T y = -1$, which is impossible. Suppose that (A) is not solvable. Then we would have $b \notin \langle a^1, \ldots, a^n \rangle$. If $r(A)$ is the rank of A, we have $r(A \,|\, b) = r(A) + 1$. For the matrix

$$A' = \left(\begin{array}{c|c} A & b \\ \hline 0^T & -1 \end{array} \right)$$

we then have $r(A') = r(A) + 1 = r(A \,|\, b)$, and so the last row $(0^T| - 1)$ is linearly dependent on the first m rows of A'. There exist therefore $y_1, \ldots, y_m$ with $\sum_{i=1}^m y_i a_i^T = 0^T$ and $\sum_{i=1}^m y_i b_i = -1$, that is, $y = (y_1, \ldots, y_m)$ is a solution of (B). $\qquad \square$

Geometrically, Theorem 15.7 is clear. If b is not in the subspace $U = \langle a^1, \ldots, a^n \rangle$, then there exists a vector y orthogonal to U forming an obtuse angle with $\boldsymbol{b}$:

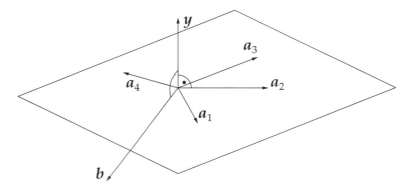

The following theorem (often called Farkas's lemma) is the fundamental result from which everything that follows is derived.

Theorem 15.8. *Exactly one of the following two statements holds:*

(A) $$A\boldsymbol{x} = \boldsymbol{b},\ \boldsymbol{x} \geq \boldsymbol{0},\ \text{is solvable,}$$

(B) $$A^T\boldsymbol{y} \geq \boldsymbol{0},\ \boldsymbol{b}^T\boldsymbol{y} < 0,\ \text{is solvable.}$$

Proof. If $\boldsymbol{x}$ were a solution of (A) and $\boldsymbol{y}$ a solution of (B), we would have the contradiction $0 \leq \boldsymbol{x}^T A^T \boldsymbol{y} = (A\boldsymbol{x})^T\boldsymbol{y} = \boldsymbol{b}^T\boldsymbol{y} < 0$. Suppose, then, that (A) is not solvable. If $A\boldsymbol{x} = \boldsymbol{b}$ were already not solvable, then we would be done by Theorem 15.7. We may therefore assume that $A\boldsymbol{x} = \boldsymbol{b}$ is solvable but possesses no nonnegative solutions. We now use induction on n. For $n = 1$ we have $x_1\boldsymbol{a}^1 = \boldsymbol{b}$, and we suppose that $x_1 < 0$ is a solution. With $\boldsymbol{y} = -\boldsymbol{b}$ we obtain $\boldsymbol{a}^{1^T}\boldsymbol{y} = -\frac{\boldsymbol{b}^T}{x_1}\boldsymbol{b} > 0$ and $\boldsymbol{b}^T\boldsymbol{y} = -\boldsymbol{b}^T\boldsymbol{b} < 0$, and so $\boldsymbol{y}$ is a solution of (B). Suppose that the assertion holds for all $k \leq n - 1$. If (A) has no nonnegative solution, then neither does $\sum_{i=1}^{n-1} x_i\boldsymbol{a}^i = \boldsymbol{b}$. Therefore, by the induction hypothesis there exists $\boldsymbol{v}$ with $\boldsymbol{a}^{i^T}\boldsymbol{v} \geq 0$ $(i = 1, \ldots, n-1)$, $\boldsymbol{b}^T\boldsymbol{v} < 0$. If also $\boldsymbol{a}^{n^T}\boldsymbol{v} \geq 0$, then we are done. We therefore suppose $\boldsymbol{a}^{n^T}\boldsymbol{v} < 0$ and set

$(\star)$
$$\overline{\boldsymbol{a}}^i = (\boldsymbol{a}^{i^T}\boldsymbol{v})\boldsymbol{a}^n - (\boldsymbol{a}^{n^T}\boldsymbol{v})\boldsymbol{a}^i \qquad (i = 1, \ldots, n-1),$$
$$\overline{\boldsymbol{b}} = (\boldsymbol{b}^T\boldsymbol{v})\boldsymbol{a}^n - (\boldsymbol{a}^{n^T}\boldsymbol{v})\boldsymbol{b}.$$

If now $\sum_{i=1}^{n-1} x_i\overline{\boldsymbol{a}}^i = \overline{\boldsymbol{b}}$ with $x_i \geq 0$ $(i = 1, \ldots, n-1)$ is solvable, then

$$-\frac{1}{\boldsymbol{a}^{n^T}\boldsymbol{v}}\left(\sum_{i=1}^{n-1} x_i(\boldsymbol{a}^{i^T}\boldsymbol{v}) - \boldsymbol{b}^T\boldsymbol{v}\right)\boldsymbol{a}^n + \sum_{i=1}^{n-1} x_i\boldsymbol{a}^i = \boldsymbol{b},$$

where all coefficients of the $\boldsymbol{a}^i$ $(i = 1, \ldots, n)$ are nonnegative, in contradiction to the assumption that (A) is not solvable. Therefore, $\sum_{i=1}^{n-1} x_i\overline{\boldsymbol{a}}^i = \overline{\boldsymbol{b}}$,

$x_i \geq 0$ $(i = 1, \ldots, n-1)$, is not solvable, and there exists by the induction hypothesis $\boldsymbol{w} \in \mathbb{R}^m$ with $\overline{\boldsymbol{a}}^{i^T} \boldsymbol{w} \geq 0$ $(i = 1, \ldots, n-1)$, $\overline{\boldsymbol{b}}^T \boldsymbol{w} < 0$. But then

$$\boldsymbol{y} = (\boldsymbol{a}^{n^T} \boldsymbol{w}) \boldsymbol{v} - (\boldsymbol{a}^{n^T} \boldsymbol{v}) \boldsymbol{w}$$

is a solution of (B), since by $(\star)$, we have

$$\boldsymbol{a}^{i^T} \boldsymbol{y} = (\boldsymbol{a}^{n^T} \boldsymbol{w})(\boldsymbol{a}^{i^T} \boldsymbol{v}) - (\boldsymbol{a}^{n^T} \boldsymbol{v})(\boldsymbol{a}^{i^T} \boldsymbol{w}) = \overline{\boldsymbol{a}}^{i^T} \boldsymbol{w} \geq 0 \quad (i = 1, \ldots, n-1),$$
$$\boldsymbol{a}^{n^T} \boldsymbol{y} = 0,$$

and

$$\boldsymbol{b}^T \boldsymbol{y} = (\boldsymbol{a}^{n^T} \boldsymbol{w})(\boldsymbol{b}^T \boldsymbol{v}) - (\boldsymbol{a}^{n^T} \boldsymbol{v})(\boldsymbol{b}^T \boldsymbol{w}) = \overline{\boldsymbol{b}}^T \boldsymbol{w} < 0.$$

$\square$

The next theorem, on nonnegative systems of inequalities, is the crucial step in the proof of the fundamental theorem in the next section.

Theorem 15.9. *Exactly one of the following two statements holds:*

(A) $\qquad\qquad A\boldsymbol{x} \leq \boldsymbol{b}, \quad \boldsymbol{x} \geq \boldsymbol{0}$, *is solvable,*

(B) $\qquad\qquad A^T \boldsymbol{y} \geq \boldsymbol{0}, \quad \boldsymbol{y} \geq \boldsymbol{0}, \quad \boldsymbol{b}^T \boldsymbol{y} < 0$, *is solvable.*

Proof. If $\boldsymbol{x}, \boldsymbol{y}$ are solutions of (A) and (B), then we have $0 \leq \boldsymbol{x}^T A^T \boldsymbol{y} = (A\boldsymbol{x})^T \boldsymbol{y} \leq \boldsymbol{b}^T \boldsymbol{y} < 0$, which is impossible. Suppose that (A) has no solution. This is clearly equivalent to $A\boldsymbol{x} + \boldsymbol{z} = \boldsymbol{b}$, $\boldsymbol{x} \geq \boldsymbol{0}$, $\boldsymbol{z} \geq \boldsymbol{0}$, having no solution. With $B = (A \mid E_m)$, where E_m is the identity matrix, this means that $B\boldsymbol{w} = \boldsymbol{b}$, $\boldsymbol{w} \geq \boldsymbol{0}$, has no solution. On account of Theorem 15.8, there exists $\boldsymbol{y} \in \mathbb{R}^m$ with $A^T \boldsymbol{y} \geq \boldsymbol{0}$, $E_m \boldsymbol{y} = \boldsymbol{y} \geq \boldsymbol{0}$, and $\boldsymbol{b}^T \boldsymbol{y} < 0$. $\square$

Example 15.10. Consider the following transportation problem. There are m factories $F_1, \ldots, F_m$ in which certain products are manufactured, and n markets $M_1, \ldots, M_n$, in which the products are sold. Every year, p_i units are produced in factory F_i, and q_j units are sold in market M_j. We would like to set up a freight plan (x_{ij}) that transports x_{ij} units of F_i to M_j in such a way that all markets are satisfied. Under what conditions is this possible?

Our inequalities are

$$\sum_{j=1}^{n} x_{ij} \leq p_i \quad (i = 1, \ldots, m),$$

$$\sum_{i=1}^{m} x_{ij} \geq q_j \quad (j = 1, \ldots, n),$$

$$x_{ij} \geq 0.$$

A necessary condition is clearly $\sum_{i=1}^{m} p_i \geq \sum_{j=1}^{n} q_j$. With Theorem 15.9 we can show at once that conversely, this condition is also sufficient. First we must bring our inequalities into the form $A\boldsymbol{x} \leq \boldsymbol{b}$, $\boldsymbol{x} \geq \boldsymbol{0}$:

$$\sum_{j=1}^{n} x_{ij} \leq p_i, \qquad \sum_{i=1}^{m} -x_{ij} \leq -q_j.$$

Let us write (x_{ij}) rowwise as

$$\boldsymbol{x} = (x_{11}, \ldots, x_{1n}, x_{21}, \ldots, x_{2n}, \ldots, x_{m1}, \ldots, x_{mn}).$$

Then A is an $((m+n) \times mn)$ matrix, and $\boldsymbol{b}$ an $(m+n)$-vector with $A\boldsymbol{x} \leq \boldsymbol{b}$:

$$A = \left.\left(\begin{array}{ccccccccc} 1 & \cdots & 1 & & & & & & \\ & & & 1 & \cdots & 1 & & 0 & \\ & 0 & & & & & \ddots & & \\ & & & & & & 1 & \cdots & 1 \\ -1 & & & -1 & & & -1 & & \\ & \ddots & & & & & & & \\ & & -1 & & \cdots & -1 & \cdots & \cdots & -1 \end{array}\right)\right\} \begin{array}{c} m \\ \\ \\ n \end{array}$$

and

$$\boldsymbol{b} = \begin{pmatrix} p_1 \\ p_2 \\ \vdots \\ p_m \\ -q_1 \\ \vdots \\ -q_n \end{pmatrix}.$$

If now $A\boldsymbol{x} \leq \boldsymbol{b}$, $\boldsymbol{x} \geq \boldsymbol{0}$, is not solvable, then there exists by Theorem 15.9 $\boldsymbol{y} \in \mathbb{R}^{m+n}$, $\boldsymbol{y} = (y_1, \ldots, y_m, y'_1, \ldots, y'_n)$, with $A^T \boldsymbol{y} \geq \boldsymbol{0}$, $\boldsymbol{y} \geq \boldsymbol{0}$, $\boldsymbol{b}^T \boldsymbol{y} < 0$. This means that

$$y_i - y'_j \geq 0 \qquad \text{for all } i, j$$

and

$$\sum_{i=1}^{m} p_i y_i - \sum_{j=1}^{n} q_j y'_j < 0.$$

Now setting $\widetilde{y} = \min y_i$, $\widetilde{y}' = \max y'_j$, we have $\widetilde{y} \geq \widetilde{y}' \geq 0$, and hence

$$\widetilde{y}\left(\sum_{i=1}^{m} p_i - \sum_{j=1}^{n} q_j\right) \leq \widetilde{y}\sum_{i=1}^{m} p_i - \widetilde{y}'\sum_{j=1}^{n} q_j \leq \sum_{i=1}^{m} p_i y_i - \sum_{j=1}^{n} q_j y'_j < 0,$$

and therefore in fact, $\sum_{i=1}^{m} p_i < \sum_{j=1}^{n} q_j$.

15.3. The Fundamental Theorem of Linear Optimization

The following result is one of the fundamental theorems in all of mathematics, theoretically elegant and applicable universally in practice.

Theorem 15.11. *Consider the standard program* (I) *and the dual program* (I*):

(I)
$$Ax \leq b, \; x \geq 0,$$
$$c^T x = \max,$$

(I*)
$$A^T y \geq c, \; y \geq 0,$$
$$b^T y = \min.$$

Then the following assertions hold:

(i) *If* (I) *and* (I*) *have admissible solutions, then they have optimal solutions, and the value of* (I) *is equal to the value of* (I*).

(ii) *If* (I) *or* (I*) *is not admissibly solvable, then neither of the two programs has an optimal solution.*

Proof. We assume that (I) and (I*) possess admissible solutions. By Theorems 15.4 and 15.5, it remains to show that the system

(15.1)
$$Ax \leq b, \quad x \geq 0,$$

(15.2)
$$A^T y \geq c, \quad y \geq 0,$$

(15.3)
$$c^T x - b^T y \geq 0,$$

has a solution. Rewritten as a system of inequalities, this is equivalent to

(15.4)
$$\begin{pmatrix} A & 0 \\ 0 & -A^T \\ -c^T & b^T \end{pmatrix} \begin{pmatrix} x \\ y \end{pmatrix} \leq \begin{pmatrix} b \\ -c \\ 0 \end{pmatrix}, \quad \begin{pmatrix} x \\ y \end{pmatrix} \geq 0,$$

having a solution. Suppose that (15.4) has no solution. Then by Theorem 15.9 there exist vectors $z \in \mathbb{R}^m$, $w \in \mathbb{R}^n$, $\alpha \in \mathbb{R}$ with $z \geq 0$, $w \geq 0$, $\alpha \geq 0$, such that

(15.5)
$$A^T z \geq \alpha c, \; Aw \leq \alpha b, \quad b^T z < c^T w.$$

We assert that $\alpha > 0$ must hold. Namely, if we had $\alpha = 0$ and if $\overline{x}, \overline{y}$ were admissible solutions of (I) and (I*), we would then have

$$0 \leq \overline{x}^T A^T z = (A\overline{x})^T z \leq b^T z < c^T w \leq (A^T \overline{y})^T w = \overline{y}^T Aw \leq 0,$$

which is impossible. Now let $x = \frac{w}{\alpha}$, $y = \frac{z}{\alpha}$. Then $Ax \leq b$, $A^T y \geq c$, $x \geq 0, y \geq 0$. From Theorem 15.4 it follows that

$$c^T w = \alpha(c^T x) \leq \alpha(b^T y) = b^T z,$$

in contradiction to (15.5). Therefore, (15.4) has a solution, and part (i) is proven.

We now assume that (I) has no admissible solution. Then (I) naturally also has no optimal solution. By Theorem 15.9, there exists, in addition to $\boldsymbol{w} \in \mathbb{R}^m$ with $A^T \boldsymbol{w} \geq \boldsymbol{0}$, also $\boldsymbol{w} \geq \boldsymbol{0}$ and $\boldsymbol{b}^T \boldsymbol{w} < 0$. If (I*) has an admissible solution $\overline{\boldsymbol{y}}$, then $\overline{\boldsymbol{y}} + \lambda \boldsymbol{w}$ is also an admissible solution of (I*) for every $\lambda \geq 0$. For the objective function of (I*) we have

$$\boldsymbol{b}^T(\overline{\boldsymbol{y}} + \lambda \boldsymbol{w}) = \boldsymbol{b}^T \overline{\boldsymbol{y}} + \lambda(\boldsymbol{b}^T \boldsymbol{w}),$$

and this expression can be made arbitrarily small, on account of $\boldsymbol{b}^T \boldsymbol{w} < 0$. The objective function of (I*) has therefore no minimum, and therefore (I*) has no optimal solution. One handles the case that (I*) has no admissible solution similarly.   □

Without using the dual program, we have the following characterization of programs with optimal solutions.

Corollary 15.12. *A standard maximum (minimum) program possesses an optimal solution if and only if there is an admissible solution and the objective function is bounded from above (below).*

We have answered the first of our questions at the end of Section 15.1. The existence of admissible solutions is characterized by Theorem 15.9, and the existence of optimal solutions by Theorem 15.11 and Corollary 15.12.

We would like to note an important consequence of the fundamental theorem, which often makes possible a quick verification of whether an admissible solution is already optimal.

Theorem 15.13. *Let programs* (I) *and* (I*) *be given as in the fundamental theorem, and* $\boldsymbol{x}$ *and* $\boldsymbol{y}$ *admissible solutions of* (I) *and* (I*). *Then* $\boldsymbol{x}$ *and* $\boldsymbol{y}$ *are optimal solutions of* (I) *and* (I*) *if and only if*

$$(15.6) \qquad y_i > 0 \implies \sum_{j=1}^{n} a_{ij} x_j = b_i \quad (i = 1, \ldots, m),$$

$$(15.7) \qquad x_j > 0 \implies \sum_{i=1}^{m} a_{ij} y_i = c_j \quad (j = 1, \ldots, n).$$

Proof. Suppose $\boldsymbol{x}$ and $\boldsymbol{y}$ are optimal solutions. By Theorems 15.11 and 15.4 we have

$$(15.8) \qquad \sum_{j=1}^{n} c_j x_j = \sum_{i,j} y_i a_{ij} x_j = \sum_{i=1}^{m} b_i y_i.$$

From the second equality in (15.8) we obtain

$$\sum_{i=1}^{m} \left(b_i - \sum_{j=1}^{n} a_{ij} x_j \right) y_i = 0.$$

Since we always have $y_i \geq 0$ and $b_i - \sum_{j=1}^{n} a_{ij}x_j \geq 0$, from $y_i > 0$ we have $\sum_{j=1}^{n} a_{ij}x_j = b_i$, and therefore (15.6). Condition (15.7) is derived similarly from the first equality in (15.8). Conversely, if $\boldsymbol{x}, \boldsymbol{y}$ satisfy conditions (15.6) and (15.7), then we have

$$\boldsymbol{c}^T \boldsymbol{x} = \sum_{j=1}^{n} c_j x_j = \sum_{j=1}^{n} (\sum_{i=1}^{m} a_{ij} y_i) x_j = \sum_{i=1}^{m} \left(\sum_{j=1}^{n} a_{ij} x_j \right) y_i = \sum_{i=1}^{m} b_i y_i = \boldsymbol{b}^T \boldsymbol{y} .$$

By Theorem 15.5 we have that $\boldsymbol{x}$ and $\boldsymbol{y}$ are optimal solutions. $\qquad \square$

Example 15.14. Consider the following program (I) and the associated dual program (I*):

(I)
$$x_1 + 3x_2 + x_4 \leq 4,$$
$$2x_1 + x_2 \leq 3,$$
$$x_2 + 4x_3 + x_4 \leq 3,$$
$$2x_1 + 4x_2 + x_3 + x_4 = \max,$$

and

(I*)
$$y_1 + 2y_2 \geq 2,$$
$$3y_1 + y_2 + y_3 \geq 4,$$
$$4y_3 \geq 1,$$
$$y_1 + y_3 \geq 1,$$
$$4y_1 + 3y_2 + 3y_3 = \min .$$

We test whether the admissible solution $\overline{\boldsymbol{x}} = \left(1, 1, \frac{1}{2}, 0\right)$ with $\boldsymbol{c}^T \overline{\boldsymbol{x}} = \frac{13}{2}$ is optimal for the program (I). If it is, then by Theorem 15.13, the first three constraints of (I*) must be equalities. We thus obtain the admissible solution $\overline{\boldsymbol{y}} = \left(\frac{11}{10}, \frac{9}{20}, \frac{1}{4}\right)$ with $\boldsymbol{b}^T \overline{\boldsymbol{y}} = \frac{13}{2}$. Therefore, $\overline{\boldsymbol{x}}, \overline{\boldsymbol{y}}$ are optimal, since the values of the programs agree.

In addition to the standard programs, there are other types of linear programs.

Definition 15.15. Let $A \in \mathbb{R}^{m \times n}$, $\boldsymbol{b} \in \mathbb{R}^m$, $\boldsymbol{c} \in \mathbb{R}^n$. A **canonical maximum program** is the problem of finding $\boldsymbol{x} \in \mathbb{R}^n$ such that

$$A\boldsymbol{x} = \boldsymbol{b}, \quad \boldsymbol{x} \geq \boldsymbol{0}, \quad \boldsymbol{c}^T \boldsymbol{x} = \max .$$

A **canonical minimum program** is defined similarly, with "maximal" replaced by "minimal."

Every canonical program can be transformed into a standard program:

$$\begin{array}{ll} A\boldsymbol{x} = \boldsymbol{b}, \quad \boldsymbol{x} \geq \boldsymbol{0}, & \qquad A\boldsymbol{x} \leq \boldsymbol{b}, \quad -A\boldsymbol{x} \leq -\boldsymbol{b}, \quad \boldsymbol{x} \geq \boldsymbol{0}, \\ \boldsymbol{c}^T \boldsymbol{x} = \max & \longrightarrow \qquad \boldsymbol{c}^T \boldsymbol{x} = \max . \end{array}$$

Conversely, a standard program with constraints $A\boldsymbol{x} \leq \boldsymbol{b}$ can be transformed into a canonical program through the introduction of so-called slack variables $\boldsymbol{z}$ (in the proof of Theorem 15.9 we have already used this idea):

$$A\boldsymbol{x} \leq \boldsymbol{b}, \quad \boldsymbol{x} \geq \boldsymbol{0}, \qquad\qquad A\boldsymbol{x} + \boldsymbol{z} = \boldsymbol{b}, \quad \boldsymbol{x} \geq \boldsymbol{0}, \quad \boldsymbol{z} \geq \boldsymbol{0},$$
$$\boldsymbol{c}^T\boldsymbol{x} = \max \qquad \longrightarrow \qquad \boldsymbol{c}^T\boldsymbol{x} + \boldsymbol{0}^T\boldsymbol{z} = \max.$$

Example 15.16. In Section 8.3 we proved the fundamental max-flow–min-cut theorem for networks, and noted in a number of places that this theorem is a special case of the fundamental theorem of linear optimization. We would like to establish this fact and prove Theorem 8.17 using the tools of linear programming. In particular, the *existence* of a maximum flow over $\mathbb{R}$ will follow, and all *without* our having to invoke continuity.

Let $\vec{G} = (V, E)$ be a network over $\mathbb{R}$ with source q, sink s, and capacity $c : E \to \mathbb{R}$. We append to $\vec{G}$ the directed edge $k^* = (s, q)$ and assign to it a very large capacity, for example, $c(k^*) > \sum_{k \in E} c(k)$. The new network will again be denoted by $\vec{G}$.

We now consider the (n, m) incidence matrix A of $\vec{G}$, where as usual, the rows are indexed by the vertices $u \in V$ and the columns by the edges $k \in E$ with

$$a_{uk} = \begin{cases} 1 & \text{if } u = k^+, \\ -1 & \text{if } u = k^-, \\ 0 & \text{if } u, k \text{ are not incident.} \end{cases}$$

A vector $(x_k : k \in E)$ is a flow, and the flow is admissible if

$$(\partial f)(u) = \sum_{k \in E} a_{uk}\, x_k = 0 \quad (u \neq q, s),$$

$$0 \leq x_k \leq c_k \quad (k \neq k^*).$$

The value of the flow is $w = \sum_{k \neq k^*} a_{sk} x_k = -\sum_{k \neq k^*} a_{qk} x_k$. Setting $x_{k^*} = w$, we also have $(\partial f)(s) = (\partial f)(q) = 0$, which was of course the reason for introducing the edge k^*.

We thereby obtain the following program:

$$A\boldsymbol{x} = \boldsymbol{0}, \quad \boldsymbol{x} \geq \boldsymbol{0},$$
$$E_m \boldsymbol{x} \leq \boldsymbol{c},$$

with

$$\boldsymbol{e}_{k^*}^T \boldsymbol{x} = \max.$$

Here E_m is the $(m \times m)$ identity matrix, and $\boldsymbol{e}_{k^*}$ is the vector of length m with 1 at the place k^* and zero otherwise. Transformed to a standard maximum program, we obtain (I) and the associated dual program (I*) in the following form:

$$(I) \quad \left(\begin{array}{c} A \\ \hline -A \\ \hline E_m \end{array}\right) x \le \left(\begin{array}{c} 0 \\ \hline 0 \\ \hline c \end{array}\right), \qquad (I^*) \quad (A^T | -A^T | E_m) \left(\begin{array}{c} y' \\ y'' \\ z \end{array}\right) \ge e_{k^*},$$

$$x \ge 0, \qquad\qquad\qquad y' \ge 0, \quad y'' \ge 0 \quad z \ge 0,$$

$$e_{k^*}^T x = \max, \qquad\qquad\qquad c^T z = \min.$$

Program (I) has the admissible solution $x = 0$, while (I*) has the admissible solution $y' = 0$, $y'' = 0$, $z = e_{k^*}$, and hence there exist optimal solutions x and (y', y'', z). The question of existence has now been answered by the fundamental theorem (and without appeal to continuity), and we have only now to show that the value of (I) corresponds precisely to the value of a maximum flow, and the value of (I*) to the capacity of a minimum cut.

Therefore let x and (y', y'', z) be optimal solutions. We set $\overline{y} = y' - y''$. The constraints in (I*) mean that

$$(15.9) \qquad A^T \overline{y} + z \ge e_{k^*}, \quad y' \ge 0, \quad y'' \ge 0, \quad z \ge 0.$$

If we had $z_{k^*} > 0$, then from (15.6) we would also have $x_{k^*} = c_{k^*} > \sum_{k \ne k^*} c_k$. But now, as we know, x_{k^*} is equal to the value of the flow induced by x on $E \setminus \{k^*\}$, and the value of a flow is restricted, by Lemma 8.16, by the capacity of every cut, and so in particular by $\sum_{k \ne k^*} c_k$. This leads then to a contradiction, and we conclude that $z_k^* = 0$, from which now follows

$$\overline{y}_q \ge 1 + \overline{y}_s$$

for the row k^* of A^T on account of $A^T \overline{y} + z \ge e_{k^*}$.

We now define the cut (X, Y) by

$$X = \{u \in V : \overline{y}_u \ge 1 + \overline{y}_s\}, \quad Y = V \setminus X.$$

Since we have $q \in X$ and $s \notin X$, we in fact obtain a cut, and furthermore, we have

$$(15.10) \qquad \overline{y}_u > \overline{y}_v \text{ for all } u \in X, v \in Y.$$

It remains to show that the capacity $c(X, Y)$ of the cut (X, Y) is exactly equal to the value of the flow x, that is, equal to x_{k^*}. Let $k = (u, v) \in E$ with $u \in X, v \in Y$, and so in particular, $k \ne k^*$. Condition (15.9) evaluated for k implies

$$-\overline{y}_u + \overline{y}_v + z_k \ge 0,$$

and from (15.10) we conclude that $z_k > 0$. From (15.6) in Theorem 15.13 it follows that $x_k = c_k$. We now assume $k = (v, u)$ with $u \in X$, $v \in Y$, $k \ne k^*$. We wish to show that then $x_k = 0$. This holds in any case when $c_k = 0$. So let $c_k > 0$. Condition (15.9) evaluated for the row k implies on account of (15.10) and $z_k \geq 0$ that

$$-\overline{y}_v + \overline{y}_u + z_k > 0 \,.$$

From (15.7) in Theorem 15.13 it follows that $x_k = 0$, as desired. We are now finished. As in Theorem 8.17 we obtain

$$x_{k^*} = \sum_{k^- \in X, k^+ \in Y} x_k - \sum_{k^+ \in X, k^- \in Y} x_k = \sum_{k^- \in X, k^+ \in Y} c_k = c(X, Y)$$

for the value of the flow, and the proof is complete.

15.4. Admissible Solutions and Optimal Solutions

We would now like to answer the second question from Section 15.1. How can we describe the set of admissible solutions and the set of optimal solutions? For this we always consider a canonical minimum program

(I)
$$A\boldsymbol{x} = \boldsymbol{b}, \quad \boldsymbol{x} \geq \boldsymbol{0},$$
$$\boldsymbol{c}^T \boldsymbol{x} = \min,$$

which represents no loss in generality by our considerations in the previous section. Let $r(A)$ be the rank of the matrix A. We may assume $r(A) = m \leq n$. In the case $n < m$, by linear algebra $m - r(A)$ equations are superfluous. Therefore this also represents no loss of generality. Let M denote the set of *admissible* solutions, and M_{opt} the set of *optimal* solutions.

Definition 15.17. A set $K \subseteq \mathbb{R}^n$ is said to be **convex** if $\boldsymbol{x}', \boldsymbol{x}'' \in K$, $0 \leq \lambda \leq 1$, implies $\lambda \boldsymbol{x}' + (1 - \lambda)\boldsymbol{x}'' \in K$.

Geometrically, this means that for every two points $\boldsymbol{x}', \boldsymbol{x}'' \in K$, the line $\overline{\boldsymbol{x}'\boldsymbol{x}''}$ joining them also lies in K (see the figure below). Clearly, the intersection of convex sets is again convex.

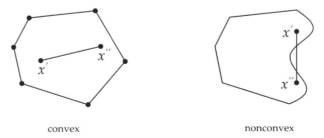

convex   nonconvex

Let $K \subseteq \mathbb{R}^n$ be convex. Then a point $\boldsymbol{p} \in K$ is called a **vertex** if $\boldsymbol{p}$ is not in the interior of a line segment lying entirely in K. That is, $\boldsymbol{p} = \lambda \boldsymbol{x}' + (1 - \lambda)\boldsymbol{x}''$, $\boldsymbol{x}', \boldsymbol{x}'' \in K$, $0 \leq \lambda \leq 1$, implies $\boldsymbol{p} = \boldsymbol{x}'$ or $\boldsymbol{p} = \boldsymbol{x}''$. Let $V(K)$ denote the set of vertices of K.

The convex set in the above figure has six vertices. A convex set can have infinitely many vertices, and it can have none at all. In the (convex) circular disk, every boundary point is a vertex, while the entire space $\mathbb{R}^n$ is clearly convex, but it has no vertices.

Theorem 15.18. *Suppose we are given the program* (I). *The set* M *of admissible solutions is convex and closed.*

Proof. Let $\boldsymbol{x}', \boldsymbol{x}'' \in M$, $0 \leq \lambda \leq 1$, $\boldsymbol{z} = \lambda \boldsymbol{x}' + (1-\lambda)\boldsymbol{x}''$. Then $A\boldsymbol{z} = \lambda A\boldsymbol{x}'$ $+(1-\lambda)A\boldsymbol{x}'' = \lambda \boldsymbol{b} + (1-\lambda)\boldsymbol{b} = \boldsymbol{b}$, whence $\boldsymbol{z} \in M$. Let $H_i = \{\boldsymbol{x} \in \mathbb{R}^n : \boldsymbol{a}_i^T \boldsymbol{x} = b_i\}$, $i = 1, \ldots, m$, and $P_j = \{\boldsymbol{x} \in \mathbb{R}^n : x_j \geq 0\}$. The sets H_i are hyperplanes and therefore convex and closed, and the same holds for each orthant P_j. Since $M = (\bigcap_{i=1}^m H_i) \cap (\bigcap_{j=1}^n P_j)$, M is also convex and closed. $\qquad\square$

Let $\boldsymbol{x} \in M$ and let $Z \subseteq \{1, \ldots, n\}$ be the set of indices k with $x_k > 0$. We then have $x_j = 0$ for $j \notin Z$. We call $\{\boldsymbol{a}^k : k \in Z\}$ the *column set associated with* $\boldsymbol{x}$, that is, $\sum_{k \in Z} x_k \boldsymbol{a}^k = \boldsymbol{b}$. In our example in Section 15.1, we saw that a (in this case *the*) optimal solution appears at a vertex. That is what we would now like to prove in general.

Theorem 15.19. *Suppose we are given the program* (I). *If* $M \neq \varnothing$, *then* M *contains vertices.*

Proof. Choose $\boldsymbol{x} \in M$ such that the index set Z of the column set associated with $\boldsymbol{x}$ is minimal (among all these index sets). If $\boldsymbol{x} \notin V(M)$, then there exist $\boldsymbol{x}' \neq \boldsymbol{x}'' \in M$, $0 < \lambda < 1$, with $\boldsymbol{x} = \lambda \boldsymbol{x}' + (1-\lambda)\boldsymbol{x}''$. From $x_j = \lambda x_j' + (1-\lambda)x_j''$ follows $x_j' = x_j'' = 0$ for all $j \notin Z$. Therefore, we have

$$\boldsymbol{b} = A\boldsymbol{x}' = \sum_{k \in Z} x_k' \boldsymbol{a}^k \,, \quad \boldsymbol{b} = A\boldsymbol{x}'' = \sum_{k \in Z} x_k'' \boldsymbol{a}^k \,,$$

that is, $\sum_{k \in Z}(x_k' - x_k'')\boldsymbol{a}^k = \boldsymbol{0}$ with $x_k' - x_k'' \neq 0$ for at least one k. Let $v_k = x_k' - x_k''$, that is, $\sum_{k \in Z} v_k \boldsymbol{a}^k = \boldsymbol{0}$, and $\rho = \min \frac{x_k}{|v_k|}$ over all $k \in Z$ with $v_k \neq 0$. Let $\rho = \frac{x_h}{v_h}$, where we may assume that $v_h > 0$. For $\overline{\boldsymbol{x}} = (\overline{x}_j)$ with $\overline{x}_k = x_k - \rho v_k$ $(k \in Z)$ and $\overline{x}_j = 0$ for $j \notin Z$ we then have

$$\overline{\boldsymbol{x}} \geq \boldsymbol{0}, \quad A\overline{\boldsymbol{x}} = A\boldsymbol{x} - \rho \sum_{k \in Z} v_k \boldsymbol{a}^k = A\boldsymbol{x} = \boldsymbol{b} \,,$$

that is, $\overline{\boldsymbol{x}} \in M$. Since $\overline{x}_h = 0$, the column set associated with $\overline{\boldsymbol{x}}$ is properly contained in Z, in contradiction to the assumption. $\qquad\square$

Theorem 15.20. *The point* $\boldsymbol{x} \in M$ *is a vertex of* M *if and only if the column set associated with* $\boldsymbol{x}$ *is linearly independent.*

Proof. Let $\{\boldsymbol{a}^k : k \in Z\}$ be the column set associated with $\boldsymbol{x}$. If $\boldsymbol{x}$ is not a vertex, $\boldsymbol{x} = \lambda \boldsymbol{x}' + (1-\lambda)\boldsymbol{x}''$, then it follows as above that $x_j' = x_j'' = 0$

for $j \notin Z$ and therefore $\sum_{k \in Z} (x'_k - x''_k) \boldsymbol{a}^k = \boldsymbol{0}$, that is, $\{\boldsymbol{a}^k : k \in Z\}$ is linearly dependent. Conversely, suppose $\{\boldsymbol{a}^k : k \in Z\}$ is linearly dependent, $\sum_{k \in Z} v_k \boldsymbol{a}^k = \boldsymbol{0}$ with $v_{k^*} \neq 0$ for some $k^* \in Z$. Since $x_k > 0$ holds for all $k \in Z$, we have for a sufficiently small $\rho > 0$ that $x_k - \rho v_k > 0$ holds for all $k \in Z$. Now let $\boldsymbol{x}', \boldsymbol{x}'' \in \mathbb{R}^n$ be determined by $x'_k = x_k + \rho v_k$, $x''_k = x_k - \rho v_k$ $(k \in Z)$, $x'_j = x''_j = 0$ $(j \notin Z)$, and so $x'_{k^*} \neq x_{k^*} \neq x''_{k^*}$, whence $\boldsymbol{x}' \neq \boldsymbol{x}$, $\boldsymbol{x}'' \neq \boldsymbol{x}$. We therefore have

$$\boldsymbol{x}' \geq \boldsymbol{0},\ \boldsymbol{x}'' \geq \boldsymbol{0}, \quad A\boldsymbol{x}' = A\boldsymbol{x}'' = \boldsymbol{b}, \quad \boldsymbol{x} = \frac{1}{2}\boldsymbol{x}' + \frac{1}{2}\boldsymbol{x}'' ,$$

and so $\boldsymbol{x}$ is not a vertex. □

From the characterization of Theorem 15.20 we can immediately draw two corollaries.

Corollary 15.21. *If $\boldsymbol{x}$ is a vertex of M, then $\boldsymbol{x}$ has at most m positive coordinates.*

We call a vertex $\boldsymbol{x}$ of M **nondegenerate** if $\boldsymbol{x}$ has *exactly* m positive coordinates. Otherwise, $\boldsymbol{x}$ is called **degenerate**.

Corollary 15.22. *If $M \neq \varnothing$, then M has finitely many vertices.*

Proof. There are only finitely many linearly independent subsets of columns, and to each of these column sets there corresponds at most one admissible solution. □

To summarize, if we are given the canonical program (I), with M the set of admissible solutions, then for M there exist the following possibilities:

(a) $M = \varnothing$ (that is, there is no admissible solution);

(b) $M \neq \varnothing$ is convex and closed and has finitely many vertices.

A set that satisfies (b) is called a *polyhedral set*. Conversely, it can be shown that a polyhedral set M that satisfies $\boldsymbol{x} \geq \boldsymbol{0}$ for all M is an admissible solution set of a canonical program. Thus the structure of the solution set of M has been satisfactorily described. There is an analogous theorem for the set M_{opt} of optimal solutions.

Theorem 15.23. *For the program (I), the set M_{opt} is convex and closed, and $V(M_{\text{opt}}) = V(M) \cap M_{\text{opt}}$.*

Proof. If $M_{\text{opt}} = \varnothing$, then there is nothing to prove. Otherwise, let w be the value of (I). Then

$$M_{\text{opt}} = M \cap \{\boldsymbol{x} \in \mathbb{R}^n : \boldsymbol{c}^T \boldsymbol{x} = w\}$$

is also closed and convex from Theorem 15.18. If $\boldsymbol{x} \in V(M) \cap M_{\text{opt}}$, then $\boldsymbol{x}$ is surely not in the interior of a line segment connecting two points in

M_{opt}, whence $\boldsymbol{x} \in V(M_{\text{opt}})$. Now let $\boldsymbol{x} \in V(M_{\text{opt}})$ and $\boldsymbol{x} = \lambda \boldsymbol{u} + (1 - \lambda)\boldsymbol{v}$, $\boldsymbol{u}, \boldsymbol{v} \in M$, $0 < \lambda < 1$. We have $w = \boldsymbol{c}^T \boldsymbol{x} = \lambda \boldsymbol{c}^T \boldsymbol{u} + (1 - \lambda)\boldsymbol{c}^T \boldsymbol{v}$, and on account of $\boldsymbol{c}^T \boldsymbol{u} \geq w$, $\boldsymbol{c}^T \boldsymbol{v} \geq w$ it follows that $w = \boldsymbol{c}^T \boldsymbol{u} = \boldsymbol{c}^T \boldsymbol{v}$, that is, $\boldsymbol{u}, \boldsymbol{v} \in M_{\text{opt}}$. From $\boldsymbol{x} \in V(M_{\text{opt}})$ it follows that $\boldsymbol{x} = \boldsymbol{u}$ or $\boldsymbol{x} = \boldsymbol{v}$, that is, $\boldsymbol{x} \in V(M) \cap M_{\text{opt}}$. □

The following theorem forms the basis for the simplex algorithm to be introduced in the next section.

Theorem 15.24. *For the program* (I), *if there exist optimal solutions, then among them is a vertex of M. In other words, $M_{\text{opt}} \neq \varnothing \Rightarrow V(M_{\text{opt}}) \neq \varnothing$.*

Proof. Let $M_{\text{opt}} \neq \varnothing$. We extend (I) by the equation $\boldsymbol{c}^T \boldsymbol{x} = w$, where w is the value of (I). The set of admissible solutions of

$$\begin{pmatrix} A \\ \boldsymbol{c}^T \end{pmatrix} \boldsymbol{x} = \begin{pmatrix} \boldsymbol{b} \\ w \end{pmatrix}, \quad \boldsymbol{x} \geq \boldsymbol{0},$$

is then precisely M_{opt}. By Theorem 15.19, M_{opt} therefore contains vertices. □

Theorems 15.20 and 15.24 make it possible in principle to determine an optimal vertex by the following method. For each vertex $\boldsymbol{x} \in M$ we call every m-set of linearly independent columns of A that contains the column set associated with $\boldsymbol{x}$ a **basis** of x. A nondegenerate vertex therefore has a basis.

We now proceed as follows: We consider in succession the $\binom{n}{m}$ submatrices $A' \subseteq A$ consisting of m columns of A, and solve $A'\boldsymbol{x} = \boldsymbol{b}$. We ignore all those A' for which $A'\boldsymbol{x} = \boldsymbol{b}$ has no solution at all or no solution $\boldsymbol{x} \geq \boldsymbol{0}$. The remaining solutions $\boldsymbol{x}$ are, according to Theorem 15.20, precisely the vertices of M, and by calculating from $\boldsymbol{c}^T \boldsymbol{x}$ the existing optimal vertex guaranteed by Theorem 15.24, we can determine the value of the program.

Example 15.25. Consider the following program:

$$5x_1 - 4x_2 + 13x_3 - 2x_4 + x_5 = 20,$$
$$x_1 - x_2 + 5x_3 - x_4 + x_5 = 8,$$
$$x_1 + 6x_2 - 7x_3 + x_4 + 5x_5 = \min.$$

We test successively the $\binom{5}{2} = 10$ submatrices

$$\begin{aligned} 5x_1 - 4x_2 &= 20, \\ x_1 - x_2 &= 8, \end{aligned} \quad \cdots \quad \begin{aligned} -2x_4 + x_5 &= 20, \\ -x_4 + x_5 &= 8. \end{aligned}$$

As vertices we have the solutions of

$$
\begin{array}{lll}
5x_1 + x_5 = 20, & -4x_2 + 13x_3 = 20, & 13x_3 + x_5 = 20, \\
x_1 + x_5 = 8, & -x_2 + 5x_3 = 8, & 5x_3 + x_5 = 8,
\end{array}
$$

$$
\boldsymbol{x'} = (3,0,0,0,5), \qquad \boldsymbol{x''} = \left(0, \tfrac{4}{7}, \tfrac{12}{7}, 0, 0\right), \qquad \boldsymbol{x'''} = \left(0,0,\tfrac{3}{2},0,\tfrac{1}{2}\right).
$$

From $\boldsymbol{c}^T \boldsymbol{x'} = 28$, $\boldsymbol{c}^T \boldsymbol{x''} = -60/7$, $\boldsymbol{c}^T \boldsymbol{x'''} = -8$ we obtain $\boldsymbol{x''}$ as the unique optimal vertex, and also the value $-\frac{60}{7}$.

For large numbers n and m this method is of course hopelessly expensive. We must therefore look for a better method for finding an optimal vertex. This is done in the following section.

15.5. The Simplex Algorithm

The simplex method for solving linear programs was proposed by George Dantzig in 1947. This algorithm is one of the most remarkable success stories in all of mathematics, equally important in both theory and practice. A reason for the unusual success was undoubtedly the simultaneous development of the first fast electronic computers. Linear programs with hundreds of constraints can now be solved. Today, the simplex algorithm is part of the basic knowledge of everyone involved in questions of optimization.

The basic idea is simple. We are given a canonical minimum program $A\boldsymbol{x} = \boldsymbol{b}$ with $r(A) = m \leq n$ and an objective function $Q(x) = \boldsymbol{c}^T \boldsymbol{x}$:

$$
A\boldsymbol{x} = \boldsymbol{b}, \quad \boldsymbol{x} \geq \boldsymbol{0},
$$

(I)

$$
Q(\boldsymbol{x}) = \boldsymbol{c}^T \boldsymbol{x} = \min .
$$

The algorithm consists of two steps:

(A) Determine an admissible vertex $\boldsymbol{x}^0$.

(B) If $\boldsymbol{x}^0$ is not optimal, determine a vertex $\boldsymbol{x}^1$ with $Q(\boldsymbol{x}^1) < Q(\boldsymbol{x}^0)$.

After finitely many steps we arrive at an optimal vertex (if one exists).

We deal first with (B): Let $\boldsymbol{x}^0 = (x_k^0) \in M$ be a vertex with basis $\{\boldsymbol{a}^k : k \in Z\}$. Then there exist unique elements $\tau_{kj} \in \mathbb{R}$ $(k \in Z, \, j = 1, \ldots, n)$ with

$$
(15.11) \qquad\qquad \boldsymbol{a}^j = \sum_{k \in Z} \tau_{kj} \boldsymbol{a}^k \qquad (j = 1, \ldots, n) .
$$

From the linear independence of $\{\boldsymbol{a}^k : k \in Z\}$ it follows that $\tau_{kk} = 1$ for $k \in Z$ and $\tau_{kj} = 0$ for $k, j \in Z$, $k \neq j$.

Let $\boldsymbol{x} \in M$ be arbitrary. Then

$$
\sum_{k \in Z} x_k^0 \boldsymbol{a}^k = \boldsymbol{b} = \sum_{j=1}^{n} x_j \boldsymbol{a}^j = \sum_{j=1}^{n} x_j \left(\sum_{k \in Z} \tau_{kj} \boldsymbol{a}^k \right) = \sum_{k \in Z} \left(\sum_{j=1}^{n} \tau_{kj} x_j \right) \boldsymbol{a}^k ,
$$

whence $x_k^0 = \sum_{j=1}^n \tau_{kj} x_j = \sum_{j \notin Z} \tau_{kj} x_j + x_k$, that is, $x_k = x_k^0 - \sum_{j \notin Z} \tau_{kj} x_j$ ($k \in Z$). This yields

$$Q(\boldsymbol{x}) = \sum_{j=1}^n c_j x_j = \sum_{k \in Z} c_k x_k + \sum_{j \notin Z} c_j x_j = \sum_{k \in Z} c_k x_k^0 - \sum_{j \notin Z} \left(\sum_{k \in Z} \tau_{kj} c_k - c_j \right) x_j .$$

Setting

(15.12) $$z_j = \sum_{k \in Z} \tau_{kj} c_k \qquad (j \notin Z) ,$$

we obtain

(15.13) $$Q(\boldsymbol{x}) = Q(\boldsymbol{x}^0) - \sum_{j \notin Z} (z_j - c_j) x_j .$$

We now distinguish the following cases:

Case 1: $z_j \leq c_j$ for all $j \notin Z$. Then we have $Q(\boldsymbol{x}) \geq Q(\boldsymbol{x}^0)$ for all $\boldsymbol{x} \in M$, and so $\boldsymbol{x}^0$ is optimal, and we are done.

Case 2: There exists $j \notin Z$ with $z_j > c_j$ and $\tau_{kj} \leq 0$ for all $k \in Z$. Let $\delta > 0$ be arbitrary. We define $\boldsymbol{x}(\delta) = (x_i(\delta))$ by

$$x_k(\delta) = x_k^0 - \delta \tau_{kj} \quad (k \in Z),$$
$$x_j(\delta) = \delta,$$
$$x_i(\delta) = 0 \quad (i \notin Z, \, i \neq j) .$$

Then $\boldsymbol{x}(\delta) \geq 0$, and by (15.11) we have

$$A\boldsymbol{x}(\delta) = \sum_{k \in Z} x_k(\delta) \boldsymbol{a}^k + \delta \boldsymbol{a}^j = \sum_{k \in Z} x_k^0 \boldsymbol{a}^k - \delta \sum_{k \in Z} \tau_{kj} \boldsymbol{a}^k + \delta \boldsymbol{a}^j = \boldsymbol{b} ,$$

that is, $\boldsymbol{x}(\delta) \in M$ for every $\delta > 0$. Since

$$Q(\boldsymbol{x}(\delta)) = Q(\boldsymbol{x}^0) - (z_j - c_j) \delta,$$

we have $Q(\boldsymbol{x})$ arbitrarily small, and therefore program (I) has no optimal solution.

Case 3: There exist $s \notin Z$, $k \in Z$ with $z_s > c_s$ and $\tau_{ks} > 0$. We set $\delta = \min x_k^0 / \tau_{ks}$ over all $k \in Z$ with $\tau_{ks} > 0$ and form $\boldsymbol{x}^1 = \boldsymbol{x}(\delta)$ as in Case 2. Let $r \in Z$ be an index such that $\delta = x_r^0 / \tau_{rs}$. Since $x_r^0 / \tau_{rs} \leq x_k^0 / \tau_{ks}$ ($k \in Z$, $\tau_{ks} > 0$), we have $x_k^1 = x_k^0 - \frac{x_r^0}{\tau_{rs}} \tau_{ks} \geq 0$, whence $\boldsymbol{x}^1 \geq \boldsymbol{0}$. That $A\boldsymbol{x}^1 = \boldsymbol{b}$ is satisfied follows as in Case 2. We now show that $\boldsymbol{x}^1$ is a vertex. The column set associated with $\boldsymbol{x}^1$ is contained in $\{\boldsymbol{a}^k : k \in (Z \setminus \{r\}) \cup \{s\}\}$. If this column set were linearly dependent, we would have

$$\sum_{k \in Z \setminus \{r\}} \mu_k \boldsymbol{a}^k + \mu \boldsymbol{a}^s = \boldsymbol{0} \qquad \text{(not all } \mu_k, \mu = 0\text{)}$$

and therefore $\mu \neq 0$, whence $\boldsymbol{a}^s = \sum_{k \in Z \smallsetminus \{r\}} (-\frac{\mu_k}{\mu}) \boldsymbol{a}^k$. On the other hand, however, we have $\boldsymbol{a}^s = \sum_{k \in Z} \tau_{ks} \boldsymbol{a}^k$, and so $\tau_{rs} = 0$, in contradiction to the choice of r. Therefore, $\boldsymbol{x}^1$ is a (possibly degenerate) vertex by Theorem 15.20.

Case 3a: $\delta > 0$. Then $Q(\boldsymbol{x}^1) = Q(\boldsymbol{x}^0) - (z_s - c_s)\delta < Q(\boldsymbol{x}^0)$.

Case 3b: $\delta = 0$. Here $Q(\boldsymbol{x}^1) = Q(\boldsymbol{x}^0)$.

This last case can occur only if $x_r^0 = 0$, that is, if $\boldsymbol{x}^0$ is a degenerate vertex. Here we have $\boldsymbol{x}^1 = \boldsymbol{x}^0$, and each time we change only a basis of $\boldsymbol{x}^0$. It could therefore transpire that we run through the bases of $\boldsymbol{x}^0$ cyclically and always keep the same value of the objective function. One must therefore establish a particular (e.g., lexicographic) order among the bases of $\boldsymbol{x}^0$ in order to run through each basis at most once (see Exercise 15.24). In practice, this case seldom arises.

In Case 3a we again determine, with the new vertex $\boldsymbol{x}^1$ with $Q(\boldsymbol{x}^1) < Q(\boldsymbol{x}^0)$, the tableau (τ'_{kj}) and test Cases 1 through 3 again. Since there exist only finitely many vertices, we must eventually end up in Case 1 (optimal solution) or Case 2 (no optimal solution exists).

For carrying out the exchange step in Case 3a, one orders the required data as usual according to the following tableau, where we set $d_j = z_j - c_j$:

		j		s			
$k \in Z$				$*$			
	k	τ_{kj}		τ_{ks}		x_k^0	x_k^0/τ_{ks}
$*$	r	τ_{rj}		$\textcircled{$\tau_{rs}$}$		x_r^0	x_r^0/τ_{rs}
		d_j		d_s		$Q(x^0)$	

The rth row and sth column used in the exchange step are indicated by $*$. The element τ_{rs} is called the **pivot element** and is circled (or shown in boldface). We now replace the basis $\{\boldsymbol{a}^k : k \in Z\}$ of $\boldsymbol{x}^0$ with the new basis $\{\boldsymbol{a}^k : k \in Z' = (Z \smallsetminus r) \cup \{s\}\}$ of $\boldsymbol{x}^1$. For the new tableau (τ'_{kj}) this means that the $*$-row now has the index s, and the $*$-column has the index r. The new tableau (τ'_{kj}) can be easily calculated using (15.11), according to which

$$(15.14) \qquad \boldsymbol{a}^r = \sum_{k \in Z \smallsetminus \{r\}} \left(-\frac{\tau_{ks}}{\tau_{rs}}\right) \boldsymbol{a}^k + \frac{1}{\tau_{rs}} \boldsymbol{a}^s,$$

and hence
(15.15)
$$\boldsymbol{a}^j = \sum_{k \in Z \smallsetminus r} \tau_{kj} \boldsymbol{a}^k + \tau_{rj} \boldsymbol{a}^r = \sum_{k \in Z \smallsetminus \{r\}} \left(\tau_{kj} - \frac{\tau_{ks}}{\tau_{rs}} \tau_{rj} \right) \boldsymbol{a}^k + \frac{\tau_{rj}}{\tau_{rs}} \boldsymbol{a}^s \quad (j \neq r).$$

If $\boldsymbol{t}_k$ denotes the kth row of (τ_{kj}) and analogously $\boldsymbol{t}'_k$ the kth row of (τ'_{kj}), then from (15.14) and (15.15) we obtain

(15.16)
$$\begin{cases} \boldsymbol{t}'_k = \boldsymbol{t}_k - \dfrac{\tau_{ks}}{\tau_{rs}} \boldsymbol{t}_r & \text{with the exception of column } r, \\[2mm] \tau'_{k,r} = -\dfrac{\tau_{ks}}{\tau_{rs}}, & (k \neq s) \\[2mm] \boldsymbol{t}'_s = \dfrac{1}{\tau_{rs}} \boldsymbol{t}_r & \text{with the exception of column } r, \\[2mm] \tau'_{s,r} = \dfrac{1}{\tau_{rs}}. \end{cases}$$

Since $x_k^1 = x_k^0 - \frac{\tau_{ks}}{\tau_{rs}} x_r^0$ $(k \neq s)$, $x_s^1 = \frac{x_r^0}{\tau_{rs}}$, the formulas (15.16) hold as well for the $\boldsymbol{x}$-columns of the tableau.

Finally, we consider the $\boldsymbol{d}$-row in our tableau. We set $\boldsymbol{d} = (d_j) = \boldsymbol{z} - \boldsymbol{c}$, where we note that $\boldsymbol{d}$ consists only of the coordinates d_j $(j \neq Z)$. From (15.12), we have $\boldsymbol{z} = \sum_{k \in Z} c_k \boldsymbol{t}_k$, $\boldsymbol{z}' = \sum_{k \in Z'} c_k \boldsymbol{t}'_k$, and therefore with (15.16) we obtain

$$\boldsymbol{d}' = \sum_{k \in Z'} c_k \boldsymbol{t}'_k - \boldsymbol{c}' = \sum_{k \in Z \smallsetminus \{r\}} c_k \left(\boldsymbol{t}_k - \frac{\tau_{ks}}{\tau_{rs}} \boldsymbol{t}_r \right) + \frac{1}{\tau_{rs}} c_s \boldsymbol{t}_r - \boldsymbol{c}'$$

$$= \sum_{k \in Z} c_k \boldsymbol{t}_k - c_r \boldsymbol{t}_r - \frac{z_s}{\tau_{rs}} \boldsymbol{t}_r + c_r \boldsymbol{t}_r + \frac{c_s}{\tau_{rs}} \boldsymbol{t}_r - \boldsymbol{c}'$$

$$= \boldsymbol{z} - \frac{d_s}{\tau_{rs}} \boldsymbol{t}_r - \boldsymbol{c}',$$

whence

(15.17)
$$\begin{cases} \boldsymbol{d}' = \boldsymbol{d} - \dfrac{d_s}{\tau_{rs}} \boldsymbol{t}_r & \text{with the exception of the } r\text{th column}, \\[2mm] d'_r = -\dfrac{d_s}{\tau_{rs}}. \end{cases}$$

From (15.13) we see that $Q(\boldsymbol{x}^1) = Q(\boldsymbol{x}^0) - d_s \delta = Q(\boldsymbol{x}^0) - \frac{d_s}{\tau_{rs}} x_r^0$, and therefore (15.17) holds also for the $\boldsymbol{x}$-column.

To sum up, we obtain the new tableau from the old one by the following calculation:

(i) Divide row $\boldsymbol{t}_r$ by τ_{rs}.
(ii) Subtract every multiple of $\boldsymbol{t}_r$ from row $\boldsymbol{t}_k$ (respectively $\boldsymbol{d}$), which gives 0 in column s, to obtain $\boldsymbol{t}'_k$ (respectively $\boldsymbol{d}'$).

(iii) Replace column s by itself multiplied by $\left(-\frac{1}{\tau_{rs}}\right)$ and the pivot element τ_{rs} by $\frac{1}{\tau_{rs}}$.

Step (B) is thereby completely described.

We now turn to step (A) to determine an admissible solution.

Case 1: The original program is of type

$$A\boldsymbol{x} \le \boldsymbol{b}, \quad \boldsymbol{x} \ge \boldsymbol{0},$$

$$\boldsymbol{c}^T \boldsymbol{x} = \max,$$

with $\boldsymbol{b} \ge \boldsymbol{0}$. By introduction of the slack variables $\boldsymbol{z}$ we obtain

$$(A \mid E_m)\begin{pmatrix}\boldsymbol{x}\\\boldsymbol{z}\end{pmatrix} = \boldsymbol{b}, \quad \begin{pmatrix}\boldsymbol{x}\\\boldsymbol{z}\end{pmatrix} \ge \boldsymbol{0},$$

$$(-\boldsymbol{c}, \boldsymbol{0})^T \begin{pmatrix}\boldsymbol{x}\\\boldsymbol{z}\end{pmatrix} = \min.$$

The admissible solution $\boldsymbol{x}^0 = \begin{pmatrix}\boldsymbol{0}\\\boldsymbol{b}\end{pmatrix}$ is by Theorem 15.20 a vertex with unit vectors from E_m as basis. It then follows that $\tau_{kj} = a_{kj}$, $z_j = 0$ ($j = 1, \ldots, n$), and hence $d_j = c_j$ and $Q(\boldsymbol{x}^0) = 0$. The starting arrangement is therefore of the form

(15.18)

	1	j	n	
$n+1$	a_{11}	a_{1j}	a_{1n}	b_1
		$\vdots$		$\vdots$
$n+m$	a_{m1}	a_{mj}	a_{mn}	b_m
	c_1	c_j	c_n	0

Case 2: Suppose the program is already given in the form

(I)

$$A\boldsymbol{x} = \boldsymbol{b}, \quad \boldsymbol{x} \ge \boldsymbol{0},$$

$$\boldsymbol{c}^T \boldsymbol{x} = \min,$$

where we may assume $\boldsymbol{b} \ge \boldsymbol{0}$. (Otherwise, we multiply the rows by -1.) We then solve the program

(II)

$$(A \mid E_m)\begin{pmatrix}\boldsymbol{x}\\\boldsymbol{z}\end{pmatrix} = \boldsymbol{b}, \quad \begin{pmatrix}\boldsymbol{x}\\\boldsymbol{z}\end{pmatrix} \ge \boldsymbol{0},$$

$$\sum_{i=1}^{m} z_i = \min.$$

For this task we already know a starting vertex, namely $\begin{pmatrix}\boldsymbol{0}\\\boldsymbol{b}\end{pmatrix}$. Since the objective function of (II) is bounded from below, there exists an optimal solution $\begin{pmatrix}\boldsymbol{x}^*\\\boldsymbol{z}^*\end{pmatrix}$. If $\boldsymbol{z}^* \ne \boldsymbol{0}$ and therefore the value of (II) is positive, then (I) has no admissible solution, since every such solution filled in with zeros would also be a solution of (II) with value of (II) equal to zero. Conversely, if $\boldsymbol{z}^* = \boldsymbol{0}$, then $\boldsymbol{x}^*$ is a (possibly degenerate) vertex of (I) whose associated column set can be extended to a basis. The starting tableau for (II) is in this case

(15.19)

	1	j	n	
$n+1$	a_{11}	a_{1j}	a_{1n}	b_1
		$\vdots$		$\vdots$
$n+m$	a_{m1}	a_{mj}	a_{mn}	b_m
	$\sum_{i=1}^{m} a_{i1}$	$\sum_{i=1}^{m} a_{ij}$	$\sum_{i=1}^{m} a_{in}$	$\sum_{i=1}^{m} b_i$

In sum, we thus obtain the following algorithmic description of the simplex method:

1. Construct the starting tableau using (15.18) or (15.19).

2. Test d_j:
   (i) If $d_j \leq 0$ for all j, then the solution is optimal.
   (ii) If there exists $d_j > 0$ with $\tau_{kj} \leq 0$ for all k, then there is no optimal solution.
   (iii) Choose $s \notin Z$ with $d_s > 0$ and determine $r \in Z$ with $\tau_{rs} > 0$ and $\frac{x_r^0}{\tau_{rs}} \leq \frac{x_k^0}{\tau_{ks}}$ for all $k \in Z$ with $\tau_{ks} > 0$.

3. Exchange $\boldsymbol{a}^r$ and $\boldsymbol{a}^s$, create the new tableau with the help of formulas (15.16) and (15.17), and go to step 2.

Example 15.26. With the following program we see that the simplex method consists in running geometrically through the edges of the convex set M of admissible solutions until an optimal vertex is found:

$$\begin{aligned}
x_1 & & & \leq 2, \\
x_1 + & x_2 & + 2x_3 & \leq 4, \\
& 3x_2 & + 4x_3 & \leq 6, \\
& & \boldsymbol{x} & \geq \boldsymbol{0}, \\
Q(\boldsymbol{x}) = x_1 + & 2x_2 & + 4x_3 & = \max.
\end{aligned}$$

The set of admissible solutions has seven vertices. The following exchange steps correspond to running through $P_1 \longrightarrow P_2 \longrightarrow P_5 \longrightarrow P_7$; P_7 is optimal. Here we get stuck at the third step at the degenerate vertex P_5 and only exchange a basis. For clarity, we have written x_j, z_i instead of the indices. The pivot elements are indicated in boldface:

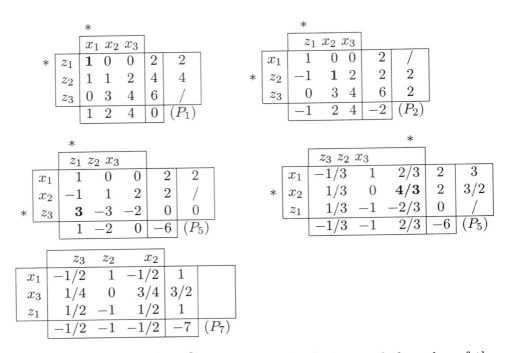

(P_1)

	x_1	x_2	x_3		
*z_1	**1**	0	0	2	2
z_2	1	1	2	4	4
z_3	0	3	4	6	/
	1	2	4	0	(P_1)

(P_2)

	z_1	x_2	x_3		
x_1	1	0	0	2	/
*z_2	−1	**1**	2	2	2
z_3	0	3	4	6	2
	−1	2	4	−2	(P_2)

(P_5)

	z_1	z_2	x_3		
x_1	1	0	0	2	2
x_2	−1	1	2	2	/
*z_3	**3**	−3	−2	0	0
	1	−2	0	−6	(P_5)

(P_5)

	z_3	z_2	x_3		
x_1	−1/3	1	2/3	2	3
*x_2	1/3	0	**4/3**	2	3/2
z_1	1/3	−1	−2/3	0	/
	−1/3	−1	2/3	−6	(P_5)

(P_7)

	z_3	z_2	x_2	
x_1	−1/2	1	−1/2	1
x_3	1/4	0	3/4	3/2
z_1	1/2	−1	1/2	1
	−1/2	−1	−1/2	−7

Therefore, $P_7 = \left(1, 0, \frac{3}{2}\right)$ is an optimal solution, and the value of the program is 7 (in that we return to the original maximum program).

With this we have answered our third question from Section 15.1 about determining solutions, but we have not performed a running-time analysis of the simplex algorithm. One can show that the simplex algorithm exhibits exponential running time. In practice, however, the algorithm is fast and is unsurpassed by other methods. Experience has shown the number of iterations to be generally linear with respect to the size of the input. This heuristic result has been verified by the fact that in a reasonable probabilistic setting, the simplex method exhibits polynomial running time on average.

In 1979, Khachiyan introduced an entirely different method, called the **ellipsoid method**, for linear programming, and this method actually runs in polynomial time with respect to the size of the input. Later, another polynomial algorithm was introduced by Karmarkar. With Kachiyan's result the problem of the complexity of linear programming was solved. Until then, linear programming over $\mathbb{Q}$ or $\mathbb{R}$ was known to be in the class NP, but one didn't know whether it was in P or perhaps was even NP complete. In contrast, integer linear programming has been shown to be NP complete. That is, the inputs A, $\boldsymbol{b}$, $\boldsymbol{c}$ are integral, and one asks whether an *integral* optimal solution exists. For most combinatorial applications (such as the Hamilton problem or the knapsack problem), one is of course looking for integer solutions. And here the problems are quite different.

15.6. Integer Linear Optimization

Let us return once more to our job-assignment problem. We are looking for an $n \times n$ matrix (x_{ij}), $x_{ij} = 1$ or 0, such that

(I)

$$\sum_{j=1}^{n} x_{ij} = 1 \quad (\forall i), \qquad \sum_{i=1}^{n} x_{ij} = 1 \quad (\forall j),$$

$$x_{ij} = 1 \text{ or } 0,$$

$$\sum w_{ij} x_{ij} = \max.$$

This is not a linear program as we know it, since the desired coordinates x_{ij} must be zero or one. In the canonical program

(II)

$$\sum_{j=1}^{n} x_{ij} = 1 \quad (\forall i), \qquad \sum_{i=1}^{n} x_{ij} = 1 \quad (\forall j),$$

$$x_{ij} \geq 0,$$

$$\sum w_{ij} x_{ij} = \max,$$

we know of course that an optimal solution $(\tilde{x}_{ij})$ exists, but we can determine a priori only that the $\tilde{x}_{ij}$ are rational numbers between zero and one. Nevertheless, we can assert that problems (I) and (II) have the same value. How so? The set of admissible solutions of (II) consists entirely of matrices (x_{ij}) with row and column sums equal to 1 and $x_{ij} \geq 0$, that is, doubly stochastic matrices. The theorem of Birkhoff and von Neumann from Section 8.2 tells us that every doubly stochastic matrix is a convex combination of permutation matrices (see Exercise 15.17). The vertices of the admissible set of (II) are therefore precisely the permutation matrices, and by Theorem 15.24, optimal solutions are always to be found among the vertices. But the permutation matrices are precisely those that are described by (I). Thus the simplex method will in fact produce a permutation matrix, that is, a $0, 1$ solution of the job-assignment problem.

To obtain $0, 1$ solutions to an integer optimization problem, we therefore proceed backward. We do not identify the vertices from the set M of admissible solutions, but the set M from the vertices. In general, the problem is as follows: Let S be a set and $\mathcal{U} \subseteq \mathcal{B}(S)$ a family of subsets over which we wish to optimize. We are given the cost function $c : S \longrightarrow \mathbb{R}$ with $c(B) = \sum_{a \in B} c(a)$, and we are seeking

$$U_0 \in \mathcal{U} \quad \text{with } c(U_0) = \min_{U \in \mathcal{U}} c(U)$$

(or analogously, $\max c(U)$).

We now make the problem an algebraic one in the following way: For $B \subseteq S$ we declare as usual the characteristic vector $\boldsymbol{x}_B = (x_{B,u})$ with

$$x_{B,u} = \begin{cases} 1 & \text{if } u \in B, \\ 0 & \text{if } u \notin B. \end{cases}$$

We then consider the closed, convex set M consisting of all convex combinations of $\boldsymbol{x}_U$, $U \in \mathcal{U}$, that is,

$$M = \sum_{U \in \mathcal{U}} \lambda_U \boldsymbol{x}_U, \quad \lambda_U \geq 0, \quad \sum_{U \in \mathcal{U}} \lambda_U = 1.$$

The vertices of the convex set M are *by construction* precisely the $0, 1$ vectors $\boldsymbol{x}_U$ ($U \in \mathcal{U}$). If we then interpret M as an admissible set of solutions, then we know from Theorem 15.24 that the objective function

$$\boldsymbol{c}^T \boldsymbol{x} = \sum_{a \in S} c_a x_a$$

assumes its minimum at a vertex $\boldsymbol{x}_U$, and this set U is the desired solution.

To be able to use the simplex algorithm, we must therefore represent the set M as the solution set of a system of inequalities $A\boldsymbol{x} \geq \boldsymbol{b}$, $\boldsymbol{x} \geq \boldsymbol{0}$. That such a description always exists is the content of a fundamental theorem of Weyl and Minkowski. The solution strategy thus reduces to the question *how* the system $A\boldsymbol{x} \geq \boldsymbol{b}$, $\boldsymbol{x} \geq \boldsymbol{0}$, can be determined. For our job-assignment problem we can give the answer with the help of the theorem of Birkhoff and von Neumann, but in general, we will face enormous difficulties. Many refined methods have been developed for special classes of $0, 1$ problems. Again, the traveling salesman problem (TSP) is a nice illustration.

We are given the cost matrix (c_{ij}) on the complete graph K_n. The underlying set S comprises the edges $\{ij : 1 \leq i, j \leq n\}$, and the distinguished sets $\mathcal{U}$ are the Hamiltonian circuits C. The convex set spanned by the vectors $\boldsymbol{x}_C$ is called the TSP-polytope M, that is,

$$M = \left\{ \sum_{C \in \mathcal{U}} \lambda_C \boldsymbol{x}_C : \lambda_C \geq 0, \sum \lambda_C = 1 \right\} \quad \text{with } \boldsymbol{c}^T \boldsymbol{x}_C = \min.$$

The bibliography will direct the reader to some of the methods by which the TSP-polytope M can be transformed into a linear program for not-too-large n. In general, however, such a rapid transformation is not to be expected, on account of the NP-completeness of TSP.

Let us return once more to the usual form $A\boldsymbol{x} \leq \boldsymbol{b}$, $\boldsymbol{x} \geq \boldsymbol{0}$. Suppose that A is an integral matrix and $\boldsymbol{b}$ an integral vector. When can we assert that all vertices of the admissible set M are integers, and therefore an optimal solution as well? In this case there is actually a complete characterization. A matrix $A \in \mathbb{Z}^{m \times n}$ is called **totally unimodular** if the determinant of every square submatrix has only the values 0, 1, and -1. The theorem of Hoffman and Kruskal now asserts that a matrix $A \in \mathbb{Z}^{m \times n}$ is totally unimodular if and only if every vertex of the admissible set $\{x : Ax \leq b, \ x \geq 0\}$ has integer coordinates, and this holds for every $b \in \mathbb{Z}^m$.

With this result we can finally derive the matching theorem $\max (|X| : X$ a matching$) = \min(|Y| : Y$ a vertex cover$)$ in a bipartite graph as a special case of the fundamental theorem of linear programming. In Section 15.2 we have already proved that the linear program $Ax \leq 1$, $x \geq 0$, $1^T x = \max$, always has an optimal solution, where A is the incidence matrix of the graph. But we can assert a priori only that among the optimal solutions there are rational vectors with $0 \leq x_i \leq 1$. However, it is easy to prove the unimodularity of A (see the exercises), and the matching theorem follows from the result of Hoffman and Kruskal. Furthermore, the incidence matrix of a graph G is unimodular *if and only if* G is bipartite. We have therefore found an algebraic confirmation that the matching problem for bipartite graphs is in a certain sense easier than for arbitrary graphs.

Exercises for Chapter 15

15.1 Construct a standard program that has an admissible solution but no optimal one.

▷ **15.2** Construct a standard program that has more than one optimal solution, but only finitely many.

15.3 Give an interpretation of the dual program of the job-assignment problem. That is, what is being minimized?

▷ **15.4** Solve the following program with the help of Theorems 15.5 and 15.13:

$$-2x_1 + x_2 \leq 2,$$
$$x_1 - 2x_2 \leq 2,$$
$$x_1 + x_2 \leq 5,$$
$$x \geq 0,$$

with $x_1 - x_2$ maximal.

15.5 Describe the linear program of the previous exercise geometrically.

15.6 The general job-assignment problem for m persons and n jobs is $\sum_{i=1}^{m} x_{ij} \leq 1 \ (\forall j)$, $\sum_{j=1}^{n} x_{ij} \leq 1 \ (\forall i)$, $x \geq 0$, $\sum_{i,j} w_{ij} x_{ij} = \max$. Solve the following problem with the matrix (w_{ij}) with the help of the dual program:

$$(w_{ij}) = \begin{pmatrix} 12 & 9 & 10 & 3 & 8 & 2 \\ 6 & 6 & 2 & 2 & 9 & 1 \\ 6 & 8 & 10 & 11 & 9 & 2 \\ 6 & 3 & 4 & 1 & 1 & 3 \\ 11 & 1 & 10 & 9 & 12 & 1 \end{pmatrix}.$$

Check the result using the method of Section 8.2.

▷ **15.7** Consider the following transport problem with cost matrix A, supply vector p, and demand vector q:

	M_1	M_2	M_3	M_4	p
F_1	4	4	9	3	3
F_2	3	5	8	8	5
F_3	2	6	5	7	7
q	2	5	4	4	

$A =$

That is, $\sum a_{ij} x_{ij}$ is to be minimized. Find an optimal solution that uses only five routes $F_i \to M_j$. Is five the minimal number for which this is possible? Hint: The value of the program in 65.

15.8 Solve the following program using the method presented at the end of Section 15.4:

$$4x_1 + 2x_2 + x_3 = 4,$$
$$x_1 + 3x_2 = 5,$$
$$x \geq 0,$$

with $2x_1 + 3x_2$ maximal. Verify the correctness of your solution with the help of the dual program.

▷ **15.9** Prove using Theorem 15.7 the following theorem from linear algebra: Let U be a subspace of $\mathbb{R}^n$ and $U^\perp = \{y : y^T x = 0 \text{ for all } x \in U\}$. Then $U^{\perp\perp} = U$ (see Section 13.4).

15.10 Let x^0 be an optimal vertex. Show that x^0 has a basis $\{a^k : k \in Z\}$ with $z_j \leq c_j$ ($j \notin Z$), where notation is as in Section 15.5.

▷ **15.11** Prove the following theorem. Exactly one of the two possibilities holds: (A) $Ax \leq b$ is solvable; (B) $A^T y = 0$, $y \geq 0$, $b^T y = -1$ is solvable. Hint: Use Theorem 15.8.

15.12 Use the previous exercise to show that the following program has no admissible solution:

$$4x_1 - 5x_2 \geq 2,$$
$$-2x_1 - 7x_2 \geq 2,$$
$$-x_1 + 3x_2 \geq -1.$$

▷ **15.13** Interpret the equilibrium theorem, Theorem 15.13, for the transportation problem. That is, why is it intuitively correct?

▷ **15.14** Describe the dual program to a canonical maximum program.

15.15 A vector $x \in \mathbb{R}^n$ is called *semipositive* if $x \geq 0$ and $x \neq 0$. Show that exactly one of the following two possibilities holds: (1) $Ax = 0$ has a semipositive solution; (2) $A^T y > 0$ is solvable.

15.16 Conclude from the previous exercise that either a subspace U contains a positive vector $a > 0$, or $U^\perp$ contains a semipositive vector. Illustrate this result in $\mathbb{R}^2$ and $\mathbb{R}^3$.

▷ **15.17** A vector x is called a convex combination of $x^1, \ldots, x^n$ if $x = \sum_{i=1}^{n} \lambda_i x^i$ with $\lambda_i \geq 0$ ($\forall i$) and $\sum_{i=1}^{n} \lambda_i = 1$. Let $K \subseteq \mathbb{R}^n$. Show that the smallest convex set that contains K (called the *convex hull*) consists of the convex combinations of vectors from K.

15.18 Given $Ax \leq b$, we know from Exercise 15.11 that this program has no solution exactly if (I) $A^T y = 0$, $b^T y = -1$, $y \geq 0$ is solvable. Now consider (II): $A^T y = 0$, $b^T y - \alpha = -1$, $y \geq 0$, $\alpha \geq 0$, $\alpha = \min$. Show that (II) has an optimal solution and that it has a positive value if and only if (I) is solvable. Let the value of (II) be positive and suppose that $(x, \beta)^T$ is a solution of the program dual to (I). Show that $-\frac{x}{\beta}$ is a solution of $Ax \leq b$.

15.19 Solve the following using the previous exercise:

$$5x_1 + 4x_2 - 7x_3 \leq 1,$$
$$-x_1 + 2x_2 - x_3 \leq -4,$$
$$-3x_1 - 2x_2 + 4x_3 \leq 3,$$
$$3x_1 - 2x_2 - 2x_3 \leq -7.$$

15.20 Solve the following program using the simplex method and illustrate it geometrically:

$$3x_1 + 4x_2 + x_3 \leq 25,$$
$$x_1 + 3x_2 + 3x_3 \leq 50,$$
$$x \geq 0,$$
$$8x_1 + 19x_2 + 7x_3 = \max.$$

Verify the solution using the dual program.

▷ **15.21** Given the standard program (I) $Ax \leq b$, $x \geq 0$, $c^T x = \max$, the following result shows how from the simplex tableau of an optimal solution of (I) one can read off at once an optimal solution of the dual problem (I*): If $\frac{t_k(k \in Z)}{d}$ is the tableau for an optimal solution of (I), then $y = (y_j)$ with

$$y_j = \begin{cases} -d_{n+j} & \text{if } n + j \notin Z, \\ 0 & \text{if } n + j \in Z, \end{cases}$$

is an optimal solution of (I*), where notation is as in Section 15.5.

15.22 Verify the previous exercise using Example 15.26.

15.23 Let $a, b \in \mathbb{R}^n$. We define the lexicographic order $a \prec b$ by $a_i < b_i$, where i is the first index such that $a_i \neq b_i$. Show that $a \prec b$ is a linear order on $\mathbb{R}^n$.

15.24 Given the program (I) $Ax = b$, $x \geq 0$, $c^T x = \min$ with notation as in Section 15.5, we give an additional rule that allows for handling degenerate vertices x^0 with $\min x_k^0 / \tau_{ks} = 0$. If $r, r', \ldots$ are the indices k with $x_k^0 / \tau_{ks} = 0$, we choose that r for which the vector $w_r = (x_r^0/\tau_{rs}, \tau_{r1}/\tau_{rs}, \ldots, \tau_{rm}/\tau_{rs})$ is as small as possible lexicographically. Show that (a) $w_r \neq w_{r'}$ for $r \neq r'$; (b) the vector $(Q(x^0), z_1, \ldots, z_m)$ is transformed by exchange into the lexicographically smaller vector $(Q(x^1), z_1', \ldots, z_m')$; (c) no basis is obtained more than once from an exchange.

▷ **15.25** Show that the incidence matrix of a bipartite graph is unimodular. Hint: If $\det A \neq 0$, then there must exist a diagonal of 1's.

Bibliography for Part 3

Boolean algebras appear as a standard topic in almost every textbook on discrete mathematics, for example in the book of Korfhage. Those wishing to delve more deeply into the theory of logical nets should consult the books of Savage and Wegener. Overviews of the field of hypergraphs are given by Berge and Bollobás. In Berge's book, the analogy with graphs is in the foreground, while Bollobás is slanted more toward the point of view of families of sets.

A lovely little book on combinatorial designs is that of Ryser. Recommended as well are the relevant chapters in the books of Biggs and Cameron. The standard text on coding theory is the book of MacWilliams and Sloane. A fine summary of information theory and coding theory can be found in McEliece. A good introduction is also offered by Welsh.

One may obtain exhaustive information on finite fields from the book of Lidl and Niederreiter. The literature on linear programming is enormous, and as representative of many books, one may recommend the book of Schrijver.

C. BERGE: *Graphs and Hypergraphs.* North Holland.

N. BIGGS: *Discrete Mathematics.* Oxford Science Publications.

B. BOLLOBÁS: *Combinatorics, Set Systems, Hypergraphs, Families of Vectors and Combinatorial Probability.* Cambridge University Press.

P. J. CAMERON: *Combinatorics, Topics, Techniques, Algorithms.* Cambridge University Press.

R. KORFHAGE: *Discrete Computational Structures.* Academic Press.

R. LIDL, H. NIEDERREITER: "Finite Fields." *Encyclopaedia of Mathematics*, volume 20. Cambridge University Press.

F. MACWILLIAMS, N. SLOANE: *The Theory of Error-Correcting Codes.* North Holland.

R. MCELIECE: *The Theory of Information and Coding.* Addison-Wesley.

H. RYSER: *Combinatorial Mathematics.* Carus Math. Monographs.

J. SAVAGE: *The Complexity of Computing.* J. Wiley & Sons.

A. SCHRIJVER: *Theory of Linear and Integer Programming.* Wiley Publications.

I. WEGENER: *The Complexity of Boolean Functions.* Teubner-Wiley.

D. WELSH: *Codes and Cryptography.* Oxford Science Publications.

Solutions to Selected Exercises

1.3 Assume that $A \subseteq N$ is a counterexample. We consider the incidence system $(A, N \smallsetminus A, I)$ with $aIb \Leftrightarrow |a - b| = 9$. For $10 \le a \le 91$ we have $r(a) = 2$; otherwise, $r(a) = 1$. Therefore, $\sum_{a \in A} r(a) \ge 92$, and on the other hand, $\sum_{b \in N \smallsetminus A} r(b) \le 90$, a contradiction. For $|A| = 54$ we take six blocks of nine consecutive numbers, with each two neighboring blocks nine apart.

1.5 Each party has between 1 and 75 seats. If the first party has i seats, then the possibilities for the second party are $76 - i, \ldots, 75$. The total number is therefore $\sum_{i=1}^{75} i = \binom{76}{2}$.

1.9 By induction, $M(i) \le M(j)$ for $1 \le i \le j \le n$. Let $2 \le k \le M(n)$. Then $S_{n+1,k} - S_{n+1,k-1} = (S_{n,k-1} - S_{n,k-2}) + k(S_{n,k} - S_{n,k-1}) + S_{n,k-1} > 0$ by induction. Likewise, one concludes that $S_{n+1,k} - S_{n+1,k+1} > 0$ for $k \ge M(n) + 1$.

1.10 For a given n let m be the number with $m! \le n < (m+1)!$ and a_m maximal with $0 \le n - a_m m!$. Hence $1 \le a_m \le m$ and $n - a_m m! < m!$. The existence and uniqueness follow by induction on m.

1.13 The trick is to consider the $n - k$ *missing* numbers. If we write them as vertical bars, we can place the k elements in the $n - k + 1$ intermediate spaces, and then read off the k-set from left to right. For part b, one shows that $\sum_k f_{n,k}$ satisfies the Fibonacci recurrence.

1.16 There are 12 ways for the first and last cards to be queens, and $50!$ ways for the remainder. The probability is therefore $12 \cdot 50!/52! = \frac{1}{221}$.

1.19 From $n + 1$ numbers, two must be adjacent, and therefore they are relatively prime. For the second question, write each integer in the form $2^k m$, m odd, $1 \le m \le 2n - 1$. There must then be two integers in a single class $\{2^k m : k \ge 0\}$. Now $\{2, 4, \ldots, 2n\}$ and $\{n + 1, n + 2, \ldots, 2n\}$ show that both assertions for n are false.

1.21 Let (k, ℓ) be the greatest common divisor of k and ℓ. For $d \mid n$ let $S_d = \{k\frac{n}{d} : (k, d) = 1, \ 1 \le k \le d\}$, that is, $|S_d| = \varphi(d)$. From $k\frac{n}{d} = k'\frac{n}{d'}$ follows $kd' = k'd$, whence $k = k', d = d'$. The sets S_d are therefore pairwise disjoint. Conversely let $m \le n, (m, n) = \frac{n}{d}$, and so $m \in S_d$, and we obtain $S = \sum_{d \mid n} S_d$.

1.23 By taking away n numbers, N decomposes into subintervals of lengths $a_0, b_1,$ $a_1, \ldots, b_s, a_s, b_{s+1}$, where the a_i denote the intervals of the remaining numbers, and b_j those of the numbers removed, and possibly $a_0 = 0, b_{s+1} = 0$. From $\Sigma b_j = n$ we have $s \le n$. To obtain the alternating property, at most one element from $a_1, \ldots, a_s$ must be removed. This yields $a_0 + \sum_{i=1}^{s}(a_i - 1) = 2n - s \ge n$ numbers with the desired properties.

1.25 Let r be as in the hint. Then $r \le \frac{n^2}{2} < n(n-1)$ on account of $n \ge 4$. The number of ones in the incidence matrix counted columnwise is therefore $(n-2)!r < n!$. There must therefore be a row of all zeros.

1.29 Let $f(n, i)$ be the number with i at the beginning. The following number is then $i + 1$ or $i - 1$ with the counts f_+, f_-. If we identify i with $i + 1$ and reduce all numbers greater than $i + 1$ by 1, we see that $f_+ = f(n-1, i)$, and analogously, $f_- = f(n-1, i-1)$. The numbers $f_{n,i}$ thus satisfy the binomial recurrence with $f(n, 1) = 1$. Therefore, $f(n, i) = \binom{n-1}{i-1}, \sum_i f(n, i) = 2^{n-1}$.

1.32 Clearly, $0 \le b_j \le n - j$. Since there are at most $n!$ sequences $(b_1, \ldots, b_n)$, it remains only to show that for a sequence $(b_1, \ldots, b_n)$ there is a permutation π. The permutation π is constructed back to front. If $b_{n-1} = 1$, then $n - 1$ comes after n, otherwise before n. Now $n - 2$ is placed according to b_{n-2}, and so on.

1.36 If we insert n into a permutation π of $\{1, \ldots, n-1\}$, we raise the number of rises by one or zero. If π has $k - 1$ rises, we increase π by one precisely when n is inserted into one of the $n - k - 1$ nonrises or at the end. A permutation π with k rises is then not increased precisely when n is inserted into one of the k rises or at the beginning.

1.39 Let R and R' be two column sets with $|R| = |R'| = r$. It is easy to construct a bijection between all paths with the r diagonals in R or in R'. We may therefore restrict to the first r columns. The paths are then uniquely determined by the r diagonals in the first r columns. If we replace the diagonals by horizontal rows, we obtain a bijection to the paths in the $n \times (n-r)$ lattice, and so for the total number we obtain $\sum_{r=0}^{n} \binom{n}{r}\binom{2n-r}{n}$.

1.41 We consider in the hint the paths from $(0, 0)$ to $(a + b, a - b)$. The permitted paths must use $(1, 1)$ as the first point. Altogether, there are $\binom{a+b-1}{a-1}$ paths from $(1, 1)$ to $(a + b, a - b)$, since they are uniquely determined by the positions of the $a - 1$ votes for A. We must now remove all paths W that intersect the x-axis. For such a W we construct a path W' from $(1, -1)$ to $(a + b, a - b)$ by reflecting all segments between points with $x = 0$ below the x-axis and leave them unchanged at the end. One sees at once that $W \to W'$ is a bijection of all paths from $(1, -1)$ to $(a + b, a - b)$, and we obtain the result $\binom{a+b-1}{a-1} - \binom{a+b-1}{a} = \frac{a-b}{a+b}\binom{a+b}{a}$.

1.43 Let n be even. From $\binom{n}{k} = \frac{n}{k}\binom{n-1}{k-1}$ it follows that $a_n = 1 + \sum_{k=1}^{n/2}[\binom{n}{k}^{-1} + \binom{n}{n-k+1}^{-1}] = 1 + \frac{1}{n}\sum_{k=1}^{n/2}[k\binom{n-1}{k-1}^{-1} + (n-k+1)\binom{n-1}{n-k}^{-1}]$ and with $\binom{n-1}{k-1} = \binom{n-1}{n-k}$, we have $a_n = 1 + \frac{n+1}{n} \sum_{k=1}^{n/2} \binom{n-1}{k-1}^{-1} = 1 + \frac{n+1}{2n} a_{n-1}$. The case of n odd proceeds analogously. For the first values we have $a_0 = 1, a_1 = 2, a_2 = \frac{5}{2}, a_3 = a_4 = \frac{8}{3} > a_5 = \frac{13}{5}$. If $a_{n-1} > 2 + \frac{2}{n-1}$, then $a_n > \frac{n+1}{2n}(2 + \frac{2}{n-1}) + 1 = 2 + \frac{1}{n}(1 + \frac{n+1}{n-1}) > 2 + \frac{2}{n}$, or $\frac{n}{2n+2} a_n > 1$. For $a_{n+1} - a_n$ we therefore have $a_{n+1} - a_n = (\frac{n+2}{2n+2} - 1)a_n + 1 = -\frac{n}{2n+2} a_n + 1 < 0$, whence $a_{n+1} < a_n$ for $n \geq 4$ by induction. The limit therefore exists, and is 2.

1.45 If $k = 2m > 2$, then 1 is in row m and column $2m$ without a circle. Therefore let k be odd. Assume that $k = p > 3$ is prime. The rows n that yield entries in column p satisfy $2n \leq p \leq 3n$, and the entry in column p is $\binom{n}{p-2n}$. From $\frac{p}{3} \leq n \leq \frac{p}{2}$ we have $1 < n < p$, and so n and p are relatively prime, and so n and $p - 2n$ as well. We now have $\binom{n}{p-2n} = \frac{n}{p-2n}\binom{n-1}{p-2k-1}$ or $(p - 2n)\binom{n}{p-2n} = n\binom{n-1}{p-2n-1}$, and it follows that $n \mid \binom{n}{p-2n}$, that is, every entry in column p is circled. Let $k = p(2m+1)$ be a composite odd number, p prime, $m \geq 1$. Then $n = pm$ satisfies the conditions $2n \leq k \leq 3n$, and one easily sees that $\binom{n}{k-2n} = \binom{pm}{p}$ is not a multiple of n.

1.46 Let the distribution be $(p_1, \ldots, p_6)$. Then $1 = (\sum p_i)^2 \leq 6 \sum p_i^2$ (prove directly or see Exercise 2.22).

1.49 For $i \in R$ let X_i be the random variable with $X_i = 1$ (rabbit lives), $X_i = 0$ (rabbit is shot). We have $EX_i = (\frac{r-1}{r})^n$, and thus $EX = (1 - \frac{1}{r})^n r$. From Markov's inequality we have $p(X \geq 1) \leq (1 - \frac{1}{r})^n r$. From $\log x \leq x - 1$ we have $\log(1 - \frac{1}{r})^n \leq -\frac{n}{r}$ and therefore $p(X \geq 1) \leq r e^{-n/r}$. Inserting $n \geq r(\log r + 5)$ yields the result $p(X \geq 1) \leq e^{-5} < 0.01$.

1.51 Let $X : \pi \to \mathbb{N}$ be the random variable with $X = k$ if the second ace comes at the kth card, whence $p(X = k) = (k-1)(n-k)/\binom{n}{3}$ with $\sum_{k=2}^{n-1}(k-1)(n-k) = \binom{n}{3}$. Let $S = \binom{n}{3}EX = \sum_{k=2}^{n-1} k(k-1)(n-k)$. Replacing the running index k with $n+1-k$, we obtain $2S = (n+1)\sum_{k=2}^{n-1}(k-1)(n-k)$, whence $EX = \frac{n+1}{2}$.

1.53 We may assume that x lies between 0 and 1. Let $kx = q + r_k$ with $q \in \mathbb{N}, 0 < r_k < 1$. We classify the integers kx by their remainders r_k. If $r_k \leq \frac{1}{n}$ or $r_k \geq \frac{n-1}{n}$, we are done. The $n - 1$ remainders therefore fall into one of the $n - 2$ classes $\frac{1}{n} < r \leq \frac{2}{n}, \ldots, \frac{n-2}{n} < r \leq \frac{n-1}{n}$. There thus exist $kx, \ell x$ ($k < \ell$) with $|r_\ell - r_k| < \frac{1}{n}$, and $(\ell - k)x$ is the desired number.

1.55 Let N be an n-set with $n = R(k-1, \ell) + R(k, \ell-1) - 1$. The only situation that causes difficulty is that in which every element forms with exactly $R(k - 1, \ell) - 1$ a red pair and with exactly $R(k, \ell - 1) - 1$ a blue pair. However, since $R(k - 1, \ell) - 1$ and $R(k, \ell - 1) - 1$ are both odd, double counting shows that this is impossible. For $R(3, 4)$ we obtain $R(3, 4) \leq 9$. The coloring on $N = \{0, 1, \ldots, 7\}$ with $\{i, j\}$ red $\iff |j - i| = 1$ or 4 $(\mod 8)$ does not have the Ramsey property, and so we have $R(3, 4) = 9$.

1.57 Let A be the event as in the hint. Then $p(A) = 2^{-\binom{k}{2}}$, that is, $\bigcup_{|A|=k} p(A) \leq \binom{n}{k} 2^{-\binom{k}{2}}$. From $\binom{n}{k} < n^k / 2^{k-1}$ follows $\binom{n}{k} < 2^{\frac{k^2}{2} - k + 1}$ with $n < 2^{k/2}$, that is, $\bigcup_{|A|=k} p(A) < 2^{-\frac{k}{2}+1} \leq \frac{1}{2}$ on account of $k \geq 4$. By symmetry we also have $p(\bigcup_{|B|=k} B) < \frac{1}{2}$, where B is the event that all persons in B are pairwise unacquainted. Therefore $R(k, k) \geq 2^{k/2}$.

1.59 Let s be as in the hint. We call a_i, $i > s$, a candidate if $a_i > a_j$ for all $j \le s$. The strategy of declaring a_i as maximum is successful if (A) a_i is maximal and (B) a_j, $s < j < i$, are not candidates. The probability for (A) is $\frac{1}{n}$ and for (B), $\frac{s}{i-1}$, and so the strategy for a_i is successful with probability $\frac{1}{n}\frac{s}{i-1}$. Summation over $i = s+1, s+2, \ldots$ yields the probability of winning $p(s) = \sum_{i=s+1}^{n} \frac{s}{n}\frac{1}{i-1} = \frac{s}{n}(H_{n-1} - H_{s-1}) \sim \frac{s}{n}\log\frac{n}{s}$. Maximizing $f(x) = \frac{1}{x}\log x$ yields $x = e$ and so $f(x_{\max}) \sim \frac{1}{e} \sim 0.37$. Rounding $x_{\max}$ to the nearest rational number $\frac{n}{s}$ changes the result by an insignificant amount.

2.1 Multiplication by $2^{n-1}/n!$ yields with $S_n = 2^n T_n/n!$ the recurrence $S_n = S_{n-1} + 3 \cdot 2^{n-1} = 3(2^n - 1) + S_0$. Hence the solution is $T_n = 3 \cdot n!$.

2.4 Let T_n be the nth number. We have $T_0 = 1$, $T_1 = 0$, and by the rule, $T_n = nT_{n-1} + (-1)^n$. This is precisely the recurrence for the derangement numbers.

2.6 With partial summation we have $\sum_{k=1}^{n-1} \frac{H_k}{(k+1)(k+2)} = \sum_{1}^{n} H_x x^{-2} = -x^{-1}H_x|_1^n + \sum_{1}^{n}(x+1)^{-1}x^{-1} = -x^{-1}H_x|_1^n + \sum_{1}^{n}x^{-2} = -x^{-1}(H_x+1)|_1^n = 1 - \frac{H_{n+1}}{n+1}$.

2.9 What is to be proved is $x^{\overline{n}} = \sum_{k=0}^{n} \frac{n!}{k!}\binom{n-1}{k-1}x^{\underline{k}}$ or $\binom{x+n-1}{n} = \sum_{k=0}^{n}\binom{n-1}{n-k}\binom{x}{k}$, but this is precisely the Vandermonde formula. The inversion formula now follows by $x \to -x$.

2.13 Let S be the set of all r-subsets, and E_i the property that i *is not* in A, for $i \in M$. Then $N(E_{i_1} \ldots E_{i_k}) = \binom{n-k}{r}$, and the formula follows.

2.17 Let p_H be the probability of not obtaining a heart. Then $p_H = \binom{39}{13}/\binom{52}{13}$, and similarly for the other suits. The probability p_{HD} of no hearts or diamonds is $p_{HD} = \binom{26}{13}/\binom{52}{13}$, and finally, $p_{HDS} = 1/\binom{52}{13}$ without heart, diamond, spade. Inclusion–exclusion takes care of the rest. Part (b) proceeds analogously.

2.18 We can move the nth disk only if the $(n-1)$-tower is in the proper order on B. This yields the recurrence $T_n = 2T_{n-1} + 1, T_1 = 1$, that is, $T_n = 2^n - 1$. In part (b) we obtain $S_n = 3S_{n-1} + 2, S_1 = 2$, that is, $S_n = 3^n - 1$. The recurrence for R_n is $R_n = 2R_{n-1} + 2, R_1 = 2$, that is, $R_n = 2^{n+1} - 2 = 2T_n$.

2.20 Let $2n$ be given. In the first round, $2, 4, \ldots, 2n$ are eliminated. The next round begins again with 1, and by the transformation $1, 2, 3, \ldots, n \to 1, 3, 5, \ldots, 2n-1$, we obtain $J(2n) = 2J(n) - 1$. The recurrence for $J(2n+1)$ is proved analogously. Let $n = 2^m + \ell$, $0 \le \ell < 2^m$. Then by induction on m, we obtain $J(n) = 2\ell + 1$.

2.22 We have $S = \sum_{j<k}(a_k - a_j)(b_k - b_j) = \sum_{k<j}(a_k - a_j)(b_k - b_j)$, whence $2S = \sum_{j,k}(a_k - a_j)(b_k - b_j) = \sum_{j,k}a_kb_k - \sum_{j,k}a_kb_j - \sum_{j,k}a_jb_k + \sum_{j,k}a_jb_j = 2n\sum_{k=1}^{n}a_kb_k - 2(\sum_{k=1}^{n}a_k)(\sum_{k=1}^{n}b_k)$. Therefore $(\sum_k a_k)(\sum_k b_k) = n\sum_k a_kb_k - S$. If $a_1 \le \cdots \le a_n$, $b_1 \le \cdots \le b_n$, then $S \ge 0$.

2.24 Induction on n. We consider the farmhouse H_1 and draw a line through it on which no further farmhouse or well lies. By rotating this line it can always be attained that in the half-plane that thus arises there are as many farmhouses as wells, and indeed at least 2 and at most $2(n-1)$.

2.26 Call a line segment $\overline{PQ}$ of length d a diameter. Suppose that three diameters, $\overline{PA}, \overline{PB}, \overline{PC}$, go out from P. The points A, B, C lie on a circle about P of radius d, and since d is maximal, on an arc of length $\le d\frac{\pi}{3}$. Let B lie between A and C

on this arc. Then no additional diameters go from B outside of $\overline{BP}$. We thus have two possibilities. Either exactly two diameters emerge from each point, or there exists a point from which at most one diameter emerges. In the first case we are done with double counting, and in the second by induction.

2.29 The equation follows from (2.8) and (2.12). In part (b), set $f(x) = \binom{n+x}{n}$.

2.30 If $n = k + (k+1) + \cdots + (\ell-1)$, then as in the hint, we have $2n = (\ell-k) \times (\ell+k-1)$. One of the two factors is even, the other odd. If $2n = xy$ is a decomposition, we obtain with $k = \frac{1}{2}|x-y| + \frac{1}{2}$, $\ell = \frac{1}{2}(x+y) + \frac{1}{2}$ a decomposition $2n = (\ell-k)(\ell+k-1)$. The desired number is therefore equal to the number of odd divisors of n, that is, $\prod(k_i + 1)$ with $n = 2^m \prod_{p_i > 2} p_i^{k_i}$ by Exercise 1.2.

2.33 We have $(-1)^{k+1}\binom{n}{k+1}^{-1} - (-1)^k\binom{n}{k}^{-1} = (-1)^{k+1}\binom{n-1}{k}^{-1}[\frac{k+1}{n} + \frac{n-k}{n}] = (-1)^{k+1}\binom{n-1}{k}^{-1}\frac{n+1}{n}$, whence $\sum(-1)^x\binom{n}{x}^{-1} = \frac{n+1}{n+2}(-1)^{x-1}\binom{n+1}{x}^{-1}$. By summation we obtain $\sum_{k=0}^{n}(-1)^k\binom{n}{k}^{-1} = \frac{n+1}{n+2}[(-1)^n + 1] = 2\frac{n+1}{n+2}$ [$n =$ even].

2.37 We have $\sum_a^b x^m = S_m(b) - S_m(a)$. From $\sum_{k=-n+1}^{-1} k^m = (-1)^m \sum_{k=1}^{n-1} k^m = (-1)^m \cdot S_m(n)$ follows

$$(-1)^m S_m(n) = \sum_{-n+1}^{0} x^m = S_m(0) - S_m(1-n) = -S_m(1-n),$$

whence assertion (a). From this follows $(-1)^m S_m(1) = 0$ and $(-1)^m S_m(\frac{1}{2}) = -S_m(\frac{1}{2})$, whence the other assertions.

2.39 From $b_k = k!a_k$ we obtain $n! = \sum_{k=0}^{n}\binom{n}{k}b_k$, and from this by binomial inversion, $b_n = D_n$ (derangement number), whence $a_n = \frac{D_n}{n!}$.

2.42 Let $p_3(n)$ be the first number and $p_{\leq 2}(n)$ the second. In the first case, let E_i be the property that $3i$ appears as a summand. With inclusion–exclusion we obtain $p_3(n) = p(n) - p(n-3) - p(n-6) - \cdots + p(n-3-6) + p(n-3-9) + \cdots$. In the second case let E_i be the property that the summand i appears three times, and the same alternating sum results. In general, we have $p_d(n) = p_{\leq d-1}(n)$.

2.45 If E_i is the property that i is a fixed point, then $N(E_{i_1} \ldots E_{i_k}) = (n-k)!$ for $k \geq t$, whence the desired number is

$$F_t = \sum_{i=0}^{n-t}(-1)^i\binom{t+i}{i}\binom{n}{t+i}(n-t-i)! = \frac{n!}{t!}\sum_{i=0}^{n-t}\frac{(-1)^i}{i!} = \binom{n}{t}D_{n-t}.$$

The equation $F_t = \binom{n}{t}D_{n-t}$ is of course also immediately clear.

3.2 We have $f(0) = 1$, $f(1) = 2$. Let A be fat. If $n \notin A$, then A is fat in $\{1, \ldots, n-1\}$. If $n \in A$, then $\{i-1 : i \in A \setminus n\}$ is fat in $\{1, \ldots, n-2\}$ or $\varnothing$ if $A = \{n\}$. Thus $f(n) = f(n-1) + f(n-2)$, that is, $f(n) = F_{n+2}$. The fat k-sets of $\{1, \ldots, n-1\}$ are precisely the k-subsets of $\{k, \ldots, n-1\}$, whence $F_{n+1} = \sum_k \binom{n-k}{k}$. Part (c) is by part (a) equivalent to $(n+1) + \sum_{k=1}^{n}\binom{n+1}{n-k}f(k-1) = f(2n)$. Let $A = \{1, \ldots, n-1\}$, $B = \{n, \ldots, 2n\}$. Then part (c) follows by classification of the fat sets X with $|X \cap B| = n - k$.

3.5   We have $L_n F_n = F_{n-1} F_n + F_n F_{n+1}$. On the other hand, $F_{2n} = F_{n-1} F_n + F_n F_{n+1}$ by repeated application of the Fibonacci recurrence (see Exercise 3.33). Inserting into $F_n = \frac{1}{\sqrt{5}}(\phi^n - \hat{\phi}^n)$ yields $L_n = \phi^n + \hat{\phi}^n$.

3.9   Consider the first number. If it is 1 or 2, we can combine it with $f(n-1)$ words of length $n-1$. If it is 0, then the second number must be 1 or 2, and we can choose the remainder in $f(n-2)$ ways. The recurrence is therefore $f(n) = 2(f(n-1) + f(n-2)) + [n=0] + [n=1]$ with $f(0) = 1$. For the generating function $F(z)$ this yields $F(z) = \frac{1+z}{1-2z-2z^2}$, and with the usual methods, $f(n) = \frac{3+2\sqrt{3}}{6}(1+\sqrt{3})^n + \frac{3-2\sqrt{3}}{6}(1-\sqrt{3})^n$.

3.12   We have $a_k = b_n = n^{\underline{k}}$, whence $\widehat{A}(z) = \sum_k \binom{n}{k} z^k = (1+z)^n$ and $\widehat{B}(z) = \sum_{n \geq k} \frac{n^{\underline{k}}}{n!} z^n = \sum_{n \geq k} \frac{z^n}{(n-k)!} = z^k \sum_{n \geq k} \frac{z^{n-k}}{(n-k)!} = z^k e^z$.

3.14   We use the recurrence $S_{n+1,m+1} = \sum_k \binom{n}{k} S_{k,m}$ from Exercise 1.35. Therefore $e^z \widehat{S}_m(z) = \sum_{n \geq 0} (\sum_k \binom{n}{k} S_{k,m}) \frac{z^n}{n!} = \sum_{n \geq 0} S_{n+1,m+1} \frac{z^n}{n!}$, whence $\widehat{S}'_{m+1}(z) = e^z \widehat{S}_m(z)$. With $\widehat{S}_0(z) = 1$ follows $\widehat{S}'_1(z) = e^z$, whence $\widehat{S}_1(z) = e^z - 1$ (since $\widehat{S}_1(0) = 0$). By induction we obtain $\widehat{S}_m(z) = \frac{(e^z-1)^m}{m!}$, $\sum_{m \geq 0} \widehat{S}_m(z) t^m = e^{t(e^z-1)}$.

3.16   We have $\sum_{n \geq 0} F_{2n} z^{2n} = \frac{1}{2}(F(z) + F(-z)) = \frac{1}{2}(\frac{z}{1-z-z^2} + \frac{-z}{1+z-z^2}) = \frac{z^2}{1-3z^2+z^4}$ by the previous exercise, whence $\sum_n F_{2n} z^n = \frac{z}{1-3z+z^2}$.

3.17   Since $G(z) - 1$ is the convolution of $G(z)$ and $\sum_{n \geq 1} n z^n = \frac{z}{(1-z)^2}$, we have $G(z) = \frac{1-2z+z^2}{1-3z+z^2} = 1 + \frac{z}{1-3z+z^2}$, and by the previous exercise, $g_n = F_{2n}$ for $n \geq 1$.

3.20   Fix x. Then $F(z)^x = \exp(x \log F(z)) = \exp(x \log(1 + \sum_{k \geq 1} a_k z^k)) = \exp(x \sum_{\ell \geq 1} \frac{(-1)^{\ell+1}}{\ell} (\sum_{k \geq 1} a_k z^k)^\ell) = \sum_{m \geq 0} \frac{x^m}{m!} (\sum_{\ell \geq 1} \frac{(-1)^{\ell+1}}{\ell} (\sum_{k \geq 1} a_k z^k)^\ell)^m$. Therefore $[z^n] F(z)^x$ is a polynomial $p_n(x)$ of degree n with $p_n(0) = 0$ for $n > 0$. The first convolution follows from $F(z)^x F(z)^y = F(z)^{x+y}$, the second by comparing the coefficient of z^{n-1} in $F'(z) F(z)^{x-1} F(z)^y = F'(z) F(z)^{x+y-1}$, since $F'(z) F(z)^{x-1} = x^{-1} \frac{\partial}{\partial z} F(z)^x = x^{-1} \sum_{n \geq 0} n \, p_n(x) z^{n-1}$.

3.24   We have $EX = \sum_{n \geq 1} n \, p_n$, $P'_X(z) = \sum n \, p_n z^{n-1}$, whence $EX = P'_X(1)$. We have $VX = EX^2 - (EX)^2 = \sum n^2 p_n - (\sum n \, p_n)^2 = \sum n(n-1) p_n + \sum n \, p_n - (\sum n \, p_n)^2$, whence $VX = P''_X(1) + P'_X(1) - (P'_X(1))^2$. Part (c) is clear.

3.26   Let k be the number of heads, $0 \leq k \leq n$. Then $p(X = k) = \binom{n}{k} 2^{-n}$, whence $P_X(z) = 2^{-n} \sum_k \binom{n}{k} = (\frac{1+z}{2})^n$. From Exercise 3.24 follows $EX = \frac{n}{2}$, $VX = \frac{n}{4}$.

3.28   We have $p(X_n = k) = \frac{I_{n,k}}{n!}$. By classifying according to the position of n we obtain $I_{n,k} = I_{n-1,k} + \cdots + I_{n-1,k-n+1}$, whence $P_n(z) = \frac{1+z+\cdots+z^{n-1}}{n} P_{n-1}(z)$, whence

$$P_n(z) = \prod_{i=1}^n \frac{1 + z + \cdots + z^{i-1}}{i},$$

since $P_1(z) = 1$. With

$$P'_n(z) = \sum_{i=1}^n \left(\prod_{j \neq i} \frac{1 + z + \cdots + z^{j-1}}{j} \right) \cdot \frac{1 + 2z + \cdots + (i-1)z^{i-2}}{i}$$

one calculates $EX = P_X'(1) = \sum_{i=1}^{n} \frac{i-1}{2} = \frac{n(n-1)}{4}$. Analogously for VX by Exercise 3.24.

3.29 Consider the left-hand 3×1 edge. Either it contains a vertical domino (above or below), in which case we may continue with B_{n-1} possibilities, or all three dominos are horizontal, and we continue in A_{n-2} ways. The recurrence is therefore $A_n = 2B_{n-1} + A_{n-2} + [n = 0]$, and by analogous reasoning, $B_n = A_{n-1} + B_{n-2}$. We thus obtain by elimination of $B(z)$, $A(z) = \frac{1-z^2}{1-4z^2+z^4}$. We see that $A_{2n+1} = 0$, which of course is clear. We treat $\frac{1-z}{1-4z+z^2}$ as usual, with the result $A_{2n} = \frac{(2+\sqrt{3})^n}{3-\sqrt{3}} + \frac{(2-\sqrt{3})^n}{3+\sqrt{3}}$.

3.31 As in Exercise 3.22, we obtain $A(z) = \frac{1}{(1-z)(1-z^2)(1-z^4)\cdots}$, whence $A(z^2) = (1-z)A(z)$. For $B(z)$ it follows that $B(z) = \frac{A(z)}{1-z}$, that is, $B(z^2) = \frac{A(z^2)}{1-z^2} = \frac{A(z)}{1+z}$. Therefore, $A(z) = (1+z)B(z^2)$, that is, $a_{2n} = a_{2n+1} = b_n$.

3.34 We know that $\sum_k \binom{n}{k} w^k = (1+w)^n$, so by Exercise 3.15, $\sum_k \binom{n}{2k+1} w^{2k+1} = \frac{(1+w)^n - (1-w)^n}{2}$. If we factor out w and set $w^2 = z$, we obtain $\sqrt{z} \sum_k \binom{n}{2k+1} z^k = \frac{(1+\sqrt{z})^n - (1-\sqrt{z})^n}{2}$, and with $z = 5$, $\sum_k \binom{n}{2k+1} 5^k = 2^{n-1} F_n$.

3.38 Let $F(x, y) = \sum_{m,n} f_{m,n} x^m y^n$. From the initial condition we get $\sum_n f_{0,n} y^n = \frac{1}{1-y}$, $\sum_m f_{m,0} x^m = \frac{1}{1-x}$, and from that $\sum_{mn=0} f_{m,n} x^m y^n = \frac{1-xy}{(1-x)(1-y)}$. For $m, n \geq 1$ we obtain the recurrence $\sum_{m,n \geq 1} f_{m,n} x^m y^n = y(F(x, y) - \frac{1}{1-y}) + (q-1)xy\, F(x, y)$, whence follows $F(x, y) = \frac{1}{1-x} \cdot \frac{1}{1-(1+(q-1)x)y}$. Therefore

$$[y^n]F(x, y) = (1 + (q-1)x)^n = \sum_{k=0}^{n} \binom{n}{k} (q-1)^k x^k.$$

Multiplication by $\frac{1}{1-x}$ (that is, summation over $[x^k]$) finally yields

$$f_{m,n} = \sum_{k=0}^{m} \binom{n}{k} (q-1)^k.$$

3.40 By convolution we have $\widehat{G}(z) = -2z\widehat{G}(z) + (\widehat{G}(z))^2 + z$, whence $\widehat{G}(z) = \frac{1+2z-\sqrt{1+4z^2}}{2}$ (since the plus sign is impossible on account of $\widehat{G}(0) = 0$). Using Exercise 1.40, we obtain $g_{2n+1} = 0$, $g_{2n} = (-1)^n \frac{(2n)!}{n} \binom{2n-2}{n-1}$. See also Section 9.4.

3.42 We set $f(0) = 1$ and have $f(1) = f(2) = 0$, whence $f(n) + f(n+1) = D_n$ for $n \leq 1$. By inserting $n+1$ into the permutations of $\{1, \ldots, n\}$ one easily proves the recurrence $f(n+1) = (n-2)f(n) + 2(n-1)f(n-1) + (n-1)f(n-2)$ $(n \geq 2)$. Rearrangement yields for $f(n) + f(n+1)$ precisely the recurrence for D_n. From Exercise 3.13 we obtain $\widehat{F}(z) + \widehat{F}'(z) = \widehat{D}(z) = \frac{e^{-z}}{1-z}$, $\widehat{F}(0) = 1$. With $\widehat{F}(z) = c(z)e^{-z}$ we obtain the solution $\widehat{F}(z) = c(z)e^{-z}$, $c(z) = -\log(1-z) + 1$. Now $f(n) = \widehat{F}(0)^{(n)}$. From $c^{(k)}(0) = (k-1)!$, $\widehat{F}(z)^{(n)} = \sum_{k=0}^{n} (-1)^k \binom{n}{k} c^{(n-k)}(z)e^{-z}$ follows $f(n) = \sum_{k=0}^{n-1} (-1)^k \binom{n}{k}(n-k-1)! + (-1)^n = n! \sum_{k=0}^{n-1} \frac{(-1)^k}{k!(n-k)} + (-1)^n$.

3.45 We have $\widehat{S}(z, n) = \sum_{m \geq 0} \sum_{k=0}^{n-1} k^m \frac{z^m}{m!} = \sum_{k=0}^{n-1} \sum_{m \geq 0} \frac{(kz)^m}{m!} = \sum_{k=0}^{n-1} e^{kz} = \frac{e^{nz}-1}{e^z-1} = \frac{e^{nz}-1}{z}\widehat{B}(z)$ from the previous exercise. Since $B_m(x)$ is the convolution of

(B_m) and (x^m), we obtain $\widehat{B}(z,x) = \widehat{B}(z)e^{xz} = \frac{ze^{xz}}{e^z - 1}$, whence $\widehat{B}(z,n) - \widehat{B}(z,0) = \frac{ze^{nz}}{e^z - 1} - \frac{z}{e^z - 1} = z\widehat{S}(z,n)$. Comparison of coefficients for z^{m+1} yields $\frac{1}{m!}S_m(n) = \frac{1}{(m+1)!}(B_{m+1}(n) - B_{m+1}(0)) = \frac{1}{(m+1)!}(\sum_k \binom{m+1}{k} B_k \, n^{m+1-k} - B_{m+1})$, $S_m(n) = \frac{1}{m+1}\sum_{k=0}^m \binom{m+1}{k} B_k \, n^{m+1-k}$.

4.1   The group of symmetries is $G = \{\, \mathrm{id}\,, i \mapsto n+1-i\}$, whence $Z(G) = \frac{1}{2}(z_1^n + z_2^{n/2})$, n even, and $Z(G) = \frac{1}{2}(z_1^n + z_1 z_2^{\frac{n-1}{2}})$, n odd. Replacing $z_1 \to 1+x$, $z_2 \to 1+x^2$ yields $m_k = \frac{1}{2}(\binom{n}{k} + \binom{n/2}{k/2}[k \equiv 0])$ for n even and $m_k = \frac{1}{2}(\binom{n}{k} + \binom{n-1/2}{\lfloor k/2 \rfloor})$ for n odd. For $n = 5$, $k = 3$, we obtain six patterns, and for $n = 6$, $k = 4$ nine patterns.

4.4   We have $|\mathcal{M}| = \binom{r+n-1}{n}$ and $|\mathcal{M}| = Z(S_n; r, \ldots, r) = \frac{1}{n!}\sum_{g \in S_n} r^{b(g)}$, whence $r^{\overline{n}} = \sum_{g \in S_n} r^{b(g)} = \sum_{k=0}^n s_{n,k} r^k$.

4.8   The symmetry group G consists of the identity, the four axes through a vertex to the opposite face, and the three axes through opposite edges, whence $|G| = 12$. Vertex and face patterns have the same cycle indicator $Z(G) = \frac{1}{12}(z_1^4 + 8z_1 z_3 + 3z_2^2)$. This can be seen from the vertex–face duality.

4.10   The group consists of id, reversion, and each exchange $0 \leftrightarrow 1$. The fixed-point sets are all $\binom{2n}{n}$ words, the $\binom{n}{n/2}$ words $a_1 \ldots a_n \, a_n \ldots a_1$ (this holds only for n even). For the fixed-point set of reversion together with the exchange we obtain the 2^n words $a_1 a_2 \ldots a_n \, \overline{a}_n \ldots \overline{a}_1$, $\overline{a}_i \neq a_i$. The lemma yields $|\mathcal{M}| = \frac{1}{4}(\binom{2n}{n} + 2^n + \binom{n}{n/2}[n \equiv 0])$.

4.11   We can choose the $k-1$ other places of the cycle in $\binom{n-1}{k-1}$ ways, whence $p = \frac{1}{n!}\binom{n-1}{k-1}(k-1)!(n-k)! = \frac{1}{n}$. If X is the length of the cycle, then $EX = \sum_{k=1}^n \frac{k}{n} = \frac{n+1}{2}$. Let Y be the number of cycles. Then $EY = \frac{1}{n!}\sum_{k=0}^n k s_{n,k} = \frac{1}{n!}(x^{\overline{n}})'_{x=1}$, whence $EY = H_n$, the harmonic number.

4.16   Since for $\mathrm{Inj}(N,R)$ we have from $f \circ \tau_g = f \circ \tau_h$ that $g = h$, every pattern contains exactly $|G|$ mappings,

$$w(\mathrm{Inj}(N,R); G) = \frac{1}{|G|}\sum_{f \in \mathrm{Inj}} w(f) = \frac{1}{|G|}\sum_{(i_1, \ldots, i_n)} x_{i_1} \cdots x_{i_n} = \frac{n!}{|G|}a_n(x_1, \ldots, x_r).$$

Setting $x_1 = \cdots = x_r = 1$, we obtain for $G = \{\,\mathrm{id}\,\}$ the number $r^{\underline{n}}$ of injective mappings and for $G = S_n$ the number $\binom{r}{n}$.

4.17   We have $G = \{\,\mathrm{id}\,, \varphi : 0 \leftrightarrow 1\}$. Let X be the set of tables. Then $|X_{\mathrm{id}}| = 16$. Writing $\overline{x} = 1 + x$, we have $x \cdot y \in X_\varphi$ if and only if $\varphi(x \cdot y) = \varphi x \cdot \varphi y$, that is, $\overline{x \cdot y} = \overline{x} \cdot \overline{y}$. From $0 \cdot 0 = a$ follows $1 \cdot 1 = \overline{a}$, and from $0 \cdot 1 = b$ follows $1 \cdot 0 = \overline{b}$. Thus there are exactly four tables

	0	1
0	a	b
1	$\overline{b}$	$\overline{a}$

in X_φ, and it follows that $|\mathcal{M}| = \frac{1}{2}(16 + 4) = 10$. For three elements one obtains $|\mathcal{M}| = 3330$.

4.19 For $G = \{\,\mathrm{id}\,\}$ all sets are inequivalent, whence $\sum_{k=0}^{n} \binom{n}{k} x^k = Z(\{\,\mathrm{id}\,\}; 1 + x, \ldots) = (1+x)^n$. For $G = S_n$ all k-sets are equivalent, and so with $x = 1$, we have $\sum_{k=0}^{n} 1 = n + 1 = \frac{1}{n!} \sum_{g \in S_n} 2^{b(g)}$. For $G = C_n$ we obtain with $x = 0$, $1 = Z(C_n; 1, \ldots, 1) = \frac{1}{n} \sum_{d \mid n} \varphi(d)$.

4.21 We have $\exp(\sum_{k \geq 1} z_k \frac{y^k}{k}) = \prod_{k \geq 1} e^{z_k \frac{y^k}{k}} = \prod_{k \geq 1} \sum_{j \geq 0} \frac{z_k^j y^{kj}}{k^j j!}$ and by multiplying out for $[y^n]$, we have $\sum_{(b_1, \ldots, b_n)} \frac{z_1^{b_1} \cdots z_n^{b_n}}{1^{b_1} \cdots n^{b_n} b_1! \cdots b_n!}$. But from Section 1.3 this is exactly $Z(S_n; z_1, \ldots, z_n)$. Setting $z_1 = z_2 = 1$, $z_i = 0$ for $i \geq 3$, then $\sum_{n \geq 0} T_n y^n = \sum_{n \geq 0} Z(S_n; 1, 1, 0, \ldots 0) y^n = \sum_{n \geq 0} \frac{i_n}{n!} y^n$. By the formula, then, we have $\sum_{n \geq 0} \frac{i_n}{n!} y^n = e^{y + \frac{y^2}{2}}$. If we wish to determine the number of permutations with only h-cycles, we set $z_h = 1$, $z_j = 0$ $(j \neq h)$ and obtain $\sum_{n \geq 0} \frac{h_n}{n!} y^n = e^{\frac{y^h}{h}}$.

4.23 Let $\overline{\mathcal{M}}$ be the set of self-dual patterns. Then $|\overline{\mathcal{M}}| = \frac{1}{|G|} \sum_{M \in \overline{\mathcal{M}}} \sum_{g \in G} |M_g|$. Let $f \in M \in \overline{\mathcal{M}}$, $G_f = \{g \in G : f \circ \tau_g = f\}$. Then f is counted in the inner sum exactly $|G_f|$ times, but f will be counted exactly as often as there is $g \in G$ with $h \circ f = f \circ \tau_g$, that is, $|\overline{\mathcal{M}}| = \frac{1}{|G|} \sum_{g \in G} |\{f \in R^N : h \circ f = f \circ \tau_g\}|$. Let $h \circ f = f \circ \tau_g$. Then f maps the cycle $(a, \tau_g a, \tau_g^2 a, \ldots)$ to $(\alpha, \beta, \alpha, \beta, \ldots)$ with $R = \{\alpha, \beta\}$, that is, the images $0, 1$ appear alternately. All cycles must therefore have even length. If the type is $t(\tau_g) = 1^0 2^{b_2} 3^0 4^{b_4} \cdots$, then there are exactly $2^{b_2} 2^{b_4} \cdots 2^{b_n}$ mappings with $h \circ f = f \circ \tau_g$, and the result follows.

5.2 If all functions are positive, then it is surely correct. In the other case, we could have, for example, $f_1(n) = n^2, g_1(n) = n^3 + n, f_2(n) = 0, g_2(n) = -n^3$.

5.4 For example, $\sqrt{|fh|}$.

5.6 From the assumption $T(\frac{n}{2}) \leq c\frac{n}{2}$ follows $T(n) \leq (c+1)n$ with a new constant $c + 1$.

5.10 An addition can at most double the previously largest element, and so $a_\ell \leq 2^\ell$ and therefore $\ell(n) \geq \lg n$. For $n = 2^\ell$ we have of course $\ell(n) = \lg n$. Let $n = 2^m + \cdots$ be the binary representation of n. With m additions we generate all powers 2^k $(k \leq m)$ and with at most m further additions, the number n.

5.12 The expression $k^2 + O(k)$ is the set of all functions $k^2 + f(k, n)$ with $|f(k, n)| \leq Ck$ for $0 \leq k \leq n$. The sum is therefore the set of all functions $\sum_{k=0}^{n} (k^2 + f(k, n)) = \frac{n^3}{3} + \frac{n^2}{2} + \frac{n}{6} + f(0, n) + \cdots + f(n, n)$. We now conclude that $|\frac{n^2}{2} + \frac{n}{6} + f(0, n) + \cdots + f(n, n)| \leq \frac{n^2}{2} + \frac{n}{6} + C(0 + 1 + \cdots + n) = \frac{n^2}{2} + \frac{n}{6} + C\frac{n}{2} + C\frac{n^2}{2} < (C+1)n^2$, and so the equation is correct.

5.15 In part (a) we have $\sum_{k \geq 0} |f(k)| = \sum_{k \geq 0} k^{-a} < \infty$ and $f(n-k) = (n-k)^{-a} = O(n^{-a})$ for $0 \leq k \leq \frac{n}{2}$. It follows that $\sum_{k=0}^{n} a_k b_{n-k} = \sum_{k=0}^{n/2} O(f(k)) O(f(n)) + \sum_{k=n/2}^{n} O(f(n)) \cdot O(f(n-k)) = 2O(f(n)) \sum_{k \geq 0} |f(k)| = O(f(n))$. In part (b) we set $a_n = b_n = a^{-n}$, and then we have $\sum_{k=0}^{n} a_k b_{n-k} = (n+1)a^{-n} \neq O(a^{-n})$.

5.18 The first recurrence contributes n. After the first round, $T(\frac{n}{4})$, $T(\frac{3n}{4})$ decomposes into $T(\frac{n}{16}), T(\frac{3n}{16}), T(\frac{3n}{16}), T(\frac{9n}{16})$ and the contribution is $\frac{n}{4} + \frac{3n}{4} = n$. At each step, n is added; altogether we have $\lg n$ rounds, and it follows that $T(n) = O(n \lg n)$.

5.20 By Theorem 5.2 we have $T(n) = n^{\lg 7}$. Consider $S(n) = \alpha S(\frac{n}{4}) + n^2$. For $\alpha < 16$, Theorem 5.2(c) yields $S(n) = \Theta(n^2) \prec T(n)$, and for $\alpha = 16$, Theorem 5.2(b) gives $S(n) = \Theta(n^2 \lg n) \prec T(n)$. In case (a), we have $\alpha > 16$ with $S(n) = \Theta(n^{\log_4 \alpha})$. Thus $S(n) \prec T(n)$ is satisfied precisely for $\alpha < 49$.

5.23 Assume $T(k) = O(k), T(k) \leq ck$. Then we have recursively

$$T(n) \leq \frac{2c}{n} \sum_{k=0}^{n-1} k + an + b = (c+a)n - c + b > cn$$

for $n \geq n_0$ since $a > 0$. The same reasoning applies to the case $T(k) = \Omega(k^2)$.

5.25 We shall prove that a calculation of $\gcd(a, b)$, $a > b$, that requires n steps implies $b \geq F_{n+1}$ and $a \geq F_{n+2}$. For $n = 1$ this is correct, and the general case follows inductively from the Fibonacci recurrence. With $a = F_{n+2}, b = F_{n+1}$ we see that the result cannot be improved. Let F_{n+2} be the smallest Fibonacci number greater than b. Then we know that the running time is less than or equal to n. From $b \geq F_{n+1} \geq \frac{\phi^n}{\sqrt{5}}$ follows $n = O(\log b)$. The last assertion says that for $b < 10^m$ the running time is at most $5m$. Now it follows easily from Exercise 3.33 that $F_{5m+2} \geq 10^m$, and thereby the assertion from the first part.

5.26 As in the hint, the waiter requires at most two flips to bring $n, n-1, \ldots, 3$ to the bottom, and then at most one additional flip. For the lower bound, we say that i neighbors $i+1$ ($i = 1, \ldots, n-1$), and n neighbors the bottom. A flip can increase the number of neighborhoods by at most 1. If the starting permutation has no neighborhoods, then we will need in any case n flips. Such permutations are easy to find for $n \geq 4$. For $n = 3$ one proves $\ell(3) = 3$ directly.

5.29 Transfer the topmost $\binom{n}{2}$ disks to B ($W_{\binom{n}{2}}$ moves), the remaining n to D (without using B, hence T_n moves), and finally the disks from B to D. For $U_n = (W_{\binom{n+1}{2}} - 1)/2^n$ we obtain with $T_n = 2^n - 1$ (Exercise 2.18) the recurrence $U_n \leq U_{n-1} + 1$, whence $W_{\binom{n+1}{2}} \leq 2^n(n-1) + 1$. There is no better algorithm known.

5.32 If $N_1 = a_k \ldots a_1$ is given in binary representation, then $N = 2N_1 + a_0$. That is, with division by 2 we obtain a_0. Altogether, we therefore need k steps (the last two places a_k, a_{k-1} are accomplished in a single step). If N has n decimal digits, then we have $2^k \leq N < 10^n$, whence $f(n) = O(n)$.

6.1 If all n vertices have different degrees, then the degrees must be the numbers $0, 1, \ldots, n-1$. But degrees 0 and $n-1$ are mutually exclusive.

6.5 For $n = 4$ we have K_4. A 3-regular graph G on $n \geq 4$ vertices contains a pair of nonincident edges $k = uv$, $k' = u'v'$. We remove k, k', join u, v with a new vertex a, u' and v' with a new vertex a' and finally a with a'.

6.8 If we remove an edge k, then by induction we obtain a graph with at least $n - q + 1$ components. Reinserting k can lower the number by at most 1.

6.10 Let A be independent with $|A| = \alpha(G)$. Then $N(A) = V \smallsetminus A$, whence $|N(A)| = n - \alpha(G) \leq \alpha(G)\Delta$.

6.13 Each of the $\chi(G)$ color classes is an independent set A_i with $\sum A_i = V$. It follows that $n = \sum |A_i| \leq \alpha(G)\chi(G)$. The graph K_n satisfies equality.

6.15 It is clear that G is connected. Let K_m on $\{u_1, \ldots, u_m\}$ be the largest complete subgraph. If $m < n$, then a $v \notin \{u_1, \ldots, u_m\}$ must be a neighbor of at least one vertex in K_m, and therefore to all, a contradiction.

6.20 A set $B \subseteq V$ meets all edges if and only if $S \setminus B$ is independent. Let A be a smallest such set, $|A| = m$, $m + \alpha(G) = n$. We have $|E| \leq \sum_{u \in A} d(u)$, since each edge is counted at least once. Since G contains no triangles, we have $d(u) \leq \alpha$, and therefore $|E| \leq \sum_{u \in A} d(u) \leq m\alpha \leq (\frac{m+\alpha}{2})^2 = \frac{n^2}{4}$. If $|E| = \frac{n^2}{4}$, then all inequalities must be equalities, and the graph $K_{n/2,n/2}$ results.

6.22 Let G be the graph with S as edge set in which two points are neighbors if $|x - y| > \frac{1}{\sqrt{2}}$. It is easy to show that G contains no K_4, and so $\lfloor \frac{n^2}{3} \rfloor$ by Exercise 6.21. The equilateral triangle shows that equality is possible.

6.25 All k-sets that contain 1 are in color class 1, all that contain 2 but not 1 in class 2, and so on up to $n - 2k + 1$. There remain all k-sets from $\{n - 2k + 2, \ldots, n\}$, and these have pairwise nonempty intersection. Incidentally, it can be proved that equality always holds. The graph $K(5, 2)$ is the Petersen graph.

6.26 The Petersen graph is $K(5, 2)$ from Exercise 6.25. Every permutation of $\{1, \ldots, 5\}$ yields an automorphism of $K(5, 2)$. Now let $\varphi : V \to V$ be an automorphism. We consider the 5-cycle $\{1, 2\}, \{3, 4\}, \{5, 1\}, \{2, 3\}, \{4, 5\}$ in $K(5, 2)$. Let $\varphi\{1, 2\} = \{a, b\}$, $\varphi\{3, 4\} = \{c, d\}$. Then without loss of generality, $\varphi\{5, 1\} = \{a, e\}$, $\varphi\{2, 3\} = \{b, c\}$. For $\varphi\{4, 5\} = \{x, y\}$ we have $x, y \notin \{a, b, c\}$ since $\{4, 5\}$ neighbors $\{1, 2\}, \{2, 3\}$, and so $\{x, y\} = \{d, e\}$. By considering an additional 5-cycle we obtain $\varphi : 1, 2, 3, 4, 5 \to a, b, c, d, e$, and hence φ induces a permutation on $\{1, \ldots, 5\}$, and $\mathrm{Aut}(\mathrm{Pet}) = S_5$.

6.28 Let S, T be the defining vertex sets with $|S| = m \leq n = |T|$. In an optimal numbering we must have 1 and $m + n$ on the same side. Let 1, $m + n$ be in T, b the bandwidth, and k and K the smallest and largest numbers in S. Then $K - k \geq m - 1$, $K - 1 \leq b$, $m + n - k \leq b$, whence $b \geq m + \lceil \frac{n}{2} \rceil - 1$ follows. The case 1, $m + n \in S$ yields analogously $b \geq n + \lceil \frac{m}{2} \rceil - 1$, and so our original choice was better. If we put $\lfloor \frac{n}{2} \rfloor + 1, \ldots, \lfloor \frac{n}{2} \rfloor + m$ in S, the remainder in T, then we obtain $b = m + \lceil \frac{n}{2} \rceil - 1$.

6.32 Number the vertices $v_1, \ldots, v_n$. Give v_1 the color 1 and inductively v_i the smallest color that does not appear among the neighbors of v_i in $\{v_1, \ldots, v_{i-1}\}$. Since v_i has at most Δ neighbors, we never need more than $\Delta + 1$ colors.

6.34 Induction on n. Assume $\chi(H) + \chi(\overline{H}) \leq n$ for all graphs on $n - 1$ vertices. Let G be a graph on n vertices $v \in V$ with $d(v) = d$. For $H = G \setminus v$, $\overline{H} = \overline{G} \setminus v$ we have $\chi(G) \leq \chi(H) + 1$, $\chi(\overline{G}) \leq \chi(\overline{H}) + 1$. The only interesting case occurs when equality holds in both cases. But then we have $\chi(H) \leq d$, $\chi(\overline{H}) \leq n - 1 - d$, and so again $\chi(G) + \chi(\overline{G}) \leq n + 1$. The second inequality follows from Exercise 6.13.

6.35 Let $C_1, \ldots, C_k$ be the color classes of a k-coloring. If we orient all the edges from left to right, that is, with increasing index of the color classes, then the condition is satisfied, since for every $k - 1$ successive edges in one direction there must always follow one in the other direction. Conversely, we fix $u_0 \in V$. For $v \neq u_0$ we consider all edge trails P from u_0 to v and supply the edges with weights 1 and $-(k - 1)$ as in the hint. Let $w(P)$ be the total weight. Since the weight of a cycle is by assumption nonpositive, there is a trail P_v with maximal weight $w(P_v)$. If now $v \to v'$, then we have $w(P_{v'}) \geq w(P_v) + 1, w(P_v) \geq w(P_{v'}) - (k-1)$, and hence $1 \leq w(P_{v'}) - w(P_v) \leq k-1$. Choosing the color 0 for u_0 and for v the color r with $w(P_v) = qk + r$, $0 \leq r \leq k-1$, we obtain a k-coloring.

6.37 (a) C_t, (b) K_{k+1}, (c) $K_{k,k}$, (d) Petersen graph ($f(3,5) \geq 10$ follows from the formula below, while the uniqueness requires some care). A vertex has k neighbors, each neighbor $k-1$ additional neighbors, and so on. Since $t = 2r + 1$, all vertices up to the $(r-1)$th iteration are distinct, and it follows that $f(k, 2r+1) \geq 1 + k \sum_{i=0}^{r-1} (k-1)^i = \frac{k(k-1)^r - 2}{k-2}$. The case $t = 2r$ is handled similarly.

6.40 The end vertices of every longest path are not cut vertices.

6.43 Let u be a vertex with maximal out-degree $d^-(u) = d$. For $v \notin N^+(u)$ we have $v \to u$ and therefore $v \to w$ for at most $d-1$ vertices of $N^+(u)$. Thus there is a path $u \to w_0 \to v$.

6.45 Suppose $d^-(v_i) < d^-(v_{i+1})$. Let $s = \#k < i : v_i \to v_k$, $s' = \#\ell > i : v_\ell \to v_i$ and analogously, t, t' for v_{i+1}. We have $d^-(v_i) = s + (n - i - s')$, $d^-(v_{i+1}) = t + (n - i - 1 - t')$, whence $s - s' < t - t' - 1$. The exchange $v_i \longrightarrow v_{i+1}$ changes $f(\pi)$ to $f(\pi) + s - s' - t + t' + 1 < f(\pi)$, and so π was not optimal.

6.46 Graphs with cut edges clearly cannot be satisfactorily oriented. For the converse, we remove an equivalence class of the relation $\approx$ (see Exercise 6.41). By induction, the remaining graph admits an orientation that is strongly connected). We orient the edges of the equivalence class cyclically and thereby obtain a desired orientation.

7.2 Let $V = \{u_1, \ldots, u_{2m}\}$ and let P_i be a path from u_i to u_{i+m} ($i = 1, \ldots, m$). Let E' contain all edges that appear in an odd number of the P_i's. Then $G' = (V, E')$ has the desired property. The graph $K_3 + K_3$ is a counterexample for disconnected graphs.

7.5 In $\sum_{u \neq v} d(u, v)$ there are $n-1$ terms equal to 1 and all others at least 2. The minimum is therefore attained when all other terms are equal to 2, and this gives the star $K_{1,n-1}$. In the other case we obtain a path of length $n-1$ by induction on n.

7.9 Suppose T and T' are distinct optimal trees and $k = uv \in E(T) \setminus E(T')$. On the unique path in T' that connects u with v lies an edge k' that joins the two components of $T \setminus \{k\}$. Now either $(T \setminus \{k\}) \cup \{k'\}$ or $(T' \setminus \{k'\}) \cup \{k\}$ has lower weight than T.

7.11 Let $n_i = |\{u \in E : d(u) = i\}|$. Then $\sum_{i=0}^{t} i n_i = 2n - 2 = 2 \sum_{i=1}^{t} n_i - 2$, $t = \max d(u)$. It follows that $n_1 = n_3 + 2n_4 + \cdots + (t-2)n_t + 2$, and so $n_1 \geq t$. Equality holds if $n_t = 1$ and all other vertices have degree 1 or 2.

7.12 Let $\mathcal{T}_k$ be the set of all trees with $d(n) = k$. For each $T \in \mathcal{T}_k$ we create $n - 1 - k$ trees in $\mathcal{T}_{k+1}$ in the following way: We consider one with n nonincident edges uv (where $d(n,v) = d(n,u) + 1$), delete uv, and attach v to n. Conversely, it is easily seen that exactly $k(n-1)$ trees from $\mathcal{T}_k$ belong to $T' \in \mathcal{T}_{k+1}$. By developing the recurrence we obtain $C(n, k) = \binom{n-2}{k-1}(n-1)^{n-1-k}$ and $t(n) = n^{n-2}$ by the binomial theorem.

7.13 From the hint we have $\det M_{ii} = \sum \det N \cdot \det N^T = \sum (\det N)^2$, where N runs through all $(n-1) \times (n-1)$ submatrices of $C \setminus \{\text{row } i\}$. The $n-1$ columns of N correspond to a subgraph of G with $n-1$ edges, and it remains to show that $\det N = \pm 1$ if G is a tree, and 0 otherwise. Suppose the $n-1$ columns produce a tree. Then there is a vertex $v_1 \neq v_i$ (v_i corresponds to the ith row) of degree 1. Let k_1 be the incident edge. We remove v_1, k_1 and again obtain a tree. There is again a vertex $v_2 \neq v_i$ of degree 1, and so on. Permuting the rows and columns in N yields a triangular matrix N' with ± 1 on the main diagonal, and so $\det N = \pm \det N' = \pm 1$. Suppose the $n-1$ columns do not yield a tree. Then there is a component that does not contain v_i. Since each of the corresponding columns contains a 1 and a -1, the rows of this component sum to zero and are therefore linearly dependent.

7.17 Every edge of K_n is in the same number a of spanning trees. Double counting yields $a\binom{n}{2} = n^{n-2}(n-1)$, whence $a = 2n^{n-3}$, and we obtain $t(G) = n^{n-3}(n-2)$.

7.18 Let a_n be the number of spanning trees in $G(2, n)$, hence $a_0 = 0$, $a_1 = 1$, $a_2 = 4$, and b_n the number of spanning forests with exactly two components of which one contains u_n, the other v_n. A case distinction as to how u_n, v_n are attached to $G(2, n-1)$ shows that $a_n = 3a_{n-1} + b_{n-1} + [n=1]$, $b_n = 2a_{n-1} + b_{n-1} + [n=1]$. Elimination of $B(z) = \sum b_n z^n$ yields $A(z) = \sum a_n z^n = \frac{z}{1 - 4z + z^2}$ and with the methods of Section 3.2 we obtain $a_n = \frac{1}{2\sqrt{3}}\left((2+\sqrt{3})^n - (2-\sqrt{3})^n\right)$.

7.21 Let A be the edge set of the tree constructed by the algorithm and let $\{b_1, \ldots, b_{n-1}\}$ be the edge set of any other tree. Let $b_i = u_i v_i$. Then there is a unique step of our algorithm in which the second of the vertices u_i, v_i is added to S and the steps are different for $b_i \neq b_j$. Since b_i is at the moment a candidate edge, it follows that $w(a_i) \leq w(b_i)$ for the corresponding A-edge.

7.24 Part (a) is clear, since all bases are the same size. Part (b) follows from axiom 3 for matroids applied to $A \setminus \{x\}$ and B. Conversely, if $\mathcal{B}$ satisfies the conditions, then one defines $\mathcal{U} = \{A : A \subseteq B \text{ for a } B \in \mathcal{B}\}$ and verifies the axioms for $\mathcal{U}$.

7.27 The minimally dependent sets in $\mathcal{M} = (E, \mathcal{W})$ are the edge sets of cycles. For $\mathcal{M}^*$ we assume that G is connected. Then A is independent in $\mathcal{M}^*$ if and only if $K \setminus A$ contains a spanning tree (see the previous exercise), and so B is minimally dependent in $\mathcal{M}^*$ if B is a minimal cut set.

7.30 Suppose $A, B \in \mathcal{U}$, $|B| = |A| + 1$, and $A \cup \{x\} \notin \mathcal{U}$ for all $x \in B \setminus A$. Let $w : S \to \mathbb{R}$ be defined by $w(x) = -|A| - 2$ for $x \in A$, $w(x) = -|A| - 1$ for $x \in B \setminus A$, $w(x) = 0$ otherwise. Let X be the solution constructed from the greedy algorithm. Then $A \subseteq X$, $X \cap (B \setminus A) = \varnothing$, whence $w(X) = -|A|(|A| + 2) > -(|A| + 1)^2 \geq w(B)$, and so the greedy algorithm does not return the optimum.

7.33 If C is a circuit with negative total weight, we can run through C as often as we like and lower the distance each time. As in the hint, suppose it has already been proved inductively that $\ell(v_{i-1}) = d(u, v_{i-1})$ after the $(i-1)$st round. In the ith round we check $\ell(v_{i-1}) + w(v_{i-1}, v_i)$ against the previous $\ell(v_i)$ and obtain $\ell(v_i) = \ell(v_{i-1}) + w(v_{i-1}, v_i) = d(u, v_{i-1}) + w(v_{i-1}, v_i) = d(u, v_i)$. Since each vertex $v \neq u$ has at most the graph distance $|V| - 1$, we are done after $|V| - 1$ iterations.

7.35 Extend the Bellman–Ford algorithm after $|V| - 1$ passes as follows: check for each edge (x, y), $\ell(y)$ against $\ell(x) + w(x, y)$. If $\ell(y) > \ell(x) + w(x, y)$ holds for an edge, print "no shortest path."

8.3 By Theorem 8.4, we must show that $|N(A)| \geq m - n + |A|$ for all $A \subseteq S$. Let $|A| = r$, $|N(A)| = s$. Then $(m - 1)n < |E| \leq rs + (n - r)n \leq n(s + n - r)$; hence $m - 1 < s + n - r$ or $s \geq m - n + r$. The graph $K_{m-1,n}$ plus an edge from a new vertex $u_m \in S$ to T shows that m cannot be improved.

8.7 Let $u_1v_1, \ldots, u_mv_m$ be a maximum matching. Assuming $m < n/2$, the remaining vertices $\{u, v, \ldots\}$ form an independent set. From $d(u) + d(v) \geq n - 1 > 2m$ it follows that there is a pair u_i, v_i to which three edges from u, v lead. Therefore, there exists a matching in $\{u, v, u_i, v_i\}$.

8.8 Clearly, mn must be even. Let m be even. Then we find a matching in each of the n paths. The number of 1-factors in $G(2, n)$ is F_{n+1} (Fibonacci number).

8.13 Let C be a Hamiltonian cycle. If we remove A, then there remain at most $|A|$ connected pieces of C.

8.15 By induction we assume that Q_{n-1} is Hamiltonian with $u_1u_2 \ldots u_{2^{n-1}}$. We have $Q_n = \{(v, 0) : v \in Q_{n-1}\} \cup \{(v, 1) : v \in Q_{n-1}\}$, and hence $(u_1, 0), \ldots, (u_{2^{n-1}}, 0)$, $(u_{2^{n-1}}, 1) \ldots, (u_1, 1)$ a Hamiltonian circuit.

8.17 Let $n \geq 3$ be odd. If we color the pieces of cheese alternately white and black like a checkerboard, then a path must alternate between white and black. The lower left corner and the middle vertex are of different colors, and so every such path has odd distance. However, since n^3 is odd, it must have even length.

8.20 The complement of a vertex cover in G is an independent set in G, hence a complete subgraph in $\overline{G}$ (see the solution to Exercise 6.20). This proves the equivalence. That, for example, the clique problem is in NP is clear.

8.22 (a) $\Rightarrow$ (b): $|S| = |T|$ is clear. Suppose $|N(A)| \leq |A|$ for $\varnothing \neq A \neq S$. Since G has a 1-factor, we must have $|N(A)| = |A|$. There is an edge k that joins $N(A)$ with $S \setminus A$, since G is connected, and this edge k is in no 1-factor. (b) $\Rightarrow$ (c): Let $A \subseteq S \setminus \{u\}$. Then $|N_{G \setminus \{u,v\}}(A)| \geq |N_G(A)| - 1 \geq |A|$. (c) $\Rightarrow$ (a): Suppose G is not connected. Let G_1 be a component with $|V(G_1) \cap S| \leq |V(G_1) \cap T|$. Let $u \in V(G_1) \cap S$, $v \in T \setminus V(G_1)$. Then $G \setminus \{u, v\}$ would have no 1-factor. Now let $k = uv$ be an edge. Then there exists a 1-factor in $G \setminus \{u, v\}$, hence with k a 1-factor in G.

8.26 If G has a 1-factor, then the second player always chooses a matching partner. In the other case, the first player chooses a maximum matching M and begins with u_1 outside of M. Player 2 must go into M, and the first player chooses the matching partner. The second player can in this way never leave M, since we otherwise would have an alternating path.

8.28 We prove the assertion more generally for all bipartite graphs $G = (S + T, E)$, $|S| = |T| = n$, that have a 1-factor with $d(u) \geq k$ for all $u \in S \cup T$. For $n = k$ we have $G = K_{k,k}$. The 1-factors in $K_{k,k}$ correspond to the $k!$ permutations. Let $n > k$. The set $A \subseteq S$ is called *critical* if $|A| = |N(A)|$. Of course, $\varnothing$ and S are always critical. Case a. $\varnothing$ and S are the only critical sets. Let $uv \in E$ and $G' = G \setminus \{u, v\}$. Since $|A| < |N(A)|$ for $A \neq \varnothing$, S in G', we always have

$|A| \leq |N(A)|$. By induction, G' has at least $(k-1)!$ distinct 1-factors, which we may extend with the edges $uv_1, \ldots, uv_k$. Case b. $A \neq \varnothing$, S is a critical set. Every 1-factor must then join precisely A with $N(A)$ and therefore $S \setminus A$ with $T \setminus N(A)$. By induction, the bipartite graph $A + N(A)$ has $k!$ 1-factors, all of which can be extended to all of G.

8.32 Let $d(G) = \min\ (|A| : A$ a vertex cover$)$. Then $d(G) + \alpha(G) = n$. By Theorem 8.5 we have $d(G) = m(G)$ and by the previous exercise $m(G) + \beta(G) = n$. For the circuit C_5 we have $\alpha = 2$, $\beta = 3$.

8.34 In the bipartite graph on $S + (\mathcal{F} \cup \mathcal{F}')$ we consider a subset $B = \{A_{i_1}, \ldots, A_{i_r}\}$ $\cup \{A'_{j_1}, \ldots, A'_{j_s}\}$ of the right side. By assumption we have $|N(B)| \geq \frac{a(r+s)}{2b} \geq r + s$. There thus exists a matching $\varphi : \mathcal{F} \cup \mathcal{F}' \to S$ from right to left. If the first player occupies $s \in S$ with $A = \varphi^{-1}(s)$, then the second counters by occupying $t \in S$ with $t = \varphi(A')$. For $n \times n$ tic-tac-toe we have $a = n$, $b \leq 4$, and so for $n \geq 3$ there is always a draw.

8.36 We consider the complete bipartite graph $K_{m,n}$ on $Z + S$ with all edges directed from Z to S, $Z = \{1, \ldots, m\}$, $S = \{1, \ldots, n\}$. The capacity c is 1 on all edges; the supply $i \in Z$ is r_i; the demand in $j \in S$ is s_j. An admissible flow then takes the values $0, 1$ (Corollary 8.18), and since $\sum r_i = \sum s_j$, the 1-values correspond to a desired matrix. We therefore must verify the condition (1) at the end of Section 8.3. Let (X, Y) be a cut with $Z \cap Y = I$, $S \cap Y = J$. Precisely the edges between $Z \setminus I$ and J contribute 1 to the capacity, so that we obtain the condition $\sum_{j \in J} s_j \leq \sum_{i \in I} r_i + |Z \setminus I| |J|$ for all $I \subseteq Z$, $J \subseteq S$. Among the k-sets $J \subseteq S$, the sharpest condition on the left is given by $J = \{1, \ldots, k\}$ since $s_1 \geq \cdots \geq s_n$, and one sees at once that for $|J| = k$ the sharpest condition on the right results for $I_0 = \{i \in Z : r_i \leq k\}$. But then we have $\sum_{i \in I_0} r_i + k |Z \setminus I_0| = \sum_{i=1}^m \min(r_i, k)$.

8.37 We number the cities such that $\ell_1 \geq \cdots \geq \ell_n$, where ℓ_i is the length of the edge that in the (NN) algorithm is attached to i. Then $c_{\text{opt}} \geq 2\ell_1$ follows at once from the triangle inequality. Let $S_k = \{1, \ldots, 2k\}$ and T_k be the tour through S_k in the same cyclic order as in the optimal tour. From the triangle inequality it follows that $c_{\text{opt}} \geq c(T_k)$. Let ij be an edge in T_k. If i was added in (NN) before j, then it follows that $c_{ij} \geq \ell_i$. Otherwise, $c_{ji} = c_{ij} \geq \ell_j$, whence $c_{ij} \geq \min(\ell_i, \ell_j)$. Summation yields $c_{\text{opt}} \geq c(T_k) \geq \sum c_{ij} \geq \sum \min(\ell_i, \ell_j) \geq 2(\ell_{k+1} + \cdots + \ell_{2k})$, since a minimum appears at most twice in the sum. In the same way it follows that $c_{\text{opt}} \geq 2 \sum_{j = \lceil n/2 \rceil + 1}^n \ell_j$. Choosing $k = 1, 2, \ldots, 2^{\lceil \lg n \rceil - 2}$ we sum all the inequalities, with the result $(\lceil \lg n \rceil + 1) c_{\text{opt}} \geq 2 \sum_{i=1}^n \ell_i = 2 c_{\text{NN}}$.

8.38 Let $\lambda = \min |A|$, $\mu = \max |\mathcal{W}|$. Then clearly we have $\mu \leq \lambda$. We consider the network $\vec{G} = (V, E)$ as in the hint. The capacity of a cut (X, Y) is $c(X, Y) = |S(X, Y)|$ on account of $c \equiv 1$ (notation as in Lemma 8.16). Since every $S(X, Y)$ is a separating set, it follows that $\lambda \leq c(X, Y)$. From Theorem 8.17, it remains to show that $w(f) \leq \mu$ for every admissible flow f. It is clear that the addition of αf_P (notation as in Theorem 8.17) corresponds exactly to a directed path from u to v, so that from $c \equiv 1$, we must have $w(f) \leq \mu$.

8.41 Edge-disjoint S, T-paths correspond to matchings; separating edge sets correspond to vertex covers; the result is Theorem 8.5.

8.43 Let $G = (V, E) \neq K_n$ be chosen as in the hint. Then $G \cup \{uv\}$ is Hamiltonian for every pair $uv \notin E$. The graph G therefore has a path $u = u_1, u_2, \ldots, u_n = v$. Let $A = \{u_i : u\, u_{i+1} \in E\}$, $B = \{u_j : u_j v \in E\}$. Hence $|A| = d(u)$, $|B| = d(v)$. Since $v \notin A \cup B$, it follows that $n \leq d(u) + d(v) = |A| + |B| = |A \cup B| + |A \cap B| < n + |A \cap B|$, that is, there is a $u_k \in A \cap B$. But now we obtain the Hamiltonian circuit $u = u_1, \ldots, u_k, v = u_n, u_{n-1}, \ldots, u_{k+1}, u$.

8.46 Someone specifies an isomorphism $\varphi : G \to H$. Then for each 1 or 0 in the adjacency matrix of G it must be verified whether the corresponding entry in the adjacency matrix of H is likewise 1 or 0. This can be carried out with $O(n^2)$ operations.

8.48 By the previous exercise, P $\subseteq$ co-NP. Let P = NP and $A \in$ co-NP. Consider A' with the truth values 1 and 0 exchanged. Then $A' \in$ NP = P, whence $A \in$ P, since P = co-P. It follows that co-NP = P = NP.

9.2 We have the search domain $S = \{1_L, 1_S, \ldots, n_L, n_S\}$ as in Section 9.1, whence $L \geq \lceil \log_3 2n \rceil$. Assume $2n = 3^k - 1, k \geq 1$. If the first test places ℓ coins on each side of the scale, then S is decomposed into the sets S_i with 2ℓ, 2ℓ, and $2n - 4\ell$ elements. Since $|S_i|$ is even, it follows that $\max |S_i| \geq 3^{k-1} + 1$, whence $L \geq k + 1$, and so $L \geq \lceil \log_3(2n + 2) \rceil$. The converse follows easily by induction.

9.4 The formula holds for $n = 1, 2$. If we remove from a tree $T \in \mathcal{T}(n, 2)$ a branch in which the leaves have length ℓ, we obtain $e(T) = e(T') + \ell + 1$, $i(T) = i(T') + \ell - 1$, whence by induction, $e(T) - i(T) = e(T') - i(T') + 2 = 2(n - 1)$.

9.8 We can at most halve the number of candidates in each round, whence $L \geq \lceil \lg n \rceil$. Conversely, if we have s candidates, we can eliminate $\lfloor \frac{s}{2} \rfloor$ candidates by disjoint comparisons.

9.11 That $L_0(n) = n - 1$ is clear; likewise $L_1(1) = 0, L_1(2) = 1$. We have $L_1(3) = 1$ by first testing $x^* = 1$. Let $n \geq 4$. If we first test $x^* = 2$, then by induction, $L \leq 1 + \max(1, \lceil \frac{n-3}{3} \rceil) = \lceil \frac{n}{3} \rceil$. Conversely, each subtree with s leaves always contains a subtree with at least $s - 3$ leaves. With an analogous approach one obtains $L_2(n) = \lceil \frac{n}{5} \rceil + 1$ for $n = 5t, 5t + 3, 5t + 4$ $(n \geq 3)$ and $L_2(n) = \lceil \frac{n}{5} \rceil$ for $n = 5t + 1, 5t + 2$ $(n \geq 12)$.

9.13 Let A be the set of 10-euro people and B that of 20-euro people. A sequence of A's and B's is inadmissible if at any point there are more B's than A's. Let $2m - 1$ be the first place where this happens. Then up to this place there are $m - 1$ of the A's and m of the B's. If we exchange these A's and B's, we obtain a sequence of $n + 1$ B's and $n - 1$ A's. Conversely, given a sequence of $n + 1$ B's and $n - 1$ A's, there is a first place at which the B's are one more than the A's. If we again exchange the A's and B's, we end up with an inadmissible sequence. One sees at once that this is a bijection of the inadmissible sequences to the $\binom{2n}{n-1}$ sequences of $n + 1$ B's and $n - 1$ A's, and we obtain $\binom{2n}{n} - \binom{2n}{n-1} = \frac{1}{n+1}\binom{2n}{n}$.

9.15 Sequence c, since after 48 the search takes place in the left subtree, while 60 is in the right one.

9.17 For $L = 1$ the graph can have at most two edges. Suppose $u \in V$ is the first test vertex. Let L be the optimal search length. If $d(u) = d > L$ and the answer is yes, then the second end vertex of k^* is among the neighbors of u, and we need

$d(u) - 1 \geq L$ additional tests. Thus $d(u) \leq L$ must hold. There is an edge not incident with u, since otherwise, $G = K_{1,d}$ and u would not be optimal. If now the answer is no, the problem is reduced to $G' = G \smallsetminus \{u\}$ with $L(G') \leq L - 1$. By induction we have $|E(G)| = d + |E(G')| \leq \binom{L+1}{2} + 1$. One proves part (b) similarly. Solving for the bounds yields $L \geq \left\lceil \sqrt{2|E| - \frac{7}{4}} - \frac{1}{2} \right\rceil, L \geq \left\lceil \sqrt{2n - \frac{7}{4}} - \frac{3}{2} \right\rceil$.

9.20 Part (a) is clear, since $k \geq \frac{n}{2}$ represents no restriction on the tests. Let $f_k(n) = L_{\leq k}(n)$. Then $f_k(n)$ is monotonically increasing in n. Namely, let $A_1, A_2, \ldots$ be the test sequence in an optimal algorithm for S. Then $A_1 \cap T, A_2 \cap T, \ldots$ forms a successful test sequence for $T \subseteq S$. Let A_1 be the first subset, $|A_1| = \ell \leq k$. Considering the subtrees rooted in A_1 and $S \smallsetminus A_1$, we thus obtain the recurrence $f_k(n) = 1 + \max(f_k(\ell), f_k(n - \ell)) = 1 + f_k(n - \ell) \geq 1 + f_k(n - k)$, whence follows $f_k(n) \geq t + \lceil \lg(n - tk) \rceil$ with $t = \lceil \frac{n}{k} \rceil - 2$. Conversely, let us take as our first test set $A_1 = \{a_1, \ldots, a_k\}$, $A_2 = \{a_{k+1}, \ldots, a_{2k}\}, \ldots, A_t = \{a_{(t-1)k+1}, \ldots, a_{tk}\}$. If $x^* \in \bigcup A_i$, then we are through with $\lceil \lg k \rceil$ additional tests. In the other case, we require $\lceil \lg(n - tk) \rceil$ additional tests.

9.23 The only alternative to elementwise search is $S = \{a, b\}$ as the first test set. We obtain $S \cap X^* = \varnothing$ with probability $(1 - p)^2$. In this case we are done, while in the other case we test $\{a\}$. Now the answer $X^* \cap \{a\} = \varnothing$ has probability $p(1 - p)$, and we obtain $X^* = \{b\}$. In the other case, test $\{b\}$. With probability p we therefore need three tests, and we obtain $\overline{L} = (1 - p)^2 + 2p(1 - p) + 3p = -p^2 + 3p + 1$, and this expression is greater than or equal to 2 precisely for $p \geq \frac{3 - \sqrt{5}}{2}$.

9.25 Replace in an optimal tree $T \in \mathcal{T}(n + 1, 2)$ a branch as in the Huffman algorithm where the leaves have length ℓ, so that $T' \in \mathcal{T}(n, 2)$ results. For $e(T), e(T')$ as in Exercise 9.4, we obtain $e(T) = e(T') + \ell + 1$. We have $h(n + 1) = \frac{e(T)}{n+1}$, $h(n) \leq \frac{e(T')}{n}$, whence $\frac{n+1}{n} h(n + 1) = \frac{e(T)}{n} \geq h(n) + \frac{\ell+1}{n}$, and thus $h(n + 1) \geq h(n) + \frac{1}{n} + \frac{1}{n}(\ell - h(n + 1))$. Equality therefore holds when $n + 1 = 2^k$ for some k.

9.29 The search domain is $S = \{y_1, \ldots, y_n, z_0, \ldots, z_n\}$, where z_i means that $y_i < x^* < y_{i+1}$. One now constructs a complete binary tree $T \in \mathcal{T}(n + 1, 2)$ with z_i in the leaves and y_i in the internal vertices and fills the vertices of T in in-order. If $x^* = y_i$, then the number of tests is $\ell(y_i) + 1$, while if $x^* = z_i$, we obtain $\ell(z_i)$, and hence $L = \lceil \lg(n + 1) \rceil$.

9.30 We carry out $\lfloor \frac{n}{2} \rfloor$ disjoint comparisons and determine the maximum of the $\lfloor \frac{n}{2} \rfloor$ larger elements and the minimum of the $\lfloor \frac{n}{2} \rfloor$ smaller ones. This yields $L \leq \lceil \frac{3n}{2} \rceil - 2$. For the lower bound, we let $\max_i$ denote the set of maximum candidates after i comparisons, with an analogous definition of $\min_i$. Let $s_i = |\max_i| + |\min_i|$. At the outset, we have $s_0 = 2n$ and at the end, $s_L = 2$. Let $x : y$ be the $(i + 1)$st comparison. If we obtain the answer $x < y$, then when $x \in \min_i$ or $y \in \max_i$ (with an arbitrary answer in the other cases), it follows that $s_{i+1} \geq s_i - 1$, except when x, y have as yet not been in any comparison. In this case we have $s_{i+1} = s_i - 2$. However, this second case can occur at most $\lfloor \frac{n}{2} \rfloor$ times, and $2 \geq 2n - L - \lfloor \frac{n}{2} \rfloor$ results.

9.32 If we always compare the minima of the current lists, then we are done in $2n - 1$ comparisons. Conversely, in the common list $\{z_1 < z_2 < \cdots < z_{2n}\}$ all comparisons $z_i : z_{i+1}$ must have been carried out, since they cannot be forced by transitivity. Therefore, $M(n, n) = 2n - 1$.

9.35 A j with $b_j > 0$ is exchanged with the largest $k > j$ that comes before j; hence we have $b'_j = b_j - 1$. Then $A = \sum_{j=1}^{n} b_j$ follows immediately, and likewise $D = 1 + \max(b_1, \ldots, b_n)$, where 1 is counted for the last round. Let (b'_j) be the inversion table after $i - 1$ rounds. If $b'_j = b_j - i + 1 \geq 0$, then j has b'_j predecessors greater than j. If j is the last element to be exchanged, then the largest k before j bubbles into the position $b'_j + j$. In the ith round, we therefore need $c_i = \max(b'_j + j - 1 : b_j - i + 1 \geq 0) = \max(b_j + j : b_j \geq i - 1) - i$ comparisons.

9.36 The probability $D \leq k$ is equal to $\frac{1}{n!}$ times the number of inversion tables with $b_i \leq k - 1$ for all i, thus equal to $\frac{k! \, k^{n-k}}{n!}$. The probability $p(D = k)$ is therefore $\frac{1}{n!}(k! \, k^{n-k} - (k-1)! \, (k-1)^{n-k+1})$ and we obtain $E(D) = n + 1 - \sum_{k=0}^{n} \frac{k! \, k^{n-k}}{n!}$.

9.39 Let $E_{n,k} = E(\ell_k(\pi))$ on $\{1, \ldots, n\}$. We first prove $E_{n,k} = E_{n-1,k} + \frac{1}{n+1-k}$ for $k < n$. The number i is the root with probability $\frac{1}{n}$. By distinguishing the cases $i < k$ and $i > k$, we obtain $E_{n,k} = \frac{1}{n} \sum_{i=1}^{k-1} (1 + E_{n-i,k-i}) + \frac{1}{n} \sum_{i=k+1}^{n} (1 + E_{i-1,k})$, whence $n \, E_{n,k} = n - 1 + \sum_{i=1}^{k-1} E_{n-i,k-i} + \sum_{i=k+1}^{n} E_{i-1,k}$. If we write the same equation for $n - 1$ instead of n and subtract the second from the first, then by induction we obtain $n \, E_{n,k} - (n-1)E_{n-1,k} = 1 + \sum_{i=1}^{k-1} \frac{1}{n-k+1} + E_{n-1,k}$, and thus $E_{n,k} = E_{n-1,k} + \frac{1}{n-k+1}$. Solving the recurrence yields $E_{n,k} = E_{k,k} + (H_{n+1-k} - 1)$. By symmetry we have $E_{k,k} = E_{k,1} = H_k - 1$ (see the previous exercise), and we obtain $E(\ell_k(\pi)) = H_k + H_{n+1-k} - 2$.

9.41 Jensen's inequality holds for arbitrary $c > 0$; we need only set $x_i = c^{y_i}$, as in Section 9.4. Carrying out the same analysis as there, we obtain $g(2) = c$, $g(n) = \frac{(n-1+2c)(n-2+2c)\cdots(2+2c)c}{n(n-1)\cdots 3} = (1 + \frac{2c-1}{n}) \cdots (1 + \frac{2c-1}{3})c$. Since $c \leq 1 + \frac{2c-1}{2}$ and $1 + x \leq e^x$ holds for every $x \in \mathbb{R}$, we end up with $g(n) \leq \exp((2c-1)(H_n - 1)) < \exp((2c-1)\log n) = n^{2c-1}$. From this we obtain, as in Section 9.4, $E(L(n)) \leq \frac{2c-1}{\log c} \log n$. The minimal value for $\frac{2c-1}{\log c}$ is attained for the solution of the equation $2 \log c - 2 + \frac{1}{c} = 0$, and for this c_0 we have $\frac{2c_0 - 1}{\log c_0} \approx 4.311$. Incidentally, it is known that $E(L(n)) = \Theta(c_0 \log n)$.

10.1 Let n be even. We divide the board into $\frac{n^2}{4}$ squares of side length 2. At most one king can be placed in each square. If we always choose the lower right corner, then we obtain $\frac{n^2}{4}$. For n odd one obtains $\frac{(n+1)^2}{4}$.

10.3 If a queen is placed in a corner, then the other three corners must remain empty.

10.5 If we count only 1 for the additions and comparisons in (10.1), then we obtain for the running time of the recurrence $T(n) \geq \sum_{k=2}^{n-1} (T(k) + T(n-k+1) + 1) + 1 = 2 \sum_{k=2}^{n-1} T(k) + (n-1)$ for $n \geq 3$, $T(2) = 0$. From this, $T(n) \geq 2 \cdot 3^{n-3}$ follows by induction.

10.8 The greedy approach takes k_m maximal with $k_m c^m \leq n$, then k_{m-1} maximal with $k_{m-1} c^{m-1} \leq n - k_m c^m$, and so on. Now let $(\ell_0, \ldots, \ell_m)$ be an optimal solution. Then we have $\ell_i \leq c - 1$ for $i < m$. Let i_0 be the largest index with $\ell_{i_0} \neq k_{i_0}$. Then $\ell_{i_0} < k_{i_0}$ by the greedy construction. There must therefore be a largest index $j_0 < i_0$ with $\ell_{j_0} > k_{j_0}$. If $I = \{i : k_i > \ell_i\}$, $J = \{j : \ell_j > k_j\}$, then $\sum_{i \in I}(k_i - \ell_i)c^i = \sum_{j \in J}(\ell_j - k_j)c^j$. On the other hand, we have $\sum_{i \in I}(k_i - \ell_i)c^i \geq c^{i_0}$ and $\sum_{j \in J}(\ell_j - k_j)c^j \leq c^{j_0+1} - 1 < c_{i_0}$ since $j_0 < i_0$. The greedy algorithm therefore actually returns the unique optimum.

10.11 In the matrix $\begin{pmatrix} 4 & 3 \\ 3 & 1 \end{pmatrix}$ the greedy algorithm constructs $x_{11} = x_{22} = 1$ with $w = 5$, while $w_{\text{opt}} = 6$.

10.15 Let A, B, C be the men and a, b, c their wives. We describe the situation in terms of the set of persons on the near side of the river. The admissible sets are $ABCabc$, $ABCab$, $ABCac$, $ABCbc$, $ABCa$, $ABCb$, $ABCc$, $ABab$, $ACac$, $BCbc$, ABC and their complements. We can describe the transport situation naturally in terms of graphs. We seek a path of odd length from $ABCabc$ to $\varnothing$. Solution: $ABCabc$, $ABCa$, $ABCab$, ABC, $ABCa$, Aa, $ABab$, ab, abc, a, ab, $\varnothing$. There is no solution for four pairs.

10.17 Illustration for $n = 3$, $m = 4$. Let $a_1 < a_2 < a_3$ be the stamps. Then we have $a_1 = 1$, $2 \leq a_2 \leq 5$, $a_2 < a_3 \leq 4a_2 + 1$. We begin the tree with $a_1 = 1$, $a_2 = 2$ and test sequentially $a_3 = 3$ to 9. We make note of each N, e.g., $N(1, 2, 3) = 12$, $N(1, 2, 4) = 14$, and return to $a_2 = 3$. The optimum is $N = 26$ for $\{1, 5, 8\}$.

10.18 Every coloring f of G^+ or G^- is also a coloring of G, by using the same colors, that is, $\chi(G) \leq \chi(G^+), \chi(G^-)$. Conversely, if f is a coloring of G, then f is a coloring of G^+ (if $f(u) \neq f(v)$) or of G^- (if $f(u) = f(v)$). Let $\omega(G)$ be the vertex number of a largest complete subgraph of G. Then $\chi(G) \geq \omega(G)$. By the branching of G into G^+, G^- we obtain the new bounds $\omega(G^+), \omega(G^-)$. At the end, we have $H = K_m$, $\chi(H) = m$, and trace back through the tree.

10.21 Let $g_1 = 1$, $g_2 = n$, $w_1 = 2$, $w_2 = n$, $G = n$. Then $w^* = 2$, $w_{\text{opt}} = n$, whence $w^* = \frac{2}{n} w_{\text{opt}} < (1 - \varepsilon)w_{\text{opt}}$ for $n > \frac{2}{1-\varepsilon}$.

10.22 Let $(x_1, \dots, x_n) \in \{0, 1\}^n$ be an optimal solution, that is, $x_i = 1$ or 0, depending on whether the ith element is or is not in the knapsack. Since $\frac{w_1}{g_1} \geq \cdots \geq \frac{w_n}{g_n}$, we have

$$w_{\text{opt}} \leq \sum_{j=1}^{k} w_j x_j + \sum_{j=k+1}^{n} \frac{w_{k+1}}{g_{k+1}} g_j x_j = \frac{w_{k+1}}{g_{k+1}} \sum_{j=1}^{n} g_j x_j + \sum_{j=1}^{k}\left(w_j - \frac{w_{k+1}}{g_{k+1}} g_j\right) x_j$$

$$\leq \frac{w_{k+1}}{g_{k+1}} G + \sum_{j=1}^{k}\left(w_j - \frac{w_{k+1}}{g_{k+1}} g_j\right) = \sum_{j=1}^{k} w_j + \frac{w_{k+1}}{g_{k+1}}\left(G - \sum_{j=1}^{k} g_j\right).$$

Let k be the largest index with $\sum_{j=1}^{k} g_j \leq G$. Then we have $0 \leq G - \sum_{j=1}^{k} g_j < g_{k+1}$ and therefore $w_{\text{opt}} \leq w^* + w_{k+1}$.

10.24 For the jobs J_i suppose $e_1 \leq \cdots \leq e_n$. Let $A = \{J_k, \dots\}$ be an optimal solution. If $k \neq 1$, then on account of $e_1 \leq e_k$ we also have that $(A \setminus \{J_k\}) \cup \{J_1\}$ is an optimal solution. Iteration shows that the greedy algorithm is successful.

11.4 By definition of the minterm we have $\overline{f(x_1,\ldots,x_n)} = \sum_{c:f(c)=0} x_1^{c_1}\cdots x_n^{c_n}$, and so by de Morgan, we have

$$f(x_1,\ldots,x_n) = \overline{\sum_{c:f(c)=0} x_1^{c_1}\cdots x_n^{c_n}} = \prod_{c:f(c)=0} (x_1^{\overline{c_1}} + \cdots + x_n^{\overline{c_n}}).$$

11.6 DNF : $\overline{x}_1\overline{x}_2x_3 + \overline{x}_1x_2\overline{x}_3 + x_1\overline{x}_2\overline{x}_3 + x_1\overline{x}_2x_3 + x_1x_2\overline{x}_3 + x_1x_2x_3$, CNF : $(x_1 + x_2 + x_3)(x_1 + \overline{x}_2 + \overline{x}_3)$. Simplification yields $f(x_1, x_2, x_3) = x_1 + x_2\overline{x}_3 + \overline{x}_2x_3$.

11.10 Let M be the number of elements in a longest chain, and m the minimal number of antichains into which P can be decomposed. Clearly, $M \le m$. Conversely, let $\ell(x)$ be the number of elements of a longest chain that ends in x. Then $A_k = \{x \in P : \ell(x) = k\}$ is an antichain, and $P = \sum_{k=1}^{M} A_k$.

11.11 The bipartite graph $G = (S + \mathcal{F}, E)$ is a tree. Therefore, $|E| = \sum_{F \in \mathcal{F}} |F| = |S| + |\mathcal{F}| - 1$.

11.13 In a minterm of the DNF we can replace $\overline{x}$ by $1 \oplus x$. Since for every x with $f(x) = 1$ only one minterm has the value 1, we can replace the OR sum $\sum$ by $\sum \oplus$. Every Boolean function therefore has an RSE representation, and since the number 2^{2^n} of Boolean functions is equal to the number of coefficient sequences (a_I), we obtain uniqueness.

11.15 We must find x with $(f \oplus g)(x_1,\ldots,x_n) = 1$. Let the RSE of $f \oplus g$ be given by $\sum c_I x_I$. If ℓ is minimal, $c_L = 1$, $|L| = \ell$, then x with $x_i = 1$ for $i \in L$, $x_j = 0$ for $j \notin L$, is a desired x.

11.16 We set x_1: white precipitate, x_2: sodium present, x_3: ammonia present, x_4: iron present. The assertions are then as follows: $A_1 : x_1 \to (x_2 \vee x_3)$, $A_2 : \neg x_2 \to x_4$, $A_3 : (x_1 \wedge x_4) \to \neg x_3$, $A_4 : x_1$. Computation yields $A_1 \wedge A_2 \wedge A_3 \wedge A_4 = x_1 \wedge x_2 \wedge \neg(x_3 \wedge x_4)$. In any case, then, sodium is present, but not iron and ammonia simultaneously.

11.18 Suppose $f(x)$ is monotone and $X = \{x : f(x) = 1\}$. With $x \in X$ we associate as usual the set $A_x \subseteq \{1,\ldots,n\}$ and set $x' = x_{i_1}\ldots x_{i_\ell}$ with $A_x = \{i_1,\ldots,i_\ell\}$. If X' is the set of minimal vectors in X, then we have $f(x) = \sum_{x \in X'} x'$.

11.20 We have $g(x,y,z,w) = \overline{z}\,\overline{w} + xyw + \overline{x}\,\overline{y}\,\overline{z}w$. A Karnaugh mapping for four variables is

11.23 An optimal logical net with fan-out 1 has as associated directed graph a tree with a unique sink $(= f)$. If we go backward, we see that the number of gates, that is, $L_\Omega(f)$, is at most $1 + r + \cdots + r^{d-1} = \frac{r^d - 1}{r-1}$. It follows that $D_\Omega(f) \ge \log_r((r-1)L_\Omega(f) + 1)$.

11.25 The number of Boolean functions in $\mathcal{B}(n)$ that depend essentially on j variables is $N(j)$, from which the formula follows. From this we compute $N(1) = 2$,

$N(2) = 10$, $N(3) = 218$, $N(4) = 64594$. As limiting value we have $\sum_{j=0}^{n-1} N(j)\binom{n}{j} \leq 2^{2^{n-1}} 2^n$, $\lim(2^{2^{n-1}+n}/2^{2^n}) = 0$, and thus $\lim N(n)/2^{2^n} = 1$.

11.27 The lower bound follows from Exercise 11.26, since $C_{\Omega_0}^*(g_j) \geq n-1$ (all variables are essential). For the upper bound we have the minterm $g_i = x_0^{c_0} \cdots x_{n-1}^{c_{n-1}}$ as in the hint. Let n be even and $f_T^{(n/2)}(x_0, \ldots, x_{n/2-1})$, $f_T^{(n/2)}(x_{n/2}, \ldots, x_{n-1})$ already realized by logical nets. Then the corresponding minterms exist, and the 2^n minterms from $f_T^{(n)}$ can be realized by AND combinations of all possible pairs. It follows that $C_{\Omega_0}^*(f_T^{(n)}) \leq 2^n + 2C_{\Omega_0}^*(f_T^{(n/2)})$. A direct realization of the minterms from $f_T^{(n/2)}$ requires $\frac{n}{2}2^{n/2} - 1$ gates, and it follows that $C_{\Omega_0}^*(f_T^{(n)}) \leq 2^n + n2^{n/2} - 2$. For odd n we replace $\frac{n}{2}$ by $\lceil\frac{n}{2}\rceil$.

11.30 Let n be even and $\mathcal{A}$ an antichain with $|\mathcal{A}| = \binom{n}{n/2}$. We set $\mathcal{A}_1 = \{A \in \mathcal{A} : n \notin A\}$, $\mathcal{A}_2 = \{A \in \mathcal{A} : n \in A\}$. Then $\mathcal{A}_1, \mathcal{A}_2' = \{A \setminus \{n\} : A \in \mathcal{A}_2\}$ are antichains in $S \setminus \{n\}$. It follows that $|\mathcal{A}_1|, |\mathcal{A}_2'| \leq \binom{n-1}{n/2}$, and so on account of $\binom{n}{n/2} = \binom{n-1}{n/2} + \binom{n-1}{n/2}$ and induction, we must have $\mathcal{A}_1 = \binom{S \setminus \{n\}}{n/2}$ or $\binom{S \setminus \{n\}}{n/2-1}$ and $\mathcal{A}_2' = \binom{S \setminus \{n\}}{n/2-1}$ or $\binom{S \setminus \{n\}}{n/2}$. One sees at once that only the combination $\mathcal{A}_1 = \binom{S \setminus \{n\}}{n/2}$ and $\mathcal{A}_2' = \binom{S \setminus \{n\}}{n/2-1}$ is possible. The proof is similar for odd n.

11.31 Let us assume the first law. Then $(x \vee y) \wedge (x \vee z) = [(x \vee y) \wedge x] \vee [(x \vee y) \wedge z] = x \vee [(x \vee y) \wedge z] = x \vee [(x \wedge z) \vee (y \wedge z)] = x \vee (y \wedge z)$.

11.34 Suppose $t < n$. Then $n - |F| > t - |F| \geq t - d(u)$ for evert pair $u \notin F$, that is, $\frac{|F|}{n-|F|} < \frac{d(u)}{t-d(u)}$. Summing these inequalities over all pairs $u \notin F$ yields $\sum_F (n - |F|)\frac{|F|}{n-|F|} < \sum_u (t - d(u))\frac{d(u)}{t-d(u)}$, whence $\sum_F |F| < \sum_u d(u)$, which is impossible, since here we have equality.

11.38 Let F be a filter. Then the minimal elements in F form an antichain. Conversely, if A is an antichain, then $F = \{x : x \geq a \text{ for an } a \in A\}$ is a filter, and the mapping is a bijection.

12.3 From $3^3 \equiv 27 \equiv 1 \pmod{13}$ follows $3^{15} = (3^3)^5 \equiv 1 \pmod{13}$. Since $15 \equiv 2 \pmod{13}$, we may restrict to 2^{83}. From $2^6 \equiv 64 \equiv -1 \pmod{13}$ follows $2^{83} = (2^6)^{13}2^5 \equiv -32 \equiv 7 \pmod{13}$.

12.6 Let T be the sum of the tens digits, and O the sum of the ones digits. Then $10T + O = 100$ and hence $T + O \equiv 1 \pmod 9$. On the other hand, $T + O = 45 \equiv 0 \pmod 9$, and so it cannot be done.

12.8 If $x^2 + ax + b$ is reducible, then $x^2 + ax + b = (x - \alpha)(x - \beta)$ with $\alpha + \beta = -a$, $\alpha\beta = b$. We have therefore only to go through the addition and multiplication tables in order to find suitable a, b for which there are no α, β, for example, $a = b = 1$. Therefore, $x^2 + x + 1$ is irreducible. The elements of GF(25) are identified with $\alpha x + \beta$ $(\alpha, \beta = 0, 1, \ldots, 4)$ with addition and multiplication modulo $x^2 + x + 1$.

12.10 Let the values be J, Q, K, A and the suits $1, 2, 3, 4$. If we normalize the first row in this sequence, then one sees that up to interchange, the only orthogonal 4×4

squares that also satisfy the diagonal condition must look as follows:

J	Q	K	A
K	A	J	Q
A	K	Q	J
Q	J	A	K

1	2	3	4
4	3	2	1
2	1	4	3
3	4	1	2

Viewed horizontally, $1, 3$ must have one color, $2, 4$ the other. However, vertically, 1 and 3, for example, collide. A checkerboard arrangement is therefore impossible.

12.12 We cannot place more than n^2 rooks, since at every level only n rooks can be placed. An arrangement with n^2 corresponds to a Latin square, whereby we place the rooks on level i in the positions of i. By cyclic rearrangement of the rook positions, one sees that on the n^d board one can place n^{d-1} rooks.

12.16 From $x \pitchfork y = x \pitchfork z$ follows $xy^{-1} = xz^{-1}$, and hence $y = z$, and likewise $y = z$ from $y \pitchfork x = z \pitchfork x$. That is, $L(x, y) = x \pitchfork y$ yields a Latin square. In a group of odd order, $x^2 = 1$ has only the solution $x = 1$ (see Exercise 12.32). The mapping $x \mapsto x^2$ is therefore a bijection. Considering the system of equations $xy^{-1} = a$, $xy = b$, we see that $x^2 = ab$, $y^2 = ba^{-1}$. The pair (x, y) is thus uniquely determined.

12.18 Since the vector addition on S forms a group, for $\boldsymbol{x} + \boldsymbol{y}$ there exists exactly one $\boldsymbol{z}$ with $\boldsymbol{x} + \boldsymbol{y} + \boldsymbol{z} = \boldsymbol{0}$.

12.20 We have $(a + b)^p = \sum_{k=0}^{p} \binom{p}{k} a^k b^{p-k}$. For $0 < k < p$ we have that $\binom{p}{k}$ is a multiple of p, and it follows that $(a + b)^p \equiv a^p + b^p \pmod{p}$. Second proof: For $a \equiv 0$, $b \equiv 0$, or $a + b \equiv 0 \pmod{p}$, the assertion is correct. If all three numbers are inequivalent to zero modulo p, then by Fermat's theorem we have $(a + b)^p \equiv a + b \equiv a^p + b^p \pmod{p}$.

12.22 For $p \equiv 3 \pmod{4}$ we have that $\frac{p-1}{2}$ is odd. From $n^2 \equiv -1 \pmod{p}$ it therefore follows that $(n^2)^{\frac{p-1}{2}} \equiv -1 \pmod{p}$, in contradiction to Fermat's theorem. For $p \equiv 1 \pmod{4}$ let $n = (\frac{p-1}{2})!$. From $k \equiv -(p - k) \pmod{p}$ for $k = 1, \ldots, \frac{p-1}{2}$ follows $n^2 \equiv (-1)^{\frac{p-1}{2}}(p - 1)! \equiv -1 \pmod{p}$ by Exercise 12.21.

12.25 From $2i \equiv 2j \pmod{n}$ follows $n \mid 2(i - j)$, whence $i \equiv j \pmod{n}$, since n is odd. The queens therefore stand on different rows and columns. The diagonal condition requires $|i' - i| \neq |j' - j|$ for two queen squares $(i, j), (i', j')$. Suppose $2i, 2i' \leq n - 1$. Then it follows at once from $i' - i \equiv 2i' - 2i$ that $i \equiv i' \pmod{n}$, which is impossible. One handles the cases $2i > n - 1$ and $2i' > n - 1$ similarly.

12.27 Let $B^2 \geq 1$. Then $f(m, n) = 2$. For $B^2 = 0$ we obtain $f(m, n) = n + 1$. In this case, $B = 0$, that is, $m(n + 1) = n! + 1$ or $n! \equiv -1 \pmod{n+1}$. It follows from Exercise 12.21 that $f(m, n) = n + 1$ is prime. Conversely, $f(1, 1) = 2$. For an odd prime p we set $n = p - 1$, $m = \frac{(p-1)!+1}{p}$ (which can be done by Exercise 12.21). With this choice we have $B = 0$ and therefore $f(m, n) = n + 1 = p$. The uniqueness follows from the observation that $f(m, n) = 2$ or $f(m, n) = n + 1$.

12.28 If the integer n has exactly k digits, then $10^{k-1} \leq n < 10^k$, that is, $k - 1 \leq \log_{10} n < k$. For $n = 4444^{4444}$ we have $\log_{10} n = 4444 \log_{10} 4444$ and from $4444 < 10^4$ we have $\log_{10} n < 4444 \cdot 4 < 20000$. For A it thus follows that $A < 9 \cdot 20000 < 199999$, whence $B < 46$. The sum S of the digits of B can thus be at most 12 (e.g., for $B = 39$). Now we have $n \equiv A \equiv B \equiv S \pmod 9$. From $4444 \equiv -2 \pmod 9$ one immediately calculates $n \equiv 7 \pmod 9$ and from that, $S = 7$ on account of $S \le 12$.

12.32 Let $a \in G$. Together with $g \in G$ we have that ag also runs through the entire group, and we obtain $\prod_{g \in G} g = \prod_{g \in G}(ag)$, whence $1 = a^n$. Let $d = \operatorname{ord}(a)$. Then we know from what we have just proved that $d \le n$. Writing $n = qd + r$, $0 \le r < d$, we have $1 = a^n = (a^d)^q a^r = a^r$, whence $r = 0$ from the minimality of d.

12.34 We remark first that $1 \ne -1$ since q is odd, whence $a \ne -a$ for all $a \ne 0$. Each pair $\{a, -a\}$ produces a square a^2. If $a^2 = b^2$, then $(a+b)(a-b) = 0$, whence $b = a$ or $b = -a$, whence $|Q| = \frac{q-1}{2}$. If we group the elements $b \ne 0, 1, -1$ pairwise as $\{b, b^{-1}\}$, we obtain $\prod_{b \ne 0} b = -1$. Choosing one element from each pair $\{b, -b\}$, this yields $(-1)^{\frac{q-1}{2}} \prod_{b^2 \in Q} b^2 = -1$. For $q \equiv 1 \pmod 4$ we have therefore $a^2 = -1$ with $a = \prod_{b^2 \in Q} b$. Conversely, if there exists a with $a^2 = -1$, then $\operatorname{ord}(a) = 4$ is in the multiplicative group of the field $\operatorname{GF}(q)$. By Exercise 12.32, it follows that $q - 1 \equiv 0 \pmod 4$. This yields the result $-1 \in Q \Leftrightarrow q \equiv 1 \pmod 4$.

12.37 Let L_0 and L_1 be orthogonal Latin squares, each filled with $0, 1, \dots, n-1$. From $M(i,j) = L_0(i,j) + n \, L_1(i,j)$ we obtain a half-magic square with the numbers $0, 1, \dots, n^2 - 1$. If we add 1 to each entry, the result is a half-magic square on $1, \dots, n^2$. If L_0, L_1 fulfill as well the diagonal condition, we obtain a magic square. Exercise 12.10 gives a pair of such Latin squares of order 4. Order 5 can be easily constructed.

12.39 The necessity results from $bk = v\lambda$. Conversely, let S be a set with v elements, $b = \frac{v\lambda}{k}$. We must construct a $v \times b$ incidence matrix M with exactly λ ones in each row and k ones in each column. Let M be an incidence matrix with k ones in each column. Assume that not all row sums are λ. Then there must exist i, j with $r_i > \lambda > r_j$. There is thus a column s with $s_i = 1$, $s_j = 0$. If we exchange this 0 and 1 in column s, then we obtain a new matrix M' with $r_i' = r_i - 1$, $r_j' = r_j + 1$. Repetition of this exchange step yields the desired matrix.

12.41 That MM^T is of the specified form follows from the definition. The determinant of MM^T is easily calculated: $\det MM^T = (r + \lambda(v-1))(r - \lambda)^{v-1}$. Since $k < v$, we have $r > \lambda$, whence $\det MM^T \ne 0$. Since the rank of a matrix is not larger than the number of columns, we conclude that $v = \operatorname{rank}(MM^T) \le \operatorname{rank}(M) \le b$.

12.45 Let $B^* = \{u_1, \dots, u_{n+1}\}$ be the line removed from the projective plane. If in the affine plane $(S, \mathcal{B})$ we define the class $\mathcal{B}_i$ as those lines that contained the point u_i in the projective plane, then the conditions are satisfied.

12.48 From Exercise 12.34 we know that $|Q| = 2n + 1$. Let $a \in Q$, $a = \alpha^2$. Then a is representable by some number r_a of differences $a = x^2 - y^2$. We must show that $r_a = r_b$ for all $b \ne 0$. Let $a = x^2 - y^2$ and $b \in Q$, $b = \beta^2$. Then $b = (\beta \alpha^{-1})^2 \alpha^2 = \gamma^2 a$, $\gamma = \beta \alpha^{-1}$, and we obtain the representation $b = (\gamma x)^2 - (\gamma y)^2$. Different representations of a correspond to different representations of b, and so $r_a = r_b$. If $b \notin Q$, then $-b \in Q$ on account of $-1 \notin Q$ (Exercise 12.34), and we conclude analogously with $-b = \beta^2$. The total number of differences from Q is $(2n+1)2n$, and so each nonzero element appears exactly n times as a difference.

12.50 Since every pair of points of a projective plane lie in exactly one block, it follows that the girth of G is at least 6. Every triangle, on the other hand, yields a cycle of length 6. The lower bound for $f(q+1,6)$ in Exercise 6.37 yields precisely the vertex number $2(q^2 + q + 1)$ of G.

13.2 We use the Kraft inequality $\sum_{i=1}^{6} 2^{-\ell_i} \leq 1$ (Theorem 9.3). The first code does not exists, while the second does.

13.5 $C = \{000000, 111000, 100110, 010101, 001011\}$. With length 5 there is no one-error-correcting code C with $|C| = 5$.

13.7 Let $C \in \mathcal{C}(n, 2d - 1)$, $|C| = M(n, 2d - 1)$. We construct C^* by appending a parity check to each codeword. If $\Delta_C(\boldsymbol{a}, \boldsymbol{b}) = 2d - 1$, then the number of 1's in $\boldsymbol{a}$ has a different parity from the number of 1's in $\boldsymbol{b}$, so that $\Delta_{C^*}(\boldsymbol{a}^*, \boldsymbol{b}^*) = 2d$ and hence $M(n + 1, 2d) \geq M(n, 2d - 1)$. Conversely, we delete a coordinate place. For part (b), we consider the last coordinate. The codewords with the same symbol in the last place form with the first $n - 1$ places a code in $\mathcal{C}(n - 1, d)$.

13.10 Since the Fano code is 1-perfect, the received word is correctly decoded if and only if at most one error has occurred. The desired probability is therefore $(1 - p)^7 + 7p(1 - p)^6$.

13.15 The code $C^\perp$ has basis $\{2210, 2101\}$. In the 4×2 matrix $\left(\begin{smallmatrix} 2 & 2 & 1 & 0 \\ 2 & 1 & 0 & 1 \end{smallmatrix}\right)^T$ each pair of rows are linearly independent, and it therefore follows from Theorem 13.5 that $d(C) \geq 3$. With $|C| = 9$, $n = 4$, $q = 3$, the Hamming bound is satisfied with equality.

13.17 Let $\boldsymbol{u}_1$ be a codeword of odd weight w. Shifting $\boldsymbol{u}_1$ n times, we obtain codewords $\boldsymbol{u}_1, \ldots, \boldsymbol{u}_n$ in which at each place, 1 appears w times altogether. It follows that $\sum_{i=1}^{n} \boldsymbol{u}_i = (1, 1, \ldots, 1)$.

13.19 Suppose we replace in the Huffman algorithm a branch with leaves u, v, $\ell(u) = \ell(v) = m$. If L is the total length of the original tree and L' the length of the new one, then $L' = L - m - 1 \geq L - n$ on account of $m \leq n - 1$. For L' we therefore have $L \leq L' + n$ and hence $L \leq n + (n - 1) + \cdots + 2 = \frac{n^2 + n - 2}{2}$. For a distribution $(p_1 \geq \cdots \geq p_n)$ with $p_i > \sum_{j > i} p_j$ we have equality.

13.21 Let $\alpha = \sum_{i=1}^{n} q^{-\ell(\boldsymbol{w}_i)}$. Then for all k we have $\alpha^k = (\sum_{\boldsymbol{w} \in C} q^{-\ell(\boldsymbol{w})})^k = \sum_{\boldsymbol{v} \in C^k} q^{-\ell(\boldsymbol{v})} = \sum_{a}^{b} N(k, \ell) q^{-\ell}$ with $a = k \cdot \min \ell(\boldsymbol{w}_i)$, $b = k \cdot \max \ell(\boldsymbol{w}_i)$. Since C is uniquely decodable, it follows that $N(k, \ell) \leq q^\ell$, and so $\alpha^k \leq k \cdot c$ with $c = \max \ell(\boldsymbol{w}_i) \geq 1$. From this we obtain $\alpha \leq \sqrt[k]{kc}$ for every k, whence $\alpha \leq \lim \sqrt[k]{kc} = 1$.

13.23 By Theorem 13.5 we must find vectors $\boldsymbol{u}_1, \ldots, \boldsymbol{u}_n \in \mathrm{GF}(q)^{n-k}$ of which each $2t$ are linearly independent. For $\boldsymbol{u}_1$ we take an arbitrary nonnull vector. Suppose that we have already constructed h vectors $\boldsymbol{u}_1, \ldots, \boldsymbol{u}_h$ with this property. We consider an i-set $U \subseteq \{\boldsymbol{u}_1, \ldots, \boldsymbol{u}_h\}$. The number of vectors that are dependent on U (but on no subset) is at most $(q - 1)^i$. As long, therefore, as

$$\binom{h}{1}(q - 1) + \binom{h}{2}(q - 1)^2 + \cdots + \binom{h}{2t - 1}(q - 1)^{2t-1} < q^{n-k} - 1,$$

[(i)] we can add an additional vector $\boldsymbol{u}_{h+1}$.

13.25 For $x, y \in \{0,1\}^n$ let $\langle x, y \rangle = |\{i : x_i = y_i = 1\}|$. Then $w(x + y) = w(x) + w(y) - 2\langle x, y \rangle$. Every $c \in C$ is a linear combination $c = g_{i_1} + \cdots + g_{i_m}$. We use induction on m. For $m = 1$ this is the assumption. Let $z = x + y$, $w(x) \equiv w(y) \equiv 0 \pmod 4$. Then it follows from the above equality that $w(z) \equiv 0 \pmod 4$, since $\langle x, y \rangle$ is even on account of $C = C^\perp$.

13.28 The sum counts the pairs $m_{ik} \neq m_{jk}$ in the $M \times n$ matrix. Since $d(C) \geq d$, it follows that $\sum \geq \binom{M}{2}d$. In a column, the contribution is $a \cdot b$, where a is the number of zeros and b is the number of ones. From $a \cdot b \leq \frac{M^2}{4}$ the upper bound follows from summation over the columns.

13.30 Let $G = \left(\begin{array}{c|c} c_1 \ldots c_d & 0 \ldots 0 \\ \hline A_1 & G_1 \end{array} \right)$. The rank of G_1 is $k-1$. Namely, if the rows of G_1 were linearly dependent, there would exist a codeword $b = (b_1, \ldots, b_d, 0, \ldots, 0)$ generated by the rows of $G \neq c$. Since $b \neq c$, there exists $b_i \neq c_i$, and $b - b_i c_i^{-1} c$ would be a codeword of weight less than d. Let C_1 be the code generated from G_1 with $d(C_1) = d_1$. It remains to show that $d_1 \geq \frac{d}{q}$. We take $u_1 \in C_1$ with $w(u_1) = d_1$. Then in C there is a codeword $u = (v_1, u_1)$. Since $\Delta(c, u) \geq d$ we must have $\Delta(v_1, (c_1 \ldots c_d)) \geq d - d_1$, from which the assertion then easily follows.

13.32 Let $A = \{0,1\}$ and $w(c)$ the weight of $c \in A^n$. From $0 \in C$ we have $w(a) \geq 2t + 1$ for $0 \neq a \in C$. Every word $c \in A^n$ with $w(c) = t + 1$ thus lies in exactly one ball $B_t(a)$, $a \in C$, $w(a) = 2t + 1$. If h_{2t+1} is the number of $a \in C$ with $w(a) = 2t + 1$, then by double counting we have $h_{2t+1}\binom{2t+1}{t+1} = \binom{n}{t+1}$. If we identify $a \in C$, $w(a) = 2t + 1$ with $U_a = \{i : a_i = 1\}$, then the U_a form a Steiner system $S_{t+1}(n, 2t + 1)$. The assertion for $\widehat{C}$ is proved analogously.

13.34 From $n^2 + n = 2^{r+1} - 2$ follows $(2n + 1)^2 = 2^{r+3} - 7$. From the observation we have the following solution pairs (n, r): $(0,0), (1,1), (2,2), (5,4), (90,12)$. The first three pairs are excluded. The pair $n = 5$, $r = 4$ yields the binary repetition code. The last pair would yield a Steiner system $S_3(90, 5)$ by Exercise 13.32. By Exercise 13.33, we have $n + 1 \equiv 0 \pmod{t + 1}$, in contradiction to $3 \nmid 91$.

13.36 $H_1 = (1)$, $H_2 = \left(\begin{smallmatrix} 1 & 1 \\ 1 & -1 \end{smallmatrix} \right)$. Let $n \geq 3$. Through multiplication of the columns by -1 (this does alter anything with respect to the Hadamard property) we may assume that the first row consists entirely of ones. From $HH^T = nE_n$ it follows that every further row contains the same number of 1's as -1's, whence $n \equiv 0 \pmod 2$. From $h_i \cdot h_j = 0$ for every pair of rows $i, j \geq 2$ it thus follows that $n \equiv 0 \pmod 4$. From $HH^T = nE_n$ follows $H^{-1} = n^{-1}H^T$, whence $H^T H = nE_n$. Parts (c) and (d) are clear.

13.38 We replace -1 by 0. Code A consists of all rows of H_n, with the first column removed. Code B consists of A together with all complements, and C consists of all rows of H together with the complements. Code B_8 is our old friend the Fano code, and $C_8 = \widehat{B_8}$ in the sense of Exercise 13.26.

13.40 By induction we have $|C(r + 1, m + 1)| = |C(r + 1, m)| \, |C(r, m)| = 2^{b+c}$ with $b = \sum_{i=0}^{r+1} \binom{m}{i}$, $c = \sum_{i=0}^{r} \binom{m}{i}$, whence $a = b + c = \sum_{i=0}^{r+1} \binom{m+1}{i}$. By Exercise 13.39 we have $d(C(r + 1, m + 1)) = 2^{m-r}$ by induction.

13.42 We have $(n + 1)(n^2 - n + 6) = (n + 1)[(n + 1)^2 - 3(n + 1) + 8] = 3 \cdot 2^{r+1}$. If $n + 1 \equiv 0 \pmod{16}$, it follows that $n^2 - n + 6 \not\equiv 0 \pmod{16}$, whence $n^2 - n + 6 \mid 24$, which because of $n \geq 15$ is impossible. Therefore, $n + 1 \mid 24$ and we obtain the possibilities $n = 7, 11, 23$. The value $n = 7$ yields the repetition code, while $n = 11$ does not satisfy the equation. So $n = 23$ remains. Here in fact we have $1 + 23 + \binom{23}{2} + \binom{23}{3} = 2048 = 2^{11}$.

13.44 Let B be the matrix as in the previous exercise. We have $B^2 = E_{12}$ and therefore $BG = B(E_{12}, B) = (B, E_{12})$. Therefore (B, E_{12}) is also a generator matrix. In other words, if $c = (c_L, c_R) \in G_{24}$, then also $\hat{c} = (c_R, c_L)$. Suppose $w(c) = 4$. Then without loss of generality we may assume $w(c_L) \leq 2$. Clearly, $c_L = 0$ implies that $c = 0$. From $w(c_L) = 1$ follows $c = g_i \in G$, in contradiction to $w(g_i) = 8$. One resolves the case $w(c_L) = 2$ similarly. We therefore have $d(G_{24}) = 8$, and by deleting the last column there results a (23,12) code with $d = 7$.

14.1 The Caesar system is cracked by considering the frequency of letters. By permutation, the frequencies of the individual letters remain the same, but letter sequences are distorted.

14.3 A permutation π can be represented by the $n \times n$ permutation matrix $A_\pi = (i, \pi_i)$ with exactly one 1 in each row and column and zeros otherwise.

14.6 If A is the matrix of the register, then t is a period if and only if $A^t s = s$. Let $p = \operatorname{per}(s)$ and $t = kp + r$, $0 \leq r < p$. Then $s = A^r (A^{kp} s) = A^r s$, whence $r = 0$.

14.8 Let $t = \operatorname{lcm}(r, s)$. Then $a_{n+t} + b_{n+t} = a_n + b_n$ for all n, whence $\operatorname{per} \leq \operatorname{lcm}(r, s)$.

14.12 Let $\mathcal{T} = \{T_1, \ldots, T_n\}$ and $k \in \mathcal{K}$. The cryptograms $C_1 = c(T_1, k), \ldots, C_n = c(T_n, k)$ are distinct, and therefore $p(C_j) = p(C_j \mid T_j) > 0$ for all j. Let $h \neq j$. Then $p(C_j \mid T_h) = p(C_j) > 0$, and so there must exist an additional key k' with $c(T_h, k') = C_j$. Since this holds for all $h \neq j$, it follows that $|\mathcal{K}| \geq n$.

14.15 A sequence of $2^n - 1$ successive vectors yields all $2^n - 1$ nonnull vectors exactly once each. We identify a vector $(x_{n-1}, \ldots, x_0)$ with the natural number $x = \sum_{i=0}^{n-1} x_i 2^i$; that is, we obtain all integers $1 \leq x \leq 2^n - 1$. Since there is one more odd number than even number, the assertion follows.

14.17 Every word $a_1 a_2 \ldots a_{n-1}$ has out-neighbors $a_2 \ldots a_{n-1} 0$ and $a_2 \ldots a_{n-1} 1$, and in-neighbors $0 a_1 \ldots a_{n-2}$ and $1 a_1 \ldots a_{n-2}$. The directed graph $\vec{G}$ thus contains an Euler cycle, and this yields the de Bruijn word.

14.19 The characteristic polynomial is $f(x) = x^5 + x + 1 = (x^2 + x + 1)(x^3 + x^2 + 1)$. Both factors are primitive (see Exercise 14.7), whence $\exp(x^2 + x + 1) = 3$, $\exp(x^3 + x^2 + 1) = 7$, $\exp(x^5 + x + 1) = 21$. The period 1 is of course realized by 0.

14.21 Let s be an input, and i the least exponent with $A^i s = \lambda s$. It then follows that the sequence $s, As, \ldots, A^{q^n - 2} s$ is of the form $b_0 = s, b_1 = As, \ldots, b_{i-1} = A^{i-1} s, \lambda b_0, \lambda b_1, \ldots, \lambda b_{i-1}, \lambda^2 b_0 \ldots$. Since $\operatorname{GF}(q)$ contains $q - 1$ nonzero elements, it follows that $i \geq \frac{q^n - 1}{q - 1}$, that is, the first $\frac{q^n - 1}{q - 1}$ vectors are linearly independent.

14.23 The encoding is exponentiation, hence easy. Decoding proceeds in two steps: First, K is calculated from $K \equiv y_j^k \equiv a^{x_j k} \equiv (a^k)^{x_j} \equiv C_1^{x_j} \pmod{p}$, and then $M \equiv C_2/K \pmod{p}$. Thus decoding with knowledge of x_j is easy. Without such knowledge it is conjectured that the problem is equivalent to that of the discrete logarithm.

14.25 $K = 9$, $C_1 = 49$, $C_2 = 9 \cdot 30 \equiv 57 \pmod{71}$, whence $M = (49, 57)$. From $2 \equiv 7^{k'} \pmod{71}$ follows $k' = 6$, $K = 19$, whence $C_2 = 2$.

14.27 Every element $a \in \mathbb{Z}_p^*$ has an order $d \mid p - 1$ (see Exercise 12.32). Let $\psi(d) = \#\{a : \mathbb{Z}_p^* : \text{ord}(a) = d\}$, whence $\sum_{d \mid p-1} \psi(d) = p - 1$. If $a \in \mathbb{Z}_p^*$ with $\text{ord}(a) = d$ exists at all, then $(a^i)^d \equiv 1 \pmod{p}$ for all a^i, that is, $a, a^2, \ldots, a^d$ are zeros of the polynomial $x^d - 1$ in $\mathbb{Z}_p$. Since $\mathbb{Z}_p$ is a field, $x^d - 1$ has precisely the elements a^i as zeros. Hence if $\psi(x) \neq 0$, then $\psi(d) = \varphi(d)$. It follows that $p - 1 = \sum_{d \mid p-1} \psi(d) \leq \sum_{d \mid p-1} \varphi(d) = p - 1$, whence $\psi(d) = \varphi(d)$ for all d, and in particular, $\psi(p - 1) = \varphi(p - 1)$.

15.2 Since M_{opt} is convex, then together with $x \neq y \in M_{\text{opt}}$, the segment $\overline{xy}$ is also in M_{opt}, and so such a program cannot exist.

15.4 Clearly, $x_1 > 0$, $x_2 > 0$, and therefore it follows from Theorem 15.13 that in the dual program the two inequalities are satisfied with equality. From this we calculate $y_2 - y_3 = \frac{1}{3}$, and it easily follows that $y_2 > 0$, $y_3 > 0$. Thus the second and third inequalities in the primal program are satisfied with equality, whence $x_1 = 4$, $x_2 = 1$, and the value of (I) is 3. Since now the first inequality is satisfied with inequality, we conclude that $y_1 = 0$, and therefore $y_1 = 0$, $y_2 = \frac{2}{3}$, $y_3 = \frac{1}{3}$, and the value of (I*) is equal to 3.

15.7 An optimal solution is $x_{14} = 3$, $x_{22} = 5$, $x_{31} = 2$, $x_{33} = 4$, $x_{34} = 1$ with value 65. It is impossible with four routes.

15.9 Clearly, we have $U^{\perp\perp} \supseteq U$. We take as columns of A a basis of U. Then $b \notin U$ means that $Ax = b$ is not solvable. Therefore, there is a solution of $A^T y = 0$, $b^T y = -1$. The condition $A^T y = 0$ implies $y \in U^\perp$ and $b^T y = -1$, and thus $b \notin U^{\perp\perp}$.

15.11 That (A) and (B) are not solvable simultaneously follows at once. Assume that (B) is unsolvable. With $C = \left(\frac{A^T}{b^T}\right)$, $c = \left(\frac{0}{-1}\right)$ this means that $Cy = c$, $y \geq 0$, is unsolvable. By Theorem 15.8 there thus exists $(z, \alpha) \in \mathbb{R}^{n+1}$ with $Az + \alpha b \geq 0$, $\alpha > 0$. But then $x = -\frac{z}{\alpha}$ is a solution of (A).

15.13 We write the transportation problem as standard minimum program (I) $-\sum_j x_{ij} \geq -p_i$, $\sum_i x_{ij} \geq q_j$, $\sum a_{ij} x_{ij} = \min$. The dual program (I*) is $y_j' - y_i \leq a_{ij}$, $\sum q_j y_j' - \sum p_i y_i = \max$. Suppose a competitor offers the planning agent in factory F_i the price y_i, transports all goods such that at least q_j units arrive at M_j, and sells everything back at price y_j' with $y_j' - y_i \leq a_{ij}$. The planner thus has to pay $\sum q_j y_j' - \sum p_i y_i \leq \sum a_{ij} x_{ij}$, which equals the value of (I). He will therefore agree, and the competitor will attempt to maximize his profit. If $y_j' - y_i < a_{ij}$, then the planning agent pays less than his costs and will close this route, $x_{ij} = 0$. The other equations are interpreted similarly.

15.14 From $\left(\frac{A}{-A}\right)x \leq \left(\frac{b}{-b}\right)$, $x \geq 0$, $c^T x = \max$ follows $A^T(y' - y'') \geq c$, $y' \geq 0$, $y'' \geq 0$, $b^T y' - b^T y'' = \min$. Setting $y = y' - y''$, we obtain the program *without* sign restriction $A^T y \geq c$, $b^T y = \min$.

15.17 If x, y are convex combinations of $x^1, \ldots, x^n$, then this holds as well for $\lambda x + (1-\lambda)y$, $0 \leq \lambda \leq 1$. Conversely, let $x = \sum_{i=1}^{n} \lambda_i x^i$ be a convex combination with $0 < \lambda_n < 1$. By induction, $y = \frac{1}{1-\lambda_n} \sum_{i=1}^{n-1} \lambda_i x^i \in K$, and therefore $x = (1-\lambda_n)y + \lambda_n x^n \in K$.

15.21 We describe (I) as usual by $(A \mid E_m)(\frac{x}{z}) = b$, $(\frac{x}{z}) \geq 0$, $-c^T x = \min$. In particular, we have $a^{n+j} = e_j$ ($j = 1, \ldots, m$), where e_j denotes the unit vector with 1 at the jth place. Let $A_Z \subseteq (A \mid E_m)$ be the $m \times m$ matrix containing exactly the columns a^k ($k \in Z$) and analogously, $c_Z = (c_k : k \in Z)$. In the simplex tableau T for the optimal solution x we therefore have $a^j = \sum_{k \in Z} \tau_{kj} a^k$, whence $e_j = a^{n+j} = \sum_{k \in Z} \tau_{k,n+j} a^k$ ($j = 1, \ldots, m$). Let $y = (y_i)$ as specified. For $n+j \notin Z$ we have $d_{n+j} \leq 0$, $c_{n+j} = 0$ and therefore $y_j = -d_{n+j} = \sum_{k \in Z} c_k \tau_{k,n+j} \geq 0$. For $n + j \in Z$ we have $\sum_{k \in Z} c_k \tau_{k,n+j} = c_{n+j} = 0$. Therefore, altogether, we have $y^T = c_Z^T(\tau_{k,n+j})$, whence $A_Z^T y = c_Z$ results. For $i \notin Z$ ($1 \leq i \leq n$) we have $a^i = \sum_{k \in Z} \tau_{ki} a^k$, whence $a^{i^T} y = \sum_{k \in Z} \tau_{ki}(a^{k^T} y) = \sum_{k \in Z} \tau_{ki} c_k = -z_i = -d_i + c_i \geq c_i$ since $d_i \leq 0$. Thus y is an admissible solution of (I*). We now apply Theorem 15.13. If $a_j^T x < b_j$, then we must have $n + j \in Z$, and $y_j = 0$ holds by hypothesis. If $x_k > 0$, then we have $k \in Z$ and therefore $a^{k^T} y = c_k$.

15.25 Let B be the $n \times q$ incidence matrix of the bipartite graph $G = (S + T, E)$ and A a square submatrix corresponding to the vertices $S' + T'$ and the edges E'. If $\det A \neq 0$, then there must be a diagonal D with 1's. If $k \in E'$ is incident only with S' but not T' (or conversely), then the corresponding 1 must be in D. It follows that $\det A = \pm \det A'$, where the edges in A' lead from S' to T'. Perhaps some 1's have already been fixed (since an end vertex already appears in D). If we continue in this way, we obtain a unique diagonal, that is, $\det A = \pm 1$, or $\det A = \pm \det C$, where the edges of C lead precisely between the vertices of C. However, in this case, because of the bipartiteness, the row sum of the vertices of C out of S is equal to the row sum of the vertices of C out of T. Therefore, the rows of C are linearly dependent, and therefore $\det C = 0$.

Index